PLUMBING TECHNOLOGY:
DESIGN AND INSTALLATION

THIRD EDITION

LEE SMITH

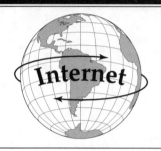

PLUMBING TECHNOLOGY:
DESIGN AND INSTALLATION

THIRD EDITION

LEE SMITH

Delmar
Thomson Learning™

Africa • Australia • Canada • Denmark • Japan • Mexico • New Zealand •
Philippines • Puerto Rico • Singapore • Spain • United Kingdom • United States

DELMAR STAFF

Business Unit Director: Alar Elken
Acquisitions Editor: Mark Huth
Executive Marketing Manager: Maura Theriault
Marketing Coordinator: Kasey Young
Project Editor: Barbara Diaz

Executive Editor: Sandy Clark
Development: Dawn Daugherty
Channel Manager: Mona Caron
Executive Production Manager: Mary Ellen Black
Art Director: Rachel Baker

For more information, contact Delmar at 3 Columbia Circle, PO Box 15015, Albany, New York, 12212-5015; or find us on the World Wide Web at http://www.delmar.com

Asia
Thomson Learning
60 Albert Street, #15-01
Albert Complex
Singapore 189969

Australia/New Zealand
Nelson/Thomson Learning
102 Dodds Street
South Melbourne, Victoria 3205
Australia

Canada
Nelson/Thomson Learning
1120 Birchmont Road
Scarborough, Ontario
Canada M1K 5G4

International Headquarters
Thomson Learning
International Division
290 Harbor Drive 2nd Floor
Stamford, CT 06902-7477
USA

Japan
Thomson Learning
Palaceside Building 5F
1-1-1 Hitosubashi, Chiyoda-ku
Tokyo 100 0003
Japan

Latin America
Thomson Learning
Seneca, 53
Colonia Polanco
11560 Mexico D. F. Mexico

Spain
Thomson Learning
Calle Magallanes, 25
28015-Madrid
España

UK/Europe/Middle East
Thomson Learning
Berkshire House
168-173 High Holborn
London
WC1V 7AA United Kingdom

Thomas Nelson & Sons Ltd.
Nelson House
Mayfield Road
Walton-on-Thames
KYT 12 5PL United Kingdom

Library of Congress Cataloging-in-Publication Data
Smith, Lee, 1935-
    Plumbing technology: design and installation/Lee Smith.--3rd ed.
        p. cm.
    Includes index.
    ISBN 0-7668-1084-4 (alk. paper)
    1. Plumbing. I. Title.

TH6123 .S57 2000
696'.1--dc21

                                                            99-057896

Contents

SECTION **4** WATER SYSTEMS

SECTION **5** HEATING WATER

SECTION **6** SPECIAL CASES

SECTION **7** FIXTURE INSTALLATION

Preface

Plumbing Technology: Design and Installation is a reflection of how plumbing as a trade has grown in complexity over the years. Our society has become increasingly dependent on skilled professionals for its health and well-being. This textbook is a vital resource in the training of these craftspeople.

The author, Lee Smith, a master plumber and a retired instructor at the Center for Arts and Technology in Chester County, Pennsylvania, has revised this benchmark textbook with a view towards modern plumbing style while retaining essential information for work on existing plumbing systems.

The following master plumbers have been of invaluable assistance in the preparation of this text: Jim Kelleher, Kelleher Plumbing Company; Bernie Feodoroff, Y.K. Mechanical Services; J.D. Witherill Plumbing Co.; Paul Pollets, Advanced Radiant Technology; Steven H. Paul, Plumbing & Heating; Al Boyarsky, Boyarsky Plumbing and Heating; Bob Murray, Yankee Professional Plumbing Service; George Slaven, George Plumbing Inc.; Karl Conley, Conley Bros. Plumbing and Heating, Inc.; Jared Mance, Kingston Mechanical; Terry Love, Love Plumbing & Remodeling; L.R. Burk, Maio Plumbing, Heating and Air Cond.; and Jim Rauer, Kish & Rauer Plumbing.

This text stresses good solid plumbing practice applicable to all areas and plumbing materials. Because our society has become so very mobile and few students will spend their whole working lives in any one area of the country, we have used a flexible approach to instruction. North America uses four major plumbing codes and numerous regional ones. Plumbing Technology uses information from all of these, along with the scientific basis for the code provisions, to provide a broad based experience to the student.

Each unit has its own set of objectives that clearly state what the student will learn in the unit. A summary is provided at the end of most chapters to enhance and review the salient points. The "Test Your Knowledge" review questions at the end of each chapter give students an opportunity to self-diagnose the extent of their knowledge. It is strongly urged that the students look up the answers in the Appendix, as additional information may often be found appended to the answer. The Instructor's Handbook has additional chapter end materials and student projects.

The units are logically organized into seven sections: Basic Knowledge, Threaded Pipe, Waste Systems, Water Systems, Heating Water, Special Cases, and Fixture Installation.

The author and publisher hope that your experience with *Plumbing Technology: Design and Installation* is positive, and that you enjoy using the text whether you are a student or an instructor.

ACKNOWLEDGMENTS

The author and Delmar would like to thank the following reviewers for the comments and suggestions they offered during the development of this project. Our gratitude is extended to:

Martin Clark-Stone
State University of New York
Canton, NY

Robert Irion
AAA Construction School
Jacksonville, FL

Frank Beatty
Pennsylvania College of Technology
Williamsport, PA

SECTION 1

BASIC
KNOWLEDGE

THE BASICS

◆ ◆ ◆ ◆ ◆ ◆ ◆ ◆ ◆ ◆ ◆ ◆ ◆ ◆ ◆ ◆

KEY TERMS

aerobic	potable
aquifer	sanitary systems
cisterns	sludge
domestic	soil
draft	water closet
effluent	water service
flue gases	

OBJECTIVES

After studying this unit, the student should be able to:

❑ Describe the progression of the plumbing art.
❑ Name three other trade areas closely related to plumbing.
❑ List the places that potable water may be obtained.
❑ Relate the breakdown of human waste within the sanitary system.

OVERVIEW

The plumber sees to the maintenance of an adequate flow of potable water, environmental comfort inside a living space, and the safe disposal of wastes produced by human habitation.

It is the rare or very small plumbing business that does not also handle heating. Depending on the kind of plumbing shop you work for, you may find that your duties overlap into a number of other trades. The trades that plumbers are involved with are always related

3

to human safety, health, and comfort.

Heating and air-conditioning control the temperature and the air quality within the living and working space.

Pipefitting is another trade that is closely related to plumbing. Many plumbers consider pipefitting a sub-genre of the plumbing trade. However, pipefitting is the skill of installing pipe of many varied materials to carry any kind of liquid, gas, or even granular materials from one location to another for any purpose. These materials could be superheated high-pressure steam, brine, nuclear cooling water, water, milk, acid, cooking oil, grain, or anything that can be pumped and transported in pipe.

When craftspeople, in any trade, work on a piping system, they must be aware of the material—what the pipe contains, its temperature, and its pressure. Knowledge protects health and sometimes even lives. In few crafts is so great an understanding of physics required than it is in the pipe trades.

The student may notice that modern plumbing is simply evolved and automated, basic sanitary practice. Be familiar with what is basic and the complex becomes more understandable.

THE BASIC WATER SUPPLY

Water is obtained from three locations: underground, surface, and less frequently from the atmosphere in the form of rain.

Underground water is generally the purest and safest. Open water, as in long rivers, lakes, and streams, should be viewed with caution. As a stream progresses and as the time for possible exposure to contaminants lengthens, the likelihood of contamination grows. Water from springs and wells is often safe to drink untreated, but almost never water from lakes, rivers, and creeks. Rain water may be safe soon after it is collected, but the necessity of holding it for

long periods in cisterns (vessels for holding rainwater) makes decontamination necessary.

> *Pure* is a relative term. There is some detectable contamination from any water source. The user's body is quite capable of handling a certain amount of contamination without harm. "City" water is delivered to the curb with a certain level of contamination. City engineers treat the water with chlorine, fluorine, bromine, or iodine. These poisons are purposely added to the water supply in concentrations that will kill bacteria but not harm humans. Nevertheless, although necessary, the additives are a source of contamination. Current practice in water supply will be explained more fully in Unit 38.

Simple Water Systems

Humans still use plumbing at its most basic level. In undeveloped countries and rural areas of developed countries, many citizens obtain their water from open streams, transport it in buckets and jars, and use it untreated.

Even in advanced countries massive power outages sometimes create severe potable water shortages. In these cases, authorities may suggest that (1) water obtained from surface sources be boiled for at least one minute plus an extra minute for each thousand feet of altitude above sea level, or (2) eight drops of 2% solution of iodine be added to each quart of water before using it.

Domestic (household or family) water obtained by lowering a bucket into a hand-dug well and then withdrawing it using muscle power is still used by great numbers of people. This is water supply at its most basic. In most places, water can be found by digging a hole. In some desert regions the aquifer, or water-bearing layer of loose material, is quite deep and essentially nonreachable or nonexistent for great stretches. Yet even there, there are places where the aquifer comes close to the surface and green plants grow and form an oasis. In these places, some of the world's most ancient yet active wells can be found.

In modern plumbing practice, the automatic delivery of water under pressure directly to the living space adds to the simple systems of the past. Pumps, whether lift or suction, or simply gravity (from a source of water higher than the valve, Figure 1–1) accomplish this. In the case of city water, the source may be a communal well or an open water supply.

From the plumber's point of view, the source determines the method that will be used to bring potable (drinkable) water into the building. In a municipality with city water, the water is supplied to a point close to the street curb line and the plumber runs a water service, or underground water supply pipe, into the house from there. In the country, a well is drilled by a commercial well driller who often installs the pump and the water service from the well to a point inside the building. The pump also may be located down in the well, in a well house, or inside the basement. Rural plumbers often install and service well pumps.

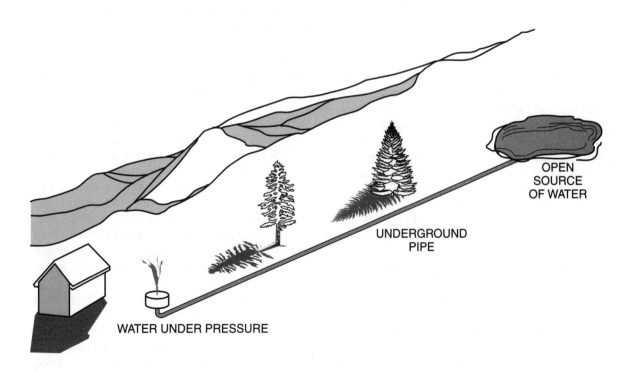

OPEN SOURCE OF WATER

UNDERGROUND PIPE

WATER UNDER PRESSURE

Figure 1–1 A gravity supplied water system

BASIC SEWAGE DISPOSAL

The sun and the air detoxify human waste admirably. What spoils this simple system is when people and their animals stay at a single place for long periods. Exposed waste creates unhealthy conditions.

A basic safe toilet for a single person or a small group traveling in the wilderness is a shallow hole in the ground, which is refilled with the excavated earth immediately after use.

A basic toilet for nomadic groups, such as backpackers or military troops, might be a shallow trench that the group uses starting at one end and backfilling with earth each day while moving toward the unused end. Aerobic (oxygen consuming) bacteria in the soil will consume and greatly reduce the quantity of the waste over time.

The same bacteria that detoxify and reduce this waste in the primitive system described above are again used in the more advanced systems of today. To allow the use of an indoor water closet, or toilet, water is mixed with the excrement so that it will flow through the pipes. In properly constructed rural systems, the destination for this soil (human waste) is beneath the earth in a septic tank as sludge, or soil that has been reduced to a nearly inert state by bacteria. The liquid portion of the waste flows out into the tile or drainage field and from there into the surrounding soil. The soil is acted upon by the sun and wind and, eventually, its excess liquid evaporates into the air. The volume of solids in the septic tank will be greatly reduced by bacteria; therefore, the septic tank should be able to operate for three or four years before being pumped out. Professionally maintained municipal disposal systems detoxify the effluent, or liquid part of sewage, to the point where it may safely be disposed of in streams. The leftover solids may be used for fertilizer if the toxic chemical content is not too high. However, this residue from larger cities invariably contains nonreducible chemical waste products like mercury and must be disposed of by dumping far out to sea.

AIR-CONDITIONING AT A BASIC LEVEL

Pioneers placed their building sites carefully. In mosquito-ridden areas, older homes would be located on a hilltop to catch all of the available breezes possible. In very cold climates, homes would be placed on a south-facing hillside or on the south side of a windbreak of thick-boughed trees like pines or hemlocks. In hot climates, breezes were maximized by placing outside walls at an angle or by placing two buildings close together to create a "breezeway" between them, Figure 1–2. It was thus logical to place a passageway room between the buildings with many screened window openings.

Reducing the wind coming from the north significantly eases the burden on the heating system. Today, while the builder may often ignore the placement of the home in a manner that will reduce heating and cooling costs, the plumber still needs to take these things into account in calculations to predict the heat load.

Air-conditioning is also heating. The object of air-conditioning is to "condition" the air of the people-space in any fashion that is required. If warmth is required, then air-conditioning should provide warmth.

At its most basic, when people were cold they might have lit a fire and warmed themselves by standing around it. The next stop in evolution was the lodge or hogan of the Native American in which a building was constructed

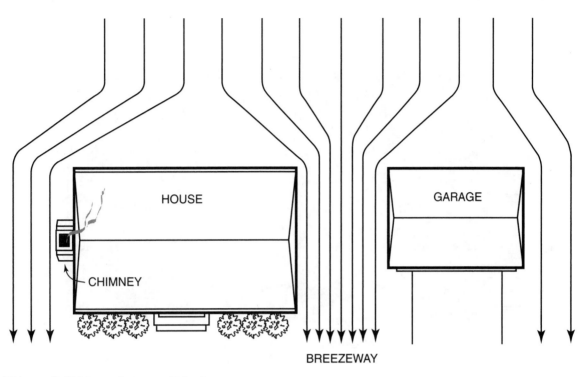

HOUSE

CHIMNEY

GARAGE

BREEZEWAY

Figure 1–2 Natural air-conditioning

of logs or adobe and a hole was left in the center of the roof, Figure 1–3. A fire pit in the center of the floor was kept supplied with wood as the fire burned. The people slept around this central fire. The building kept the cold winds out and protected the fire from anything but a vertical deluge of rain. European settlers were quick to adapt this technique.

This technique worked without a chimney because warm air from the fire rises. When some of the warm air found the escape hole in the roof, it rushed out. Nature tends to allow no vacuum, so other air must replace the air that had just escaped through the "chimney" hole. This process took place at a faster and faster rate until a draft (the tendency of rising warm air to form a column) was created. The air inside the lodge quickly cleared of the initial smoke produced when the fire was first lit. What would have happened if the fire was built too high and sparks began to come into contact with the bark or thatching of the roof?

Homes with fireplaces and brick chimneys are simply a modernization of this very ancient lodge design and have the same design problem. The fire must have air to burn and that air must come from somewhere. In early times that air came rushing through the cracks in the building's construction.

The heated air containing the flue gases

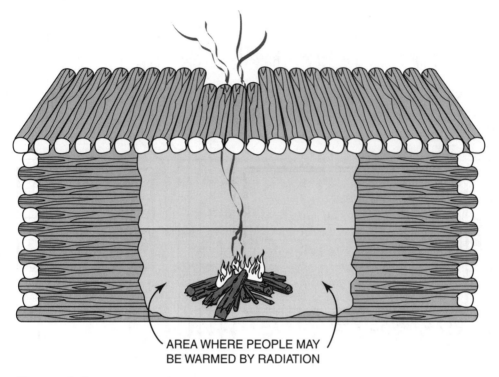

AREA WHERE PEOPLE MAY
BE WARMED BY RADIATION

Figure 1–3 An early lodge

(waste products of combustion) must be allowed to escape safely, that is, without poisoning the occupants or setting the building on fire. These flue gases are not used to heat the home, but in some cases they may be used to heat domestic water. If the heating is accomplished by a warm-air heating system, the hot flue gas is used to heat the air that does not actually come in contact with it, and that uncontaminated air is used to heat the living space.

It might seem that a fireplace would be rather dangerous, and it is true that a blocked chimney might cause flue gases to contaminate the air in a building. Normally, the flame uses some local air to burn the fuel and then that air goes directly up the chimney. The heating actu-

ally takes place through a process called radiation. The sun radiates heat, and the earth absorbs its heat. The bricks and the flame of the fireplace radiate heat, and our bodies and everything surrounding the fireplace absorb it.

Today's heating systems using open flame for warmth do the task even more efficiently by exchanging the heat of the flue gases with captive air, steam, or water systems and distributing it separately through the house. Combustion requires air. This combustion air may be brought in from the outside of the house, reducing the use of inside air to maintain the fire. This will reduce the flow of outside air from doors and windows into the house to replace air used up by the burner gun.

SUMMARY

At its most basic, the art of plumbing is simple. The plumber uses a basic knowledge of chemistry, biology, and physics to maintain safe and comfortable living spaces, but it is this very simplicity that dictates the most careful applications of science. As much as is possible, drainage and venting systems must operate by gravity and atmospheric pressure. A moving part in any system is subject to breakdown and failure. For the most part, sanitary (drainage, waste, and vent) systems are silent, hidden within the structure, and expected to last for years without service. They must have very few or no moving parts.

Gravity-operated pipe systems depend on careful balances. A pipe going directly from the kitchen sink to the septic tank will eventually carry the waste water from what sink to the tank, but it will also allow the smells and toxic gases from the septic tank to waft back into the kitchen. Therefore, the plumber constructs the waste line with a bend in it so that a small quantity of water is trapped within that pipe under the sink. Not surprisingly, this is called a "trap." This water provides a seal so that gases from the sewage system may not find their way back up through the sink and into the house. The existence of that small but vital quantity of water hangs in the balance and is easily upset. In order to ensure its existence, the plumber installs another pipe whose sole purpose is to supply air to the trap at the precise point that will allow the last remaining glassful of water to remain in the trap and not be drawn down the waste pipe with the water preceding it.

Excess pressure below the trap causes gas to bubble up through it. Excess pressure above the trap may push the water out and cause the precious seal to be lost. A perfect balance of pressures must exist.

◆ ◆

TEST YOUR KNOWLEDGE

1. Plumbing shops usually do _heating_ also.

2. Ordinarily the safest source of drinking water is from _underground_ _____ sources.

3. An aquifer is a/an _____ _water bearing layer of loose material_.

4. Human waste buried in the ground is reduced to almost nothing by _aerobic_ pb _____.

5. Two buildings in close proximity often have a natural _breezeway_ _____ between them.

6. The smoke from a fire in a lodge-type house will find its way safely through a hole in the roof because the rising heat will create a _draft_ _____.

7. A fireplace heats mainly through _radiation_ _____

8. The maintenance of the trap in a drainage system depends on a perfect balance of _pressure_ _____ above and below that trap.

UNIT 2

◆ ◆ ◆ ◆ ◆ ◆ ◆ ◆ ◆ ◆ ◆ ◆ ◆ ◆ ◆ ◆ ◆ ◆ ◆

SAFETY

OBJECTIVES

After studying this unit, the student should be able to:

❏ Discuss the importance of safety on the job.
❏ Know where the latest safety information may be obtained.
❏ Avoid some of the hazards that are likely to be encountered in shops and on building sites.

IMPORTANCE OF SAFE WORK HABITS

This unit touches on some safety considerations. The subject of on-the-job safety would fill a very large book. For a detailed, authoritative source, the student is referred to OSHA regulations, which may be viewed at http://www.osha.gov/ on the Internet.

No single chapter, unit, or even book can touch all of the bases concerning safety. This unit, and the many safety tips to be found throughout this book, are simply a general first exposure to the subject.

It is wise to observe all the safety rules. Aside from the brain, the human body is the worker's most valuable tool. This tool is no different from any other in at least one respect: if it is to last for as long as it is needed, it must be maintained. This unit is about using the brain to keep the body in good working condition for life. It is all about developing safe working attitudes so that safety considerations become an automatic response.

The underlying thought behind the phrase "working with your hands" is that working with the hands is wonderful because it does not involve any deep thinking or great philosophies. Wrong! The hands do their wonderful work, whether it is painting a scene, playing a sonata, writing a manuscript, or installing a water softener, only because an alert and intelligent brain tells those hands what to do. Still, those eyes and hands and feet are vitally important. The brain keeps them safe from harm. Make safety a habit. Learn the rules, accept them and enjoy working with your hands for a long time.

How to Develop Safe Work Habits

❏ Take all safety rules seriously.
❏ Be alert to potential danger. If danger were always obvious, accidents would seldom happen.
❏ Keep tools clean, sharp, and in good condition.
❏ Handle chemicals used in the trade carefully and with knowledge.
❏ Regard all revolving machinery with extreme caution.
❏ Accept responsibility for the safety of others.
❏ Only operate unfamiliar equipment after learning the safety rules concerning it.
❏ Do not allow work areas to become cluttered.
❏ Wear safety equipment when it is required.
❏ Set up the job with safety in mind at the beginning.

POTENTIAL DANGERS

In most cases, accidents occur because a worker did not recognize a potential danger. A high percentage of the time, operator error rather than equipment malfunction is the cause of an accident. Sleepy, drug-affected, inattentive, or unknowledgeable workers cause most of the on-the-job injuries and deaths every year. A dangerous situation is not always easy to recognize. Safety engineers learn all of the hazards and how to avoid them, then they carefully craft the rules to fit specific work situations. These safety rules must be obeyed to the letter in order to avoid accidents.

Tools

A dull, poorly maintained tool is more dangerous than a sharp, clean, well-maintained one. Dull tools slip, spin off the work, overheat, bend, break, and cause needless injuries. Keep tools in good repair, clean, free of rust, and sharp. Do not use damaged tools. They could break and possibly cause an injury. For example, a loose hammer handle may come off at any time injure the user or a bystander. If a tool is dirty, oily, or rusty, it is difficult to tell whether it is truly safe to use.

Ladders. The plumber needs to use ladders often. Misuse of ladders is a frequent cause of injury. Some important safety considerations concerning ladders follow:

1. Whenever ladders are used in one place on a long-term basis, they should be tied into place or fastened in such a way that they cannot be displaced. A ladder fastened or held is said to be secured, Figure 2–1.
2. When ladders are being moved throughout the workday and a stepladder is not available, one person should be assigned to hold the ladder while another works from it. While someone is on the ladder, at no time should the ladder be unsecured. The person holding the ladder in place is required to wear an approved hard hat.
3. Ladders that do not conduct electricity are much safer to use when there is the slightest danger of the ladder coming into contact with overhead electrical sources. Ladders that conduct electricity must be clearly marked.
4. Wooden ladders should be painted with a clear coating so that defects can be seen.
5. Never stand on the top platform of a stepladder.

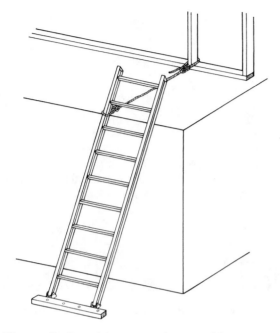

Figure 2-1 Ladder secured top and bottom

6. The landing area at the base and the top of the ladder must be kept clear at all times.

Scaffolds. A scaffold is a temporary platform often made right on the job site to support workers and materials while construction is in progress. Improperly and hastily constructed scaffolding causes many accidents every year. Some rules established by the Occupational Health and Safety Administration (OSHA) are:

1. Scaffolds over 10 feet in height must have *back rails* and *toe rails*. A back rail is a board as big as or bigger than 2" × 4", placed at a height of 42 inches. A toe rail is a board placed vertically at the platform level to keep tools and materials from working their way off the platform and injuring someone below.

2. If a scaffold is between 4 feet and 10 feet in height and is less than 45 inches wide or long, it must have a back rail.
3. The legs of the scaffold must be placed on a firm foundation so that no sinking will occur during use. Bricks, boards, and cement blocks are *not acceptable.*
4. Scaffolds can only be erected, altered, or moved under the supervision of experienced and competent persons.
5. Scaffolds must be designed to support at least *four* times the total weight that will actually be on them.

Chemicals

Learn the characteristics of the chemicals to be used. Store and use them safely. Many chemicals used by the plumber can be hazardous. Cutting oil can create a slipping or falling hazard. Gasoline or naphtha can cause a fire hazard. Drain-cleaning chemicals can burn skin and eyes seriously. Oil-soaked materials can catch on fire without being exposed to flame. This is called spontaneous combustion. Clean up all oil spills and place oily rags in a self-closing metal container. All flammable liquids should be kept in a safe place specially designed for this type of storage, Figure 2–2. This precaution also includes propane cylinders that may leak and fill an enclosed space with explosive fumes.

Machinery

Revolving machinery has a high accident potential. The student plumber should operate machinery only after receiving instruction for the safe operation of the tool. Loose clothing should not be worn when operating machinery. Long hair catching in pipe lathe can also cause injuries. Rings and other jewelry are a hazard with revolving machinery and also in welding and soldering operations. Revolving machinery

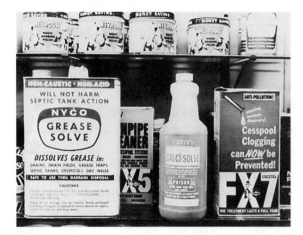

Figure 2–2 Chemicals and flammable liquids should be stored in a safe place.

Figure 2–3 Do not use gloves or loose clothing while using the grinder. Safety goggles must be worn while grinding.

includes pipe lathes of all kinds, power-driven drills, and grinders. Grinders should have the tool rest adjusted up to within 1/16 inch of the grinding wheel. Grinding goggles must be worn. Gloves should not be worn when operating a grinder, Figure 2–3. Gloves are sometimes caught between the tool rest and the grinding wheel, resulting in a serious hand injury. Soft metals, such as copper or aluminum, should never be ground on a grinder. The swift expansion of soft metal under the frictional heat of grinding can result in an explosion.

RESPONSIBILITY

The student plumber must always remember the responsibility to other workers. Workers must warn people around them when they may be causing a potential danger. For example, if a worker is pouring molten lead, persons nearby should be warned of splattering molten lead.

Clean Work Areas

Work areas cluttered with tools, materials, and equipment are serious hazards. Safety engineers call these tripping, falling, or slipping hazards. Someone whose vision is limited when carrying a large object is a likely target for such accidents. Do not leave things lying about; pick them up.

Job Setup

When setting up a job, always think of safety. If most of the work is being done above the ground, use a scaffold (a temporary platform) rather than a ladder. If there is going to be an open trench, erect barricades to prevent people from falling in and being hurt. Check for proper lighting on the job. If the equipment or materials are heavy, obtain help carrying them.

People Working Below

Every worker has a responsibility to the people who work in the vicinity. When working overhead, the area underneath should be roped off if possible so that others are not injured by falling objects. But when this is not practical or possible, great care must be assumed to set things away from edges so that there is no danger of them falling on someone below.

Horseplay

Horseplay is always dangerous on the job or in the shop. When playing, you cannot pay full attention to danger. Many accidents happen every year because of horseplay.

PERSONAL SAFETY EQUIPMENT

Work Shoes

Plumbers move heavy tubs and other heavy equipment. When handling these, cuts through thinly covered shoes can occur, causing serious injury. But even lightweight fittings and tools can cut an improperly protected foot and cause loss of work. Lifting heavy objects is more comfortable and physically safer when sturdy shoes are worn. Invest in a good quality pair of leather work-type shoes. Work shoes with steel toes provide even better protection.

Safety Glasses

Molten solder and lead, concrete spatter and wood chips, and metal sparks from the grinding operation are all serious dangers to the plumber.

Figure 2–4 Your safety is increased with the use of safety glasses. (*Courtesy of Eastern Safety Equipment Co.*)

The great majority of accident-related blindness could be prevented with an inexpensive pair of safety glasses, Figure 2–4. Many people refuse to wear eye protection because of some initial discomfort. Adjust your glasses so that they are as comfortable as possible, while still being firmly in place. Then, persevere. In a few days, it will feel strange to be without them.

Buy a pair of safety glasses that have side shields. The blinking mechanism of the eye does not seem to react as well to particles approaching from the sides. Keep the glasses on at all times when on the work site. You will lose them and break them less often; but more importantly, you will be protected from the most dangerous flying object of all: the unexpected one.

Gloves

Gloves can be a great comfort in cold weather and a protection from minor cuts, scrapes, and burns. However, beware of the loose-fitting glove when working around revolving machinery. Many injuries occur when loose finger ends on gloves are caught up in machinery and the whole hand is drawn into the works.

Using leather gloves to pick up very hot materials will quickly render the gloves stiff and useless. Use a pair of pliers to pick up hot pipe assemblies and other hot items.

Ear Protection

Continual loud noise can cause permanent hearing loss. When working at the job site, machinery and vehicles that make loud noises are unavoidable, so wear ear plugs. They are inexpensive and effective.

Masks

When working at a job site where the quality of the air is questionable, air filter masks are a must. Several types can be used: from the paper variety to the air-flowing variety, which blows fresh air over a plexiglass mask. Your choice of mask should depend on the air contamination present at the time. Dangerous air quality would be evident, for example, if asbestos insulation was used to insulate a pipe.

Clothing

It is a good idea to have your own personal safety equipment if it is not provided by your employer; for example, a hard hat, Figure 2–5. The clothes that we wear on the job take a lot of punishment. The temptation here is to wear old and raggedy clothing. However, clothing that is oily can cause skin irritation and skin disease. Clothing that has ragged holes is more susceptible to fire and is, therefore, dangerous. Our work clothing is our first line of defense against cuts, burns, and abrasions. It is not wise to wear worn out or dirty clothing.

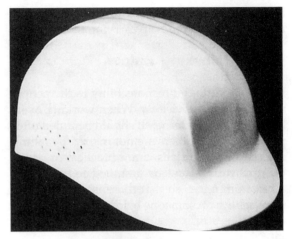

Figure 2–5 A hard hat is a good idea for some jobs. *(Courtesy of Eastern Safety Equipment Co.)*

FIRST AID

Occasionally, injuries will occur. When they do there is no substitute for having them looked after by a medically trained person; that is, a doctor or nurse. It is simply a good professional procedure to have even small injuries treated as soon as possible. This will prevent a small injury from becoming infected and causing a loss of earnings. It is also wise from the viewpoint of workmen's compensation or other medical insurance to have the date and time of the injury recorded. This will certify that the injury occurred on the job.

The student who intends to work at any of the construction trades should take an approved course in CPR (cardio-pulmonary resuscitation) and first aid at the earliest opportunity, Figure 2–6. These are occasionally given by your local Red Cross chapter. It is usually necessary to take a short refresher course at least once a year to maintain your Red Cross certification in CPR. You will find opportunity to use this training for either your own benefit or a fellow worker's benefit.

Figure 2–6 A well-stocked first aid kit. *(Courtesy of Eastern Safety Equipment Co.)*

SAFETY TIPS FOR THE PLUMBING SHOP

- ❑ Wear safety glasses and leather shoes.
- ❑ Pick up objects with the knees bent and the back straight.
- ❑ Clean up after completing a job.
- ❑ Obtain help to move a length of pipe.
- ❑ Use tools only for the purpose intended.
- ❑ Grind burrs off of chisels.
- ❑ Report all injuries no matter how small.
- ❑ Secure ladders that will be used in one spot at both the top and bottom.

- ❑ Request help when lifting heavy objects.
- ❑ Never run in the shop.
- ❑ Know where the fire equipment is.
- ❑ Never move things by throwing them.
- ❑ Do not wear gloves when grinding.
- ❑ Do not look at the arc on the electric welding machine.
- ❑ Tools should not be carried in the back pocket.
- ❑ Never strike hardened metal surfaces together.

◆ ◆

TEST YOUR KNOWLEDGE

1. When should gloves not be worn?

2. Can you make your own scaffolding?

3. Is it all right to paint a ladder with varnish?

4. Can the legs of a scaffold be supported on cement blocks?

5. How should you store paint thinner?

6. Are shatterproof sunglasses good enough for eye protection?

7. When should you make your first safety survey?

8. If you have only a short job to do, can you get away with someone just holding the ladder as you use it?

9. What concern should you have about aluminum ladders?

10. What is wrong with work pants that have a little hole in the knee?

FITTING SPECIFICATION AND IDENTIFICATION

KEY TERMS

adapters
branched fittings
bullhead tee
bushings
close nipple
compression joint
couplings
elbows
female end
flanges
hexagon bushing
long nipple
long-turn pattern tee
lug elbow, drop-eared
   elbow, eared elbow,
   drop elbow

male end, spigot end
pipe nipple
reducers
reducing elbow
return elbow
short-turn pattern tee
shoulder nipple
side outlet, three-way elbow
slip coupling
straight tee
street, service elbow
tucker or slip-and-caulk
union

OBJECTIVES

After studying this unit, the student should be able to:

❏ Identify the various patterns of fittings.
❏ Specify fittings correctly.

Learning about fitting specification and identification is somewhat like learning a language. Today there are a number of ways by which fittings can be specified. The elements are mentioned in different order in different manufacturers' catalogs. The system described in this unit is very common. However, there are variations within this system. There are different names for fittings from one job area to another. A wye may be called a Y-branch, a lateral, or simply a Y. It is important to learn as many of the names for the fittings as possible.

Fittings may change the direction of the pipe as in the case of an elbow, Figure 3–1. Fittings may also provide a branch to the main run of pipe. These include wyes, Figure 3–2, tees, Figure 3–3, and crosses, Figure 3–4. Some fittings, such as bushings and reducers, may change the size of a fitting or a pipe, Figure 3–5. Other fittings, such as couplings and unions, may simply connect pipe to pipe, Figure 3–6 and Figure 3–7. Fittings that change from one type of pipe material to another are called **adapters.** Figure 3–8 shows an adapter that goes from a screwed fitting to copper tubing.

Regardless of the material from which a fitting is made, it is generally called by the same name. Figure 3–9 shows Y-pattern fittings: the one on the left is for galvanized steel pipe; the middle one is for DWV copper tubing; and the

Figure 3–1 Elbow

Figure 3–2 Wye

Figure 3–3
Reducing tee

Figure 3–4 Cross

Figure 3–5 Bushing used
to reduce size of fitting

Figure 3–6 Coupling

Figure 3–7 Union

Figure 3–8 Adapter

one on the right is for ABS plastic pipe.

Union-type fittings, which make the pipe portable, come in many different forms. Figure 3–10, shows flanges. A flange is a type of union joint using a gasket and bolts. Figure 3–11 shows a compression joint taken apart. A compression joint is a fitting used to join tubing by means of pressure or friction. This is also a union-type fitting. Figure 3–12 shows a slip coupling. A slip coupling is used on traps and other easy to get at locations.

Figure 3–9 Y-pattern fittings

Figure 3–10 Flanges

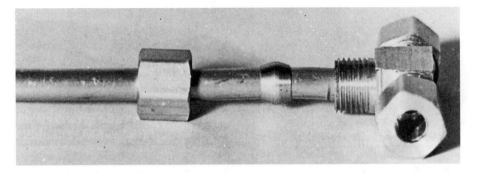

Figure 3-11 Compression joint

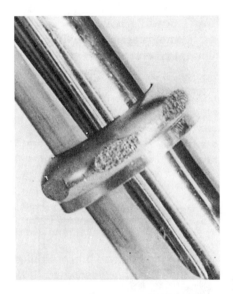

Figure 3-12 Slip coupling

MALE AND FEMALE END PREPARATIONS

When specifying fittings, the end of a fitting or pipe that enters into a fitting opening is called the **male** or **spigot end**. The fitting opening that accepts the male or spigot end is said to have a **female end** preparation.

FITTING SPECIFICATION

1–2″, 90°, elbow, galvanized malleable iron, screw, standard.

The quantity is given first on the specification. The size or sizes are given second. The number of degrees, if applicable, is given third. The fourth item is the fitting pattern name. The fifth is the material. Sixth, is the type of end preparation (screwed, sweat, butt weld, etc.). The weight of the fitting desired is the last item listed. In actual practice, some of the items are abbreviated:

1–2″, 90°, ell, galv. m.i., scr., std

When specifying a fitting with more than one size, the largest size is generally mentioned first. When more than one size is involved and different end preparations are also needed, the end preparation is given with the size. For example, 2 3/4″ FPT × 1/2″c × 1/2″c tee means that the plumber wants two tees with a 3/4-inch female pipe thread (FPT) on one end of the run. The other two outlets are to have 1/2-inch copper soldered into them. Notice that the fitting mentioned is not only a tee but also an adapter. Many fittings do more than one job.

A tee has three outlets. It is specified by giving the large size of the run first, the opposite end next, and the branch last. The run is measured straight through the tee; the branch is on the side.

A tee in which all three outlets are the same size is known as a **straight tee** and may be specified as one size, as a 1 1/4-inch tee.

A tee in which both ends are the same and the branch is smaller might be specified as a 1 1/2 inch × 1 1/4 inch tee.

A tee having a branch that is larger than the run is called a **bullhead tee,** and might be specified as a 3/4 inch × 3/4 inch × 1 inch tee. Notice that the larger size comes last in this case.

A cross has four outlets. Otherwise, it is specified the same way as a tee.

A wye has a branch on a 45-degree angle to the run and is read the same way as the tee.

Specifications for several different fittings follow:

1 – 1 1/4" 90° elbow, cast iron, drainage
1 – 1" × 3/4" × 1/2" tee, cast iron, scr., xh
1 – 2" × 1 1/2" reducing coupling, brass, scr., xh

In the case of the elbow above, the word "drainage" makes elements like galvanized or standard weight unnecessary. This happens frequently in fitting specifications.

PIPE NIPPLES

A **pipe nipple** is a short piece of pipe threaded at each end. Nipples are made from black iron, galvanized steel, or brass pipe. The inside diameters of pipe nipples measure the same as long pipe, that is, from 1/8 inch to 12 inches. Three types of nipples are now in use.

The **close nipple,** or all-thread nipple, Figure 3–13, has two standard pipe threads. One

pipe thread is cut from each end. A close nipple cannot be made by making a long thread on a piece of pipe and cutting off the necessary amount of thread. This nipple would not have the taper needed to make a tight joint.

The **shoulder nipple** has a short space between the threads, Figure 3–14.

The **long nipple,** Figure 3–15, is made in graduations of 2", 2 1/2", 3", 3 1/2", 4", 4 1/2", 5", 5 1/2", 6", 7", 8", 9", 10", 11", and 12" in length. The lengths from 7 inches to 12 inches long are not widely used.

To specify a pipe nipple, the inside diameter is given first, the length next, and the weight and kind last; for example:

1/2" × 3" galvanized steel nipple
2" × 4 1/2" black wrought-iron nipple
3" × 5 1/2" extra-heavy, black steel nipple

Figure 3–13 Close nipple

Figure 3–14 Shoulder nipple

Figure 3–15 Long nipple

Brass nipples may have a chrome or tin-plate finish.

Remember the following points regarding specification of pipe nipples:

❏ The pipe size is stated first.
❏ The length of the nipple is given next.
❏ The type of material is given last.
❏ If the nipple is of any weight other than standard, this information should be included before the listing of type of material. See the specifications for the 3 inch × 5 1/2 inch extra-heavy, black steel pipe nipple.
❏ For further information, consult the manufacturer's catalog.

ELBOWS

Elbows are pipe fittings made to change the direction of a pipeline. It is often necessary to offset a pipeline to avoid an obstruction. By using two elbows of the same angle, the pipe may be returned to the original direction.

Elbows are usually made of malleable iron. Black iron is used on air pipes, oil lines, or railings. Galvanized iron is used on water lines or outside piping to prevent rusting. Black cast-iron elbows, made with a heavy bead, are used on heating systems.

Brass fittings are made in standard and extra-heavy types. The standard fitting is similar to the malleable fitting in appearance. The extra-heavy fitting is similar to the cast-iron fitting with a large bead, Figure 3–16.

Most changes are made with 45-degree or 90-degree elbows. The 90-degree elbow shown in Figure 3–17 changes the direction of the pipe by 90 degrees, or places the pipe at right angles to its original direction. Elbows at a 45-degree angle are preferred because they cause less friction in the line.

The size of the elbow refers to the inside

Figure 3–16 Extra-heavy 90-degree elbow (Courtesy Stanley G. Flagg & Co.)

Figure 3–17 Standard 90-degree elbow

diameter of the pipe on which it fits. Elbows are made in the following sizes:

1/8"	1/4"
3/8"	1/2"
3/4"	1"
1 1/4"	1 1/2"
2"	2 1/2"
3"	4"
5"	6"
8"	

Figure 3–18 shows a 45-degree elbow. Notice that because this change of direction is only half as great as the 90-degree elbow, water flows more easily through such a line.

Reducing elbows, Figure 3–19, are made in sizes from 1/4″ × 1/8″ to 6″ × 2″. The reducing elbow not only changes the direction of the pipeline but also changes the size of the pipe used.

A street or a service elbow has one outside thread and, on the opposite end, an inside thread, Figure 3–20. It is made in 45- and 90-degree angles in sizes from 1/8 inch to 4 inches, and is also obtainable in reducing sizes in the 90-degree pattern. This elbow is made for close work because it is shorter than a close nipple and elbow.

Figure 3–21 shows a side outlet or a three-way elbow. This has an additional outlet 90 degrees from each of the others. It is mostly used on railings.

Figure 3–22 shows a lug elbow (also known as a drop-eared elbow, eared elbow, or drop elbow), which has two lugs for fastening. It is used most frequently in the installation of faucets for automatic washers and the riser pipe terminal for shower arms, but it can be used in any location where it is necessary to fasten an elbow securely.

The return elbow is a 180-degree bend used on pipe coils, Figure 3–23. It is made in three patterns: close, medium, and open. The difference is in the *spread*, or in the center-to-center distance, of the openings. The spread of a 1-inch elbow is:

Figure 3–18 45-degree elbow

Figure 3–19 Reducing elbow

Figure 3–20 Service elbow (*Courtesy Stanley G. Flagg & Co.*)

Figure 3–21 3-way elbow (*Courtesy Stanley G. Flagg & Co.*)

Figure 3–22 Lug elbow

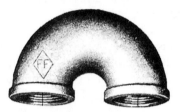

Figure 3–23 Return elbow

Close—1 1/2 inches
Medium—1 7/8 inches
Open—2 1/2 inches

COUPLINGS, REDUCERS, AND BUSHINGS

Couplings are straight fittings designed to join two pieces of pipe of the same size end to end. They are used when a full length of pipe is not enough. They may be obtained in all pipe sizes up to 12 inches. However, pipe with an inside diameter of over 6 inches is rarely used in threaded pipe work.

Bushings are used to reduce the outlets of boilers, tanks, etc. They have both inside and outside threads.

Black cast-iron (BC) bushings are used on heating, oil, or air lines. Galvanized malleable or brass bushings are used whenever there is a chance that rusting may occur.

There are three types of bushings: hexagon, face, and eccentric. The hexagon bushing has a six-sided collar. The face bushing is used when appearance is important. The eccentric or off-center bushing is used on radiators because the lowered center of one end of the bushing permits condensation to drain away, Figure 3–24.

Hexagon and flush bushings, Figure 3–25 and Figure 3–26, are made in a variety of sizes, starting with a reduction of 1/8 inch in the 1/4 inch × 1/8 inch size and going up to 10 inches × 2 inches with a reduction of 8 inches in size.

Reducers are used to reduce pipelines to smaller sizes. They are produced in sizes from 1/4 inch × 1/8 inch to 6 inches × 2 inches in standard malleable fittings, with a variety of sizes between. Consult the manufacturer's catalog for the sizes available. They may be obtained with one of several pipe size reductions.

Figure 3–24 Eccentric bushing

Figure 3–25 Hexagon bushing

Figure 3–26 Flush bushing

For steam lines, eccentric reducing couplings are used, Figure 3–27.

Caution is advised before using either bushings or reducers on gas installations. Local utility companies have different regulations regarding their use and may refuse service if the gas piping does not conform to their regulations.

UNIONS

Practically all pipe threads are right hand and must be turned in a clockwise direction to be tightened. When a pipe must be connected to a stationary fixture or boiler, it cannot be tightened into the fixture and the fitting at the same time without the use of a union. With a union, the fixture and fitting may be connected and disconnected without cutting the pipe.

Figure 3–27 Eccentric coupling

A union consists of three parts. Two parts are screwed to pieces of pipe, and a hexagon nut or collar is used to draw the two pieces together.

There are two types of unions. The gasket type requires a washer, Figure 3–28. The ground joint type has an iron-to-iron, Figure 3–29A, or a brass-inserted seat, Figure 3–29B, to make a watertight joint. Both types are made in sizes

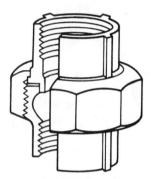

Figure 3–28 Gasket-type union

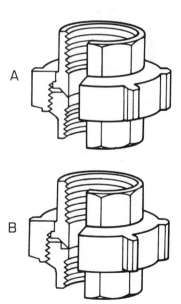

Figure 3–29 Ground joint-type union

from 1/8 inch to 4 inches and in different weights to withstand 150 to 300 pounds of pressure per square inch. Washers for the gasket-type union are made of various materials, such as rubber, leather, copper, or asbestos.

Black malleable iron unions are used for air, oil, and gas lines. Galvanized iron or brass unions are used where rusting is a problem. Some gas companies require the use of galvanized unions in their gas lines. Check local code requirements. Unions should be located close to the fixture or equipment so that as little pipe as possible is disturbed.

The three parts of a union are carefully machined to fit. Therefore, the parts from one manufacturer are not interchangeable with those from another. Always keep all three parts of a union together.

FITTINGS WITH BRANCHES

Branched fittings are used to bring two pipelines together into a single line or to install a branch line into an existing main line. These are called by a wide variety of names, depending on how they are used. But regardless of their specific name, they may be classified as a tee, a wye, or a combination of the two. A *tee* provides an outlet at a right angle (90 degrees) to the main run of pipe. A *wye* provides a branch outlet at a 45-degree angle to the main run of pipe.

Tees

Drainage tees are made in sizes from 1 1/4 inches to 8 inches. Reducing sizes range from 1 1/2 inches × 1 1/4 inches to 8 inches × 6 inches. All reductions are made in the branch opening. There are three patterns of tees; straight, short turn, and long turn or TY.

The straight tee, Figure 3–30, is permitted

Figure 3–30 Straight tee

only on vent lines. Otherwise, the abrupt turn would interfere with drainage.

The short-turn pattern tee, Figure 3–31, is used on vertical pipes. It is usually used in partitions and walls.

The long-turn pattern tee, Figure 3–32, is used on either horizontal or vertical lines where space permits and may also be used at 90-degree turns. The top opening in this case is used as a cleanout (an opening through which cleaning of the pipe may be done) and is left plugged.

All drainage tees are made in double patterns, such as the short-turn double pattern, Figure 3–33. Special tees are constructed with a

Figure 3–31 Short-turn tee

Figure 3–32 Long-turn tee or TY

Figure 3–33 Short-turn double pattern

Figure 3–34 Single Y

Figure 3–35 Double TY. (*Courtesy Stanley G. Flagg & Co.*)

bell on the top opening for caulking; this tee is frequently referred to as a **tucker** or a **slip-and-caulk** fitting.

Wyes

Because branches should enter drains at a 45-degree angle, wyes are used when the first fitting is some distance from the drain.

Drainage wyes and TYs are made in single and double patterns and in straight and reduced sizes. Notice that only the branch is reduced, Figure 3–34 and Figure 3–35.

Figure 3–36 shows how wyes are used. The 45-degree elbow could be turned either horizon-

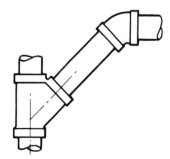

Figure 3–36 Wye connection

tally or vertically. A plug could be screwed in at the top of the wye for use as a cleanout.

VALVES, PLUGS, AND CAP

Fittings that stop or control the flow within pipelines are valves, plugs, and caps. Plugs and caps simply stop the flow. A plug threads into a fitting to stop the flow. A cap threads onto a pipe to stop the flow. Both of these fittings have one size specification; for example, 1/2" or 3/4", and so forth. The plug has a male thread and the cap a female one.

When a plug or cap is installed, it is meant to remain for an indefinite period of time. A valve, on the other hand, is installed to control the flow within the pipeline whenever it is called for. The valve may do this automatically or manually, depending on the kind of valve that it is.

Valves may have many names applied to them and the student of plumbing should be aware of many of the names. Some names for different, and sometimes the same, valves are:

Pave
Hydrant
Bib
Shut-off
Stop
Spigot
Valve
Cock

Even when they are called valves, there are globe valves, gate valves, needle valves, pressure relief valves, and check valves, to mention a few. Some of these will be more closely described in later units.

TEST YOUR KNOWLEDGE

1. What are fittings that reduce the size of the opening in another fitting called?

2. What are fittings that add a branch to an existing line of pipe called?

3. Two pieces of pipe may be joined by the use of what type of fitting?

4. Different pipe materials are joined together by the use of what kind of fitting?

5. Name a fitting type that joins pipe together but that also makes it easy to take the joint apart again.

6. What kind of fitting is bolted together to join pipe?

7. To identify a fitting is to name and perhaps elaborate on its type. To specify a fitting is to do what?

8. How long may a pipe nipple be and still be classified as a nipple?

9. Name two fittings that stop or control the flow in a line of pipe.

10. What is a bib?

UNIT 4

◆ ◆ ◆ ◆ ◆ ◆ ◆ ◆ ◆ ◆ ◆ ◆ ◆ ◆ ◆ ◆ ◆

ELEMENTARY SKETCH MAKING AND READING

KEY TERMS

dimensioning
isometric
orthographic
single-line technique

OBJECTIVES

After studying this unit, the student should be able to:

❏ Define the purpose of a simple piping sketch and its important elements.
❏ Define the different types of sketches.
❏ Explain how a sketch is made.
❏ Understand dimensioning and correct piping symbols.

A sketch is a simple drawing with a minimum of detail. Its purpose is to preserve information. Sometimes the information is for the maker of the sketch. Often it is passed on to someone else to make the actual installation. The maker of the sketch must always make it so that someone else will be able to make, or at lease understand, the installation from the sketch. It seems at the time that details of the job will easily be remembered when it is time to install the system. Frequently the time between sketch and installation is longer than anticipated. The information that was omitted from the drawing because one was in a hurry or was sure that one would remember often must then be rediscovered.

If the maker of the sketch, the draftsperson, always makes the assumption that someone else will be required to make the installation from the sketch, even if he/she knows otherwise, missing information and obscure drawings will be avoided.

Sketches are important to the pipe trades. A simple, clearly executed drawing can convey more information on one page than many pages of written words. The plumber often works from sketches.

TYPES OF DRAWINGS

Sketches may be orthographic or isometric.

Orthographic

An orthographic drawing, as it applies to building plans, separates horizontal piping from vertical piping. Piping which runs horizontally, for the most part is best represented with a *plan* view. For those who have studied *mechanical drawing* the plan view corresponds to the *top* view. Piping which runs mostly up and down (vertically), would be shown with an *elevation* view. A drawing of a room with the viewer's eye at the ceiling looking down toward the floor would be a *plan* view. If the viewer's eye was in the center of the room looking at pipe arrayed against one wall then this would be an *elevation* view. Again, for people with mechanical drawing this would correspond to a *front* or perhaps a *side* view.

Isometric

An isometric drawing shows piping both horizontally and vertically in the same view. In order to accomplish this the angles are distorted as in Figure 4–1. All the corners of the box which Figure 4–1A represents are actually 90 degree or right angles. But if a protractor were laid on this figure it would be discovered that none of the visible corners of the box are actually drawn at 90 degrees.

By distorting the angles the draftsperson can represent pipe systems which run both horizontally and vertically on one view. Because of its resemblance to a picture, this is often the drawing which is easiest for the reader to visualize, if the system is not complicated.

The isometric drawing is, however, the more difficult to draw. Keep in mind a few basic facts about isometric lines when doing this kind of a drawing.

1. *All lines which are vertical on the object being drawn are drawn vertically on the drawing.* Note: If the drawing is on a flat surface like a table, the term *vertical* would be indicated on the drawing by lines drawn toward

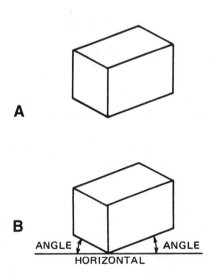

Figure 4–1 An isometric drawing of a box

the viewer and away from the viewer as it sits on the table.

2. *All lines which are horizontal on the object being drawn are drawn with a 30-degree slope on the paper.* Note: The lines which are meant to represent horizontal lines on the box in Figure 4–1B, are represented by lines which slope up from the lowest point in two directions. Which 30-degree angle used is determined by the direction the edge is running. If the box was positioned so that some horizontal lines ran in a north/south direction and some in an east/west direction, then, depending on the drawing's perspective, the north/south lines might slope up to the left and the east/west lines would slope up to the right.

Compare the two sketches Figures 4–2A and 4–2B. Figure 4–2A is an isometric sketch of a simple lavatory pipe layout, without measurements. Sketch 4–2B is an orthographic sketch of the same thing. Notice how much more clearly the isometric sketch shows the orientation of the 1 1/4" "P" trap. The orthographic drawing is more precise especially when dimensions or measurements are included. However, in order to show all of the dimensions necessary for installation the orthographic sketch requires another view. In the orthographic sketch in Figure 4–2B, would you be able to tell how far out from its horizontal waste line the outlet of the trap would be placed? Could you tell how far back into the wall the holes must be cut in the floor for the vertical supply lines to pass through?

The isometric drawing technique can create confusion if *crosses* are not supplied by the sketch maker where the piping penetrates walls and floors (these are called *pipe penetrations*). Notice that in the isometric sketch, Figure 4–2A, the draftsperson even makes imaginary lines, the crosses and the wall surface lines conform to the principle that edges which are vertical are

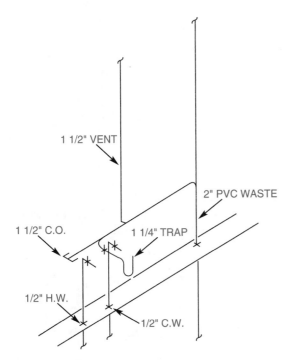

Figure 4–2A

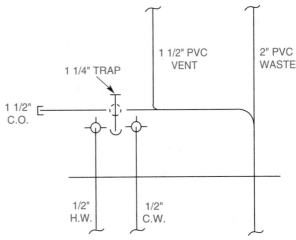

Figure 4–2B

represented by "vertica" drawing lines, and edges which are horizontal slope up to the left or up to the right on a 30-degree angle, depending on whether they run north/south or east/west. Dimension lines are also imaginary lines. They would also run according to the isometric drawing rules.

ELEMENTS

Sketches are ideally made on graph paper which aids neatness. However, some sketches will be made on opened-up lunch bags and other emergency drawing materials. Nevertheless a sketch must be neat and readable enough so that anyone can understand it. It must contain enough information so that the job can be accurately identified even if the maker of the sketch is not present. While sketches are rarely made to scale (the drawing being precisely proportional to the object being drawn), measurements placed on the drawing must be accurate and carefully recorded.

Drawings are subject to rough usage, and measurements can sometimes become difficult to read. Lettering style and the methods of entering dimensions should be given special attention. Building drawings which have feet and inch measurements will show the feet and inches with a hyphen in between, i.e., 7'-9 3/8". This is not a "minus" sign but a separator of feet and inches. Figure 4–2C shows a way to write numbers which results in maximum readability

for drawings even when faded or partially smudged. Experience has shown that this method results in consistently legible numbers even when applied with a paintbrush over concrete, wood, and steel. The small numbers and the arrows on this figure indicate the order and the direction the strokes should take for right-handed draftspersons. Left-handers should take note that pencil strokes are always toward the body or toward the palm of the hand. The student is urged to practice making numbers in this fashion until he/she is proficient.

HOW A SKETCH IS MADE

Sometimes the sketch is made from a larger, more complex drawing. The overall dimensions of a particular area, such as a cellar, are laid out on the sketch sheet. Then, the particular piping system is sketched in and a few important dimensions are noted. This becomes the working sketch for the plumber installing the systems. This usually occurs on new construction.

At other times, the dimensions of the space to be worked are obtained at the job site. This often occurs when the home has already been constructed and the plumbing will be some type of alteration.

The drawing is usually made on graph paper with a lead pencil. The lead pencil makes erasing and correcting mistakes easy.

A title block, Figure 4–3, is always included at the bottom of the sketch. All important infor-

Figure 4–2C

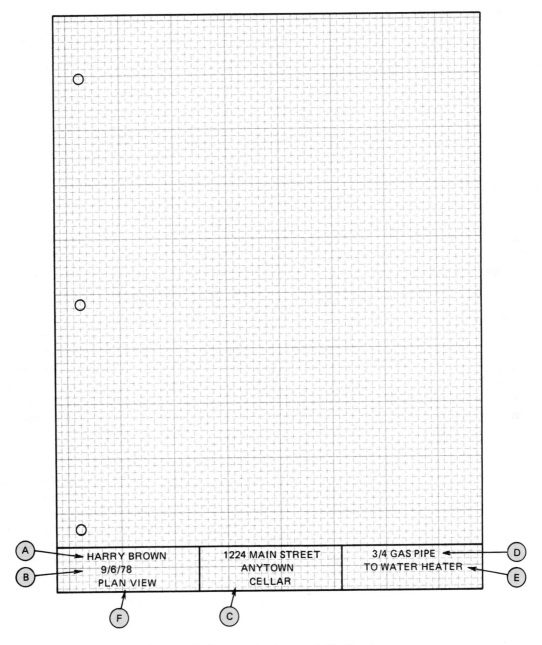

Figure 4–3 8 1/2 × 11 quadrille lined pape

mation is then located in one area where it can be easily picked out. To make a title block, draw a straight heavy line across the page about 1 inch to 1 1/2 inches up from the bottom. The title block can be divided further with short vertical lines to keep the information separated.

The title block should contain the following points:

A. the name of the person who made the sketch.
B. the date that the sketch was made.
C. the address of the building and the particular area of the building involved.
D. the material to be used.
E. the system being altered or installed.
F. the particular view shown.

There are four other points a good drawing must have, Figure 4–4.

❏ It must be readable by anyone familiar with sketches.
❏ It must contain all of the necessary measurements.
❏ A direction should be supplied to orient the reader in the building. An arrow placed on the floor plan pointing north is helpful, Figure 4–4.
❏ On elevation views, a notation should be made as to which direction is shown, for example, "The northeast laundry room wall."

Pipe sketches are made by using the single-line technique. This means that the pipeline is represented by a single pencil line. This line depicts the centerline of the pipe. Fittings are indicated by short crossing lines, Figure 4–5. On a plan view, a vertical pipe is depicted by a circle. When the pipe is facing up, the full circle is visible, Figure 4–6. When the pipe is facing down, part of the circle is obscured by the centerline of the pipe, Figure 4–7. On an elevation view, the circle is used for pipes that face toward or away from the viewer.

Figure 4–8 gives the single-line symbols for a wide variety of pipe fittings.

DIMENSIONING

Dimensioning refers to the placement of measurements on the drawing. Dimensions are usually taken from the centerline of the pipe or a fitting to the centerline of another pipe or fitting. They may also be from a pipe or fitting to a part of the building. It is often advisable to first make a rough sketch with the dimensions of the space being worked in on it. Then only the essential space dimensions should be put in the working sketch. Dimensions are indicated by the measurement being written on the drawing, Figure 4–9. The measurement is placed between two arrows as in Figure 4–9. It is understood that the measurement indicated goes from the point of one arrow to the point of the other arrow. Often a light extension line is drawn to clarify the point from which the measurement is to be taken. The extension line should not touch the object being measured. Observe that in Figure 4–9 the extension line does not quite touch the 90-degree elbow.

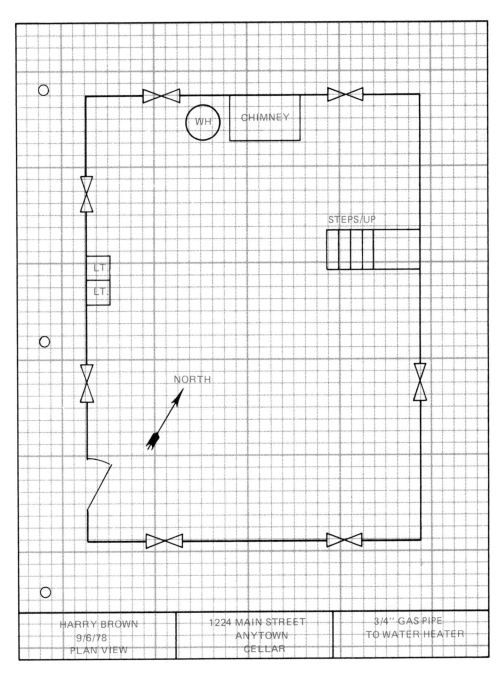

Figure 4–4

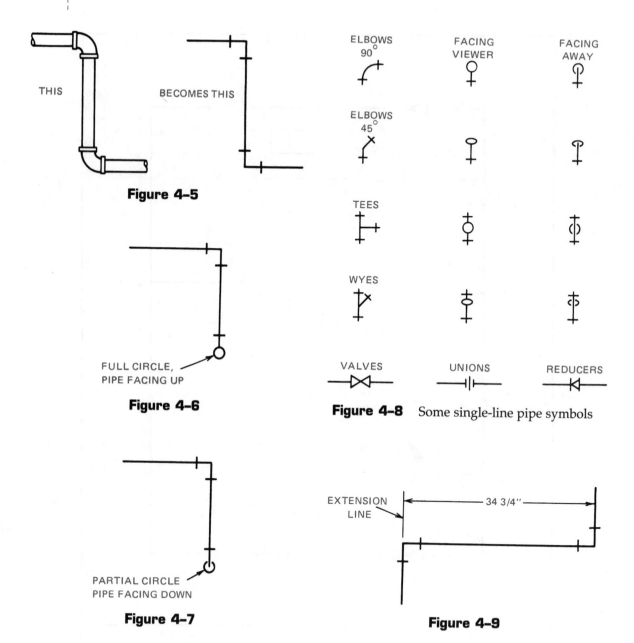

THIS

BECOMES THIS

Figure 4–5

FULL CIRCLE,
PIPE FACING UP

Figure 4–6

PARTIAL CIRCLE
PIPE FACING DOWN

Figure 4–7

ELBOWS 90°   FACING VIEWER   FACING AWAY

ELBOWS 45°

TEES

WYES

VALVES   UNIONS   REDUCERS

Figure 4–8   Some single-line pipe symbols

EXTENSION LINE

34 3/4"

Figure 4–9

TEST YOUR KNOWLEDGE

1. Who needs to understand a sketch?

2. What is the advantage of an isometric sketch?

3. The terms *top view, front view,* and *side view* in mechanical drawing correspond to which views in architectural drafting?

4. Aside from the location of the job and its orientation, what element on the sketch should be given the greatest care?

5. What do dimension, extension, and hidden lines have in common?

6. What should you include with the sketch's title?

7. What is a single-line pipe drawing?

UNIT 5

◆ ◆ ◆ ◆ ◆ ◆ ◆ ◆ ◆ ◆ ◆ ◆ ◆ ◆ ◆ ◆

BUILDING CONSTRUCTION

KEY TERMS

balloon construction
frost line

OBJECTIVES

After studying this unit, the student should be able to:

❏ Describe the reason for various construction practices.
❏ Identify the names of construction members.
❏ Define the parts of a building's construction.

RESIDENTIAL CONSTRUCTION PRACTICES

Much has been learned about making buildings that will be safe, dry, and warm, Figure 5–1. Many years ago it was felt that the proper home had to be constructed only from stone or masonry. Of course, vacation or summer houses and peasant cottages might be of rough log or sawn lumber construction. But, a real home was usually made with stone walls. The stone was very solid and safe but lent itself to a damp, cold interior environment in winter months. Modern frame-type construction was a result of a scarcity of craftsmen and materials. It was originally thought that frame construction would have a short life span. It had the advantage, however, of being inexpensive and quick. It has been discovered that, with the proper building practice, this type of home will last indefinitely. Indeed, many homes built before this country was called the United States of America are still in use

Figure 5-1

today. The term **balloon construction** was given to this construction method early in the country's development. This was a disparaging term used to describe the speed with which a house of this type could be built. While some still refer to all frame construction as balloon construction, strictly speaking, the term only applies when the corner posts of the building run from the base of the first floor to the roof line. Many barns are built in this manner, Figure 5–2.

FOUNDATIONS

It has been discovered through trial and error over many years that a building, if it is to last, must have a solid foundation that will not heave or shift. Many soil types will shift when damp. All soil types will heave when frozen. The power of freezing water to expand is considered limit-

less by some authorities. To avoid the problems associated with freezing and shifting soil, building codes specify that foundations must extend down below the frost line. The **frost line** is the depth below the surface to which the soil can be expected to freeze in winter, Figure 5–3. On deep shifting type soils, many times the only acceptable foundation is the floating type. This means that the foundation will be a solid, poured concrete pad upon which the house is built.

For the house to have a firm foundation, the soft upper soil is removed. This layer must be removed until (1) the depth is below the local frost line and (2) firm rock or undisturbed virgin subsoil has been reached. A concrete footing is then poured. The width of the footing is determined by the solidity of the soil at that point. A common ratio between the width of the footing and the width of the foundation wall is about 1 1/2 to 1. An 18-inch-wide footing

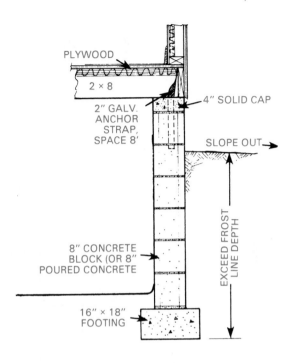

Figure 5-3

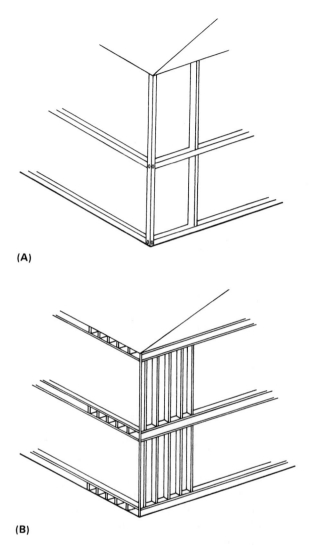

(A)

(B)

Figure 5-2 (A) Early balloon or barn construction, and (B) modern frame construction

might be used to support a 12-inch-wide foundation wall. The job of the footing is to distribute the weight of the house or building over a wider area. When installing pipelines, this footing should never be disturbed. If a pipeline must be run under a footing, an absolute minimum of earth should be removed beneath the footing. Even though great care in the excavation may have been exercised when the original foundation trench was dug, some settling of the foundation usually takes place. Therefore, the pipe should be run inside an iron pipe. This *sleeve* should be two pipe sizes larger than the line being run, Figure 5-4. The same sleeve arrangement should be used when running through a foundation wall. The pipe should be sealed inside of the sleeve with bituminous compound or a similar material. Remember that it is a responsibility of the tradesman that *the building will be as strong after any installation as it was before.*

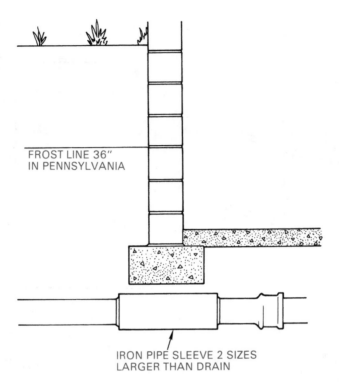

FROST LINE 36"
IN PENNSYLVANIA

IRON PIPE SLEEVE 2 SIZES
LARGER THAN DRAIN

Figure 5–4

THE PAD OR SLAB FOUNDATION

If a slab of concrete is poured for the weight of the house to rest upon, the weight of the house is distributed over a very large area. The system works well over soft soil types. The homeowner, of course, must live without a cellar. This type of foundation presents its own problems for the plumber. The main house drain and its branches must be installed and inspected before the slab is poured, Figure 5–5. Many times, water supply lines and even heat-circulating lines must be installed beneath the concrete foundation slab. These pipes will be very difficult to get to in the future. Therefore, the plumber must pay close attention to the following considerations:

❑ Is the material of high enough quality to last for the life of the building?

❑ Are the rough-in measurements exactly right? There can be no adjustment after the slab is poured.

❑ Will the various joint materials withstand the lime that will leach from the concrete?

❑ Are the pipes touching any of the steel-reinforcing rods in the concrete? Remember, dissimilar materials will cause electrolysis, which will corrode the pipes.

❑ Are there enough cleanouts? Are they properly placed?

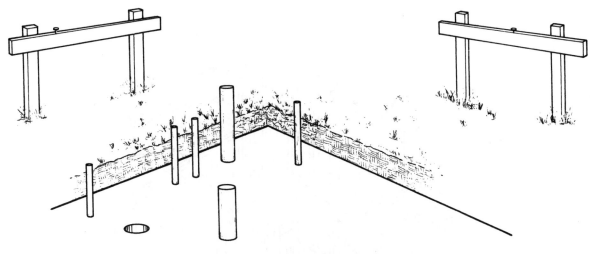

Figure 5–5

DECKS, ROOF, AND FRAMING

The student should examine the drawings carefully. In this manner, he/she will become familiar with the names of the parts of the construction. There are many names for the various members of a building, Figure 5–6. A few commonly used names that the plumber should be familiar with are presented here.

Base: The finish of a room where the wall joins the floor.

Bridging: Pieces of wood fitted between each joist to keep them from twisting.

Board Foot: The equivalent of a board 1 foot square and 1 inch thick.

Casing: The trimming around a door or window opening.

Concrete: An artificial building material consisting of sand, cement, gravel, broken stone or other aggregate, and enough water to cause the cement to set.

Conductor: Pipes that carry rainwater from the roof to the ground.

Fascia Board: This is where the rain gutter is fastened.

Flashing: This is usually a soft metal strip used to seal the roof from rain. Plumbers place a roof flashing on their vent pipes where they go through the roof.

Form: This is usually a temporary plywood structure in which the concrete is poured to form the footing and foundation walls.

Framing: The rough wooden structure of the house.

Furring: Thin wooden strips often used to make a wall thicker to accommodate pipes.

Header: A short joist that supports the ends of other longer joists. Also, timber used across the top of a window or door to form a lintel.

Joists: Timbers that support the flooring and the ceiling materials.

Knee Brace: This piece of timber goes from the joists in the attic floor to the rafters. It helps to support the roof.

Lath: This is most often a steel mesh that is laid over the studs to support the plaster in a wall. In older homes, this may be made from thin

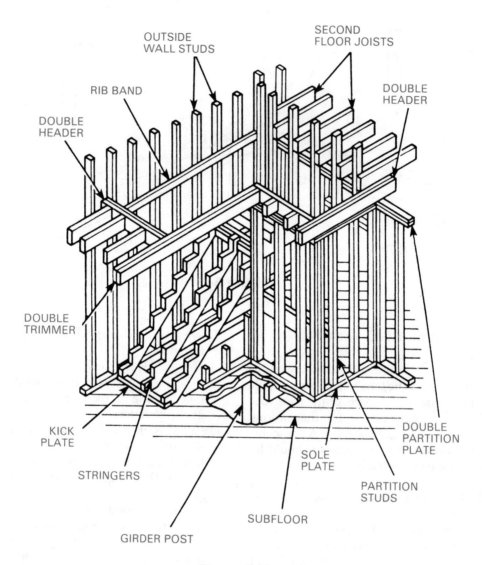

Figure 5–6

wood strips.

Ledgerboard: A horizontal timber fastened in place to provide a ledge. Some plumbers secure a ledgerboard to the studs in the bathroom wall to support the bathtub.

Molding: These are wood strips of various widths. They are used to give a finished appearance around doors, windows, and any rough joining.

Piers: Masonry supports.

Piles: Poles that are treated and driven into soft, swampy ground to provide a firm base for a footing.

Plaster: A mixture of cement, sand, and lime used to cover walls.

Plumb: This means on a vertical line.

Rafters: These timbers support the roof.

Scab: A short piece of lumber used to splice or support two pieces of wood.

Scaffold or Staging: A temporary platform erected to work at heights.

Sheathing: Any material used as a covering over wall studs. Sheathing is also placed over rafters to provide a base for roofing material.

Sills: These are horizontal timbers that support the framing above.

Studs: These are vertical timbers used to form walls and partitions.

Subfloor: This is nailed directly to the joists. The finished flooring is laid on top of this.

These are a few of many terms that the student will have to become familiar with to function in the residential plumbing trade.

◆ ◆

TEST YOUR KNOWLEDGE

1. Houses using frame-type construction use _____ as a primary building material.

2. Balloon construction was an early derogatory term for _____ _____.

3. Barn construction can be distinguished from modern frame construction by _____.

4. Foundations prevent buildings from settling in soft soil and _____ when the soil freezes.

5. Because a wider base gives greater support, a _____ is at least one and a half times as thick as the stone, block, or concrete wall it supports.

6. To prevent frost heave, a footing must be deeper than the local _____ _____.

7. Pipes running through foundation walls or under footings must be protected by a/an _____ two pipe sizes larger than the pipe.

8. To distribute the weight of a house evenly over a large area in soft soil, a concrete _____ may be used as a foundation.

ELEMENTS OF BLUEPRINT READING

KEY TERMS

bid
elevation
front elevation
plan view
scale

OBJECTIVES

After studying this unit, the student should be able to:

❏ Explain the different kinds of drawings.
❏ Visualize the completed building.
❏ Find architects' specifications.
❏ Make pipe sketches from building plans.
❏ Describe the material requirements from a blueprint.

INTRODUCTION

Any building craftsman must be able to read blueprints. Much of the information necessary to accomplish the job must be gleaned from sketches or blueprints. As is often the case, the plumber may find that the only direction that the draftsman has supplied for the plumber is the location of fixtures and equipment. It is left up to the plumber to properly route and install the rough-in piping. The plumber may then take his or her copy of the drawings and sketch in pencil where the pipes must go. This might seem more simple than it really is.

Many young craftspeople are dismayed upon opening a blueprint and finding so much information there. Many times their first impression is that they will never be able to read blueprints. It takes time to understand a building drawing. It takes careful inspection and study to get all of the information necessary to do the plumbing in a house. To make quick judgments about the placement and routing of piping can be disastrous. A particular wall may be too narrow to contain the proposed vent line. Or perhaps there will be insufficient fall if a pipe is run in the wrong location. Blueprint reading requires great concentration and methodical search methods.

The plumber needs to know a lot about the construction of the building so that he or she can plan the kinds of installations that would be appropriate. Some questions that the plumber would have in mind are as follows:

- ❏ How much space is in the walls?
- ❏ What are the construction materials?
- ❏ How is the building situated on the site?
- ❏ How thick will the wall and floor covering materials be?
- ❏ Where will the fixtures and equipment be located?
- ❏ In which direction do the joists run?
- ❏ Where will the service piping be located?

These are just a few of the questions that need answering. The blueprint can answer many of these questions for the person who can read it. Reading drawings is not difficult, but it takes a good deal of patience.

Throughout this unit we will be using some terms that may be unfamiliar. For the sake of clarity, we will define some of them now.

Architectural Drawing: The drawings, or the blueprints, for the building. This includes a number of sheets of plans and elevations.

Blueprint: At one time, blueprints were actually white lines on a blue background or blue lines on a white background. The copying machines that turned the draftsman's inked drawings into "blueprints" were responsible for this. Now, "blueprints" can also be black or white or any combination. Any formal drawing made to build or manufacture a product can be called a blueprint.

Commercial Building: A structure designed or sales, rental, or manufacture of a product. A building meant to provide office or warehousing space. An apartment building or condominium. In other words, a building to either house a business or to provide a service to the public.

Detail: An area of the drawing set aside to make a part of the main drawing more clear. It is usually enlarged and shown from a different angle.

Elevation Views: This is a portion of the drawing that shows a vertical surface of the building.

Notes: Notes are distributed throughout the drawing. They often contain vital information for the craftsman.

Plan Views: Technically, a plan shows a horizontal surface of the building, for instance, a floor plan. However, the term "plans" is often used to refer to all of the architectural drawings.

Residence: The place where people live. An apartment house could answer this description. But the term "residentia" when referring to blueprints usually designates a one- or two-family home.

Rough-in Sheets: These are small but formal drawings of fixture layouts. They are provided by the manufacturer of the fixture. They show the craftsman all of the essential measurements that are needed to install the pipelines prior to the arrival of the fixtures themselves.

Scale: This is a ratio between the size of the drawing and the size of the building. A scale of 1/8" = 1'-0" means that a distance of 1/8 inch on the drawing is equal to 1 foot in the actual building.

Schedule: One of a number of charts placed on the drawing to display information about materials or design loads to make this information easily available.

Section: This is a detailed drawing as in "cross-section." A portion of the main drawing is "sliced" and then turned at a 90-degree angle to make its construction clearer.

Specifications: Usually these are found in a separate report that is meant to accompany the drawings. They provide detailed information about materials, requirements, and workmanship. The specifications, applicable codes, and the blueprints provide the basis for the contract between the builder or owner and the tradesman. Specifications can also be found on some drawings under "General Notes."

Working Drawings: These are all of the plans and elevation sheets that make up the "blueprint."

The specifications and the blueprints form the basis upon which estimates or job bids are made. If the tradesman overlooks a specification, he or she may very well be called upon to correct the job to conform to specifications out of his or her own pocket. Careless reading of the blueprints can also create problems. The plumber must make a materials list by looking at a blueprint that may not have the actual piping drawn on it. From this list and his or her experience with time requirements for different plumbing jobs, the plumber comes up with a bid. This is submitted to the builder or general contractor with bids from other plumbers. The plumber who submits the lowest bid price and who will also conform to specifications will usually get the job. If the plumber submits a bid that is too high he or she will not get the job. If on the other hand the bid is too low, he or she will get the job but will not make a profit or perhaps even lose money. Errors of blueprint reading are almost always costly.

Plumbers must submit drawings to the inspecting authority. Many architectural drawings do not have the piping systems indicated. Therefore, the plumber is called upon to do the drawings. Drawings must also be provided for the craftsmen who are to work on the job. There are computer programs available to aid in the production of accurate drawings. Advantages of using computer-aided drafting programs include:

❑ Neat, clean drawings.
❑ Information to produce a drawing is stored on a magnetic disk. Therefore, a fresh unmarked drawing can be produced quickly and easily.
❑ Changes are automated by the program and are easily accomplished.

VISUALIZATION

To visualize something, so far as the craftsman is concerned, is to understand enough about an object or a design to make an exact copy; in other words, to know everything about it structurally. Listening to a spoken description of a plumbing job will only do for the smallest projects. Having an accurate drawing gives us more information about an object or a building than if we had the object or building before us.

To properly visualize a building, some skills must be learned. First, let us define two terms: vertical and horizontal. A vertical line goes up and down. A wall is constructed on a vertical plane. Horizontal is parallel to the earth's surface. A floor is constructed on a horizontal plane. When you are examining a plan view, you are looking at a drawing of something that lies mostly horizontally. Looking down at a roof or a floor drawing would be looking at a plan view. Elevations are drawings of objects in a mostly vertical plane. Looking at a drawing of the front wall of a house might be described as looking at the front elevation. Drawings of buildings are composed of plans, elevations, and details.

SKETCHING

A sketch is an informal drawing often made directly onto the blueprint using a 6-foot folding rule as a straightedge. Or if more than this is needed, a piece of tracing paper might be placed over the blueprint and a penciled sketch made on that. The drawing on the tracing paper can then be copied on common office duplicating equipment. Often, a sketch is not more than a line drawing on a paper bag. While the proportion and the angles of a sketch may be incorrect, the measurements indicated must be as accurate as possible.

DRAWING

Building plans are generally drawn in what is called *orthographic projection*. This kind of drawing shows plans and elevations in separate views in the manner that has been presented here. There is another type of drawing called *isometric projection*. This type distorts the angles of the object in order to show three views in one drawing. The isometric drawing, while being harder to draw, is generally much easier to visualize. Complex drawings will show more detail if done in orthographic projection.

Essential tools needed for doing presentable drawings are not very complicated or expensive. Some basic drawing tools are:

1. Straightedge. This can be a 12-inch rule. In this way it can serve two purposes: to measure on the drawing and to draw straight lines.
2. Pencils. A couple of pencils of different softness.
3. Triangles. 30, 60, 90 triangles and a 45, 45, 90 triangle.
4. Tee square, for drawing parallel lines.
5. Large, soft eraser for erasing and cleaning the drawing.
6. Erasing shield. So that lines that are to remain are not inadvertently erased.
7. Scaling ruler or architect's rule.
8. Pencil sharpener.
9. Drawing compass, for drawing circles.
10. Protractor, for laying out angles.

Isometric projection, Figure 6–1, is used widely for schematically showing small piping areas and also for large systems. Many plumbing inspectors ask for an isometric sketch before issuing the plumbing permit. As has been noted, three views of an area can be shown with one isometric drawing. When making isometric drawings, the student should keep in mind lines that are actually vertical are drawn vertically. However, *every* horizontal line on the object or in the space being drawn is going to be on a 30-degree angle from the true horizontal of the drawing. This includes imaginary lines like extension lines; dimension lines; lettering lines; x's, which denote wall or floor penetrations, and any other line that would be horizontal as we visualize our drawing. A little thought will show that horizontal lines from any corner of the room can run in two directions that are at 90 degrees to each other. On the drawing this will be shown as slanting the 30-degree line up from

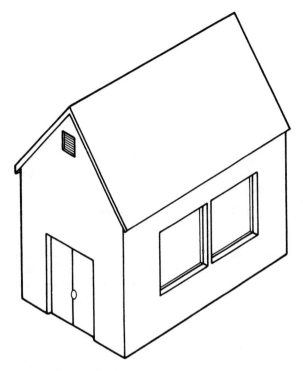

Figure 6–1 An isometric drawing

the left or up from the right, depending on whether the line is running east and west or north and south.

A handy way to start an isometric drawing is to start with an inverted "Y" figure. The stem of the "Y" should be vertical and the two branches should be drawn 30 degrees from the horizontal in two directions. Now when making the drawing over this figure *all* horizontal and vertical lines should be parallel to one of these lines. It helps to draw the x's denoting pipe going through a wall or a floor before drawing the actual pipe line.

SCALE

When the term scale is used in blueprint reading, it is a reference to the size of the drawing compared to the size of the object or building, which the drawing represents. While scale can be expressed as a ratio, i.e., 1:12, on architectural drawings, this would be described as 1″ = 1′-0″. The reader of this blueprint would then know that an actual measured distance on the drawing of 1 inch would be equal to 1 foot or 12 inches in the building. The dimensions or measurements that are written on the drawing are measurements of the actual building. When necessary dimensions are not on the drawing, the reader can calculate the distance that would be in the actual building by placing a rule or tape measure on the drawing itself. A *scaling rule* or *architect's scale* is marked in such a way that when it is placed on the drawing the scale will indicate actual building dimensions.

VIEWS

Building plans are given in a number of sheets. Ordinarily, each *view* is placed on a sheet by itself. There will usually be floor plans for each story of the building, and at least one front elevation and one side elevation. The side of the building facing the street is usually given the designation front elevation. When a right side or a left side elevation is given, the reader must imagine standing at the front of the house. Then, if he or she walked to the right side of the building and faced that wall, he or she would be looking at the right elevation. Three views are the minimum number in orthographic projection drawings to safely indicate all necessary measurements. The student should take note that at least three dimensions are required to locate any given point. These are:

Up and Down: Above or below a reference point that may be the earth's surface or the floor level of the space being worked in or even down from the ceiling.

Side to Side: This could be represented by the directions east and west or north and south.

Nearer and Farther: This could also be represented by the directions east and west or north and south. However, if side to side is east and west, this must be north and south. Or, if side to side is north and south, this must be east and west.

The student should also take note that all of these directions are at right angles, or at 90-degree angles, to each other. When making a sketch, a question should be constantly asked: "Have I indicated the location of this point in three directions?"

SYMBOLS

Some line symbols used in drafting are:

Centerline: This is used to indicate the center of an object.

Hidden line: This indicates that the line cannot be seen from the vantage point taken but is behind the surface.

Cutting plane line: This indicates where the drawing has been cut in order to show a detail.

Break line: This indicates that the object is longer than what the drawing shows. The part between the break lines has been removed to make the drawing clearer.

Dimension line: This indicates a measurement of the object. It is understood that the measurement is from the point of the arrow at one end to the point of the arrow at the other end.

1-11/16"

Extension or witness line: When it is impossible or inconvenient to have the dimension arrow terminate at a line representing the outline of the object drawn, the draftsman will draw a witness or extension line as an imaginary extension of the outline. The arrow may terminate at this line. There must always be a space between the end of this extension line and the object to demonstrate that it is not a part of the object.

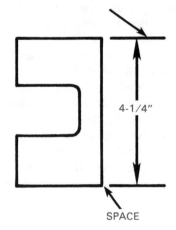

EXTENSION LINE

4-1/4"

SPACE

DIMENSIONS

Because it is understood that the dimension indicated is always from the point of one arrow to the point of another arrow, dimensions can be indicated in a number of ways.

GETTING INFORMATION

Because all notes on the blueprint as well as the architect's specifications are a part of the contract, the plumber must have a firm understanding of the requirements as well as the applicable plumbing code before making the materials list. As the *take-off sheets,* or materials list, are prepared, constant reference to the specifications and codes and blueprints may need to take place. Every wall or floor that is to be penetrated by pipe or is to support pipe must be completely understood. When a trench is to be dug, the question must be asked: "How deep?" Is the grade line, or the surface of the ground, in evidence going to be the finished grade line? Will there be danger of the pipes freezing? The answers to these questions and similar questions about every line of pipe in the installation can be found on a properly prepared set of architectural drawings.

ARCHITECTURAL SYMBOLS

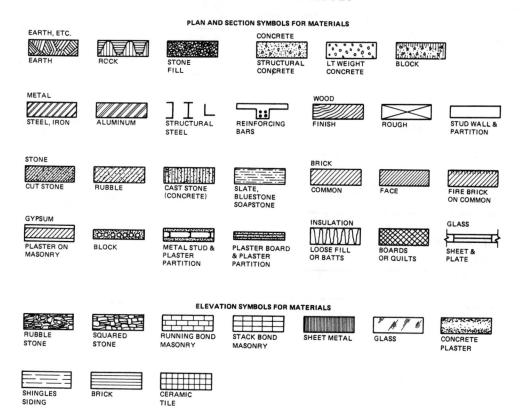

ELECTRICAL SYMBOLS

SWITCH OUTLETS
S - SINGLE POLE SWITCH
S_2 - DOUBLE POLE SWITCH
S_3 - THREE WAY SWITCH

S_4 - FOUR WAY SWITCH
S_D - AUTOMATIC DOOR SWITCH
Scb- CIRCUIT BREAKER

CONVENIENCE OUTLETS
DUPLEX OUTLET
WEATHERPROOF
RANGE OUTLET
SPECIAL PURPOSE

LIGHTING PANEL
POWER PANEL
POWER TRANSFORMER
PUSH BUTTON
TELEPHONE

GENERAL OUTLETS
CEILING   WALL
OUTLET DROP CORD
PULL SWITCH
JUNCTION BOX

HVAC SYMBOLS

EXPOSED RADIATOR
RECESSED RADIATOR
THERMOSTAT
THERMOSTATIC TRAP

THERMOMETER
PRESSURE GAUGE
RELIEF VALVE

FLOAT TRAP
BOILER RETURN TRAP
REDUCING PRESSURE VALVE
AUTOMATIC AIR VENT

PIPING SYMBOLS

VENT
COLD WATER
HOT WATER
HOT WATER RETURN
SOIL OR WASTE (ABOVE GRADE)
SOIL OR WASTE (BELOW GRADE)

PIPE FITTING SYMBOLS

VALVE
HOSE BIBB — HB
90° ELBOW
45° ELBOW
TEE
CLEANOUT — CO
FLOOR DRAIN — FD

PLUMBING FIXTURE SYMBOLS

RECESSED TUB		CLOTHES DRYER	D
ROLL RIM TUB		WASHING MACHINE	WM
ANGLE TUB		WALL – TYPE DRINKING FOUNTAIN	
SHOWER STALL		DRY WELL	DW
WATER CLOSET		WATER HEATER	WH
BIDET		KITCHEN SINK R & L DRAIN BOARD	
		COMBINATION SINK AND DISHWASHER	DW
		COMBINATION SINK AND LAUNDRY TRAY	S T
URINAL STALL TYPE		HOT WATER TANK	HWT
LAVATORY		WATER METER	M

SUMMARY

Blueprints, or architectural drawings, can be intimidating. At first glance, there is a welter of information. Occasionally, the plumbing and pipefitting are only partially illustrated, and from experience, the craftsperson is expected to deduce a great amount of information. It is not unusual that the "plans" are drawings representing a "typical" home or a particular building and the actual layout of the services must adapt to the local building site and conditions.

The apprentice is advised to borrow a set of plans and to take them home after work. Follow out a single system on the plans. Starting with the lowest plan view that shows your system of choice (usually the basement), for example, choose the soil pipe. Determine where the terminal ends are on the particular plan view that you are following, then determine where the pipe becomes vertical to accommodate the next higher story. Go to the next higher plan view, perhaps the first floor plan view, and do the same thing. Do not be distracted by the large amount of information shown; simply follow the system that you have elected to learn. Occasionally, you will discover that the pipelines increase in number as the floors get higher in elevation. This is because they branch out in the vertical spaces between the floors or stories. This can be seen on the plan views but sometimes the elevation views show it better. When you have located the terminal ends, you might go back and look at the elevations anyway to discover how and where branching occurred.

Locate the notes, schedule, and specification areas to discover the materials that should be used on this system. If you are patient and trace out one system at a time, you will learn how to read blueprints quickly as well as learn much about plumbing.

◆ ◆

TEST YOUR KNOWLEDGE

1. What blueprint views correspond to the mechanical drawing top, front, and side?

2. What kind of projection is used to make a traditional set of building blueprints?

3. How many dimensions are needed to locate anything in a building?

4. Can you name three different kinds of invisible lines?

5. The relationship between lengths on the drawing and lengths in the actual building is known as _____.

6. What drawing in the packet of blueprints would you first check to find the dimensions of the home's front entryway?

7. Name a kind of drawing that can show all of the required dimensions in a single view.

8. Architectural drawings are also called _____.

◆ ◆ ◆ ◆ ◆ ◆ ◆ ◆ ◆ ◆ ◆ ◆ ◆ ◆ ◆ ◆ ◆ ◆

PLUMBING
MATH

KEY TERMS

chalk line   6-foot folding rule
English style   soapstone
level   square
metric style   steel tape
plumb bob

OBJECTIVES

After studying this unit, the student should be able to:

❑ Take measurements accurately.
❑ Add and subtract measurements expressed in feet, inches,
  and fractional parts of an inch.
❑ Multiply and divide feet and inches by a decimal constant.
❑ Express all answers in on-the-job usable form.

A thorough understanding of basic mathematics as it applies to plumbing is necessary. Some of the methods used by the plumber may vary from arithmetic as it is learned in school. Most of the changes are the result of the plumber's measuring tool, the 6-foot folding rule.

Measuring accurately is essential in producing high-quality work. Jobs proceed in three steps.

1. The space in which the system is to be installed is measured.

2. Calculations are made to find the actual pipe lengths.
3. Measuring is used again in preparing to cut the pipe and install the system. Because errors occur as the job proceeds, it is important that the first measurements are taken as skillfully as possible.

TOOLS

The tools most commonly used in measuring are the 6-foot folding rule, the steel tape, the level, the framing square, the plumb bob, and soapstone or a marking pencil.

Six-Foot Folding Rule

The measuring tool used most frequently is the 6-foot folding rule. The student plumber must learn to use this tool to the limit of its accuracy. To use the rule, it is simply laid on, or held along the distance to be measured. The end of the rule is lined up with one end of the object and the length is read off to the closest division on the rule.

The smallest graduation on the rule is 1/16 inch. The fractions 1/8, 1/4, and 1/2 are also included. The rule is usually numbered from 1 to 72 inches. Some 6-foot folding rules are marked off in feet and inches. This unit will use the more common rule, marked off in inches.

The inch lines of the rule are drawn completely across the rule, Figure 7–1. The 1/2-inch lines have a small break in the center, Figure 7–2. The 1/4-inch lines divide the inch into quarters. The 1/4-inch and the 3/4-inch lines have a bigger break in them than the 1/2-inch lines. The 1/2-inch line is also a 1/4-inch line (2/4), Figure 7–3. The 1/8-inch lines are shorter and the 1/16-inch lines are the shortest of all, Figure 7–4. Every line on the rule represents 1/16. Only the shortest lines are called 16th frac-

tions. A measurement of 2/16 would be called 1/8. A measurement of 4/16 would be called 1/4, and so on. Look at Figure 7–4 and read from left to right including all fractional lines, 1", 1-1/16", 1-1/8", 1-3/16", 1-1/4", 1-5/16",

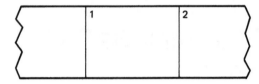

Figure 7–1

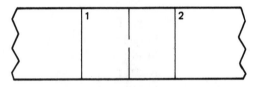

Figure 7–2

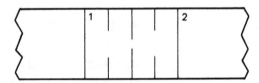

Figure 7–3

Figure 7–4

1-3/8", 1-7/16", 1-1/2", 1-9/16", 1-5/8", 1-11/16", 1-3/4", 1-13/16", 1-7/8", 1-15/16", and 2". Notice that the top numbers of all the fractions are odd numbers. An even number on the top of a fraction means the bottom number is too large.

Steel Tape

The **steel tape** is a light flexible blade with the measurements marked on it, Figure 7–5. It has the advantage of containing the full range of feet and inches in a small light tool and it may be hooked to the end of a board or pipe and read off at the other end quickly and easily. It has the disadvantage of being unable to support its own weight. Therefore, a measurement in midair can be difficult.

Level

The **level,** Figure 7–6, is used by the plumber to make surfaces either vertical or horizontal.

Some levels may also indicate a 45-degree angle. The level is placed upon or against the object being leveled. The plumber observes a bubble in a liquid-filled glass chamber. The position of the bubble tells the plumber how well the surface is aligned.

The 2-foot level is an ideal tool for making vertical or horizontal reference lines on a wall. A measurement from the floor is marked using the 6-foot rule. The level is placed under this mark, the bubble is centered, and a perfectly horizontal line is drawn along the top of the level.

Framing Square

The **square**, Figure 7–7, forms an accurate 90-degree angle. Although it is used more by the pipe fitter, it is also used by the plumber. It may be used for making square cuts on pipe and lumber, setting flanges, and welding fittings. Making pipe hangers for the larger pipe sizes is made easier by the steel square. It is marked in inches on both the inside- and outside-right angle. This may be used along with the 6-foot folding rule to find the long side of a right triangle.

Figure 7–5 Steel tape

Figure 7–6 Level

Figure 7–7 Steel square

Plumb Bob

The plumb bob is, in its simplest form, a string with a weight on one end, Figure 7–8. To use the plumb bob, one end is fastened to a solid object and the weight is allowed to hang. The string will be a vertical reference line. The weight will also indicate a spot directly below the fastening point.

Chalk Line

The chalk line, Figure 7–9, is used to make a straight line on a vertical or horizontal surface. It consists of a strong string that rolls up into a container called a chalk box. The chalk box is filled with powdered chalk. The string, which is dusted with chalk, is stretched tightly against the surface. The plumber then plucks the string using the thumb and forefinger. An accurate line of chalk will be left upon the surface.

Soapstone

Soapstone is a flat marking tool somewhat like chalk. It leaves a white mark. It may be sharp-

Figure 7–9 Chalk line

ened by being rubbed against a rough surface. The plumber should carry a soft pencil for marking on light surfaces and soapstone for marking on dark surfaces.

ENGLISH STYLE VS METRIC STYLE

The 6-foot folding rule produces what is called English style measurements. Throughout most of the world, the metric style measurement is used. Americans have demonstrated a great reluctance to change from the familiar English measurement despite much pressure to do so. Because most trades and fellow workers, architects, suppliers, and building codes express distances this way, and the individual plumber needs to be able to communicate measurements, there is little advantage to be gained, at present, from learning *metric* measurements.

Insofar as applying mathematics to our measurements, the metric system is far superior to the English system. It would almost be worthwhile to learn the metric system and to measure with a metric rule when doing on-the-job calculations. If it is needed to then communicate our answers back to someone else, our answers would need to be converted to English style measurements.

Figure 7–8 Plumb bobs

But conversions are the problem. Conversions are what cause the difficulties and most of the mistakes.

A COMPARISON OF METRIC AND ENGLISH UNIT CONVERSION

The *metric* measurement is based on divisions of *ten,* and makes all *fractions* decimal-type fractions. So 1 and 3/10 *meters* will be expressed as 1.3 meters. This makes difficult conversions easy. A *decimeter* is 1/10th of a *meter* (again, always in multiples of 10).

Converting a metric measurement

Problem: How many decimeters would there be in that 1.3 meter measurement?

Solution: Multiply by 10. And to multiply by ten, all that we must do is move the decimal point one place to the right. Therefore 1.3 meters becomes 13 decimeters. Conversions are quite simple. The following is a comparable English style conversion.

Converting an English Measurement

Problem: Convert 1 and 3/10 feet to inches.

Known: There are 12 inches to 1 foot.

First conversion: 1 foot is equal to 12 inches. So there are at least 12 inches.

Second conversion: How many inches are there in 3/10 feet?

$12 \times 3/10 = 3\ 6/10$ inches.
Add these inches to the 12 that is known,
$12 + 3\ 6/10 = 15\ 6/10$ inches.

Problem: We know that 1 3/10 feet is equal to 15 6/10 inches but 6/10 of an inch can't be found anywhere on our *English measure rule!* It's divided into 1/16s, 1/8s, and 1/4s and 1/2s. So ...

Problem: Convert 6/10 inches to the nearest measurement on the rule.

Known: Because 1/16s are the smallest fractions, it is easy to convert 1/16s to any of the other fractions on the rule. It is known that there are 16 sixteenths in an inch. So ...

Third conversion: Find out how many sixteenths there are in 6/10 of an inch.
$16 \times 6/10 = 9\ 6/10$ sixteenths.

Solution: In the second conversion we discovered that 1 3/10 feet was equivalent to 15 6/10 inches. But the fractional part of the answer could not be used. So put down **15 whole inches.** In the third conversion we discovered that the nearest sixteenth fraction to 6/10 of an inch is **10/16.** Why 10/16? Because the real answer 9 6/10 sixteenths is closer to 10/16 than it is to 9/16 and so we must take the higher fraction of 10/16. So now we have the answer, right? Well, yes and no. Even though 15 10/16 inches is technically correct to the nearest sixteenth, it is proper usage to reduce all fractions to their "lowest term." So ...

Fourth conversion: Convert fraction to its lowest term. 10/16 is equal to 5/8.

The final answer: 1 3/10 feet is equal to 15 5/8 inches.

> **HINT:**
> To reduce a fraction to its lowest term (for fractions that are found on the English rule), keep dividing the top and then the bottom by 2 until the top number is **odd.**

We have become used to these conversions, but as you can see the *metric* system conversion is much easier. Because the English system grew out of such things as the length of a man's foot and the distance between two knuckles on the thumb, for every conversion the craftsperson had to memorize a fairly complex unit relationship. There are 36 inches in a yard, 5280 feet in a mile and so forth. With the metric system, which is based on the length of a meter, we only need to know the order of the units. As the unit gets larger it increases in size by 10 times.

The user of the English system still has to frequently mix the two styles when a constant has to be used. Constants are almost always in decimal (using 10 for a base) form. The formulas for a circle frequently use 3.14159 (represented by the Greek letter π). Sometimes trigonometric functions must be used. These are always in decimal form. The craftsperson must convert the measurement to metric form, multiply by the decimal constant, and then convert back to English style.

A QUICK OVERVIEW OF A COMMON CONVERSION PROBLEM

Problem: Calculate the length of a piece of pipe which will offset the pipeline 6'-11 3/4". This offset must be at a precise angle of 45 degrees. The secant or cosecant of 45 degrees must be used; multiply this by the amount of the offset. So the problem becomes
6'-11 3/4" × 1.414

First, convert feet to inches and add it to the rest of the inches.
6'-11 3/4" becomes 72" + 11 3/4".

Second, do the addition;
72" + 11 3/4" becomes 83 3/4".

Third, convert the fraction to a decimal fraction by dividing the bottom of the fraction into the top;
3/4" becomes .75".

Fourth, replace the English-style fraction with the decimal-style fraction.
83 3/4" becomes 83.75 inches.

Now the English-style measurement is in a form that can be multiplied or divided by a decimal constant. Let's say that that constant is 1.414 and multiplication is being used. The problem then becomes 1.414 × 83.75 = 118.4225.

The decimal-style answer, 118.4225, is in inches. To be accurate to the nearest 1/16", three decimal places are plenty; 118.4225 yields 118.422.

An easy method to solve the required English-style fractions from a decimal answer is:

GIVEN A DECIMAL ANSWER OF: 118.422"

.422   = *just* the fractional part
× 16   = use 16 for 1/16ths
‾‾‾‾‾‾
2532
422
‾‾‾‾‾‾
6.752   = The number of 1/16ths

Round using one place to the right of the decimal point. 6.752 rounds to 7/16". An answer of 3,499 would be equated with 3/16". An answer of 3.500 would be equated with 4/16″ or 1/4″.

Now put the fraction back with the whole number part of the old decimal answer, i.e. 118.422″ becomes 118 7/16″. Convert inches to feet and inches.

DECIMAL MEASURE: 118.422 becomes
ENGLISH MEASURE: 9′ -10 7/16″

Rationale: Remember that to find the decimal equivalent of a fraction like 5/16, *divide* the denominator, 16, into the numerator, 5, yielding 0.3125. Having the decimal equivalent, 0.3125, wouldn't it then make sense to now *multiply* the decimal by 16 to get back to the fraction if necessary? When the decimal fraction is not precisely equivalent to a 1/16ths fraction the real number answer has significant digits to the right of the decimal point these are used to either accept the number of 1/16ths indicated by the number to the left of the decimal or to increase it by one. Look at only the first decimal place to the right. If that digit is 5 or more, increase the number to the left. If it is 4 or less leave the number of 1/16ths indicated by the number to the left of the decimal point the way it is. Of course, 8 or 4 could be used for the nearest 8th fraction or 4th fraction. But since the smallest length generally indicated on the rule is the 1/16″ distance, why not use 16 and then derive all other fractions from that?

If the student has followed the discussion up to this point with ease, he/she is probably ready to go on to the unit end *summary*. If there is still some difficulty, the information which follows is a very condensed review of basic arithmetic.

Measurements taken by the plumber are usually expressed in feet, inches, and fractional parts of an inch. The plumber must be able to use these measurements to do plumbing problems. The answers obtained must be in a form applicable to the 6-foot folding rule. Answers like 7 1/5 inches are useless in plumbing. To apply a measurement to a pipe cut, the fractional part must be in 1/2s, 1/4s, 1/8s, or 1/16s of an inch.

It is often necessary to change the form of a measurement to make the mathematics easier.

CHANGING INCHES TO FEET

Example:

Change 113 1/4 inches to feet and inches.
There are 12 inches in one foot.
Divide 12 in to 113 1/4 inches.

$$
\text{A. } 12\,\overline{)\,113\ 1/4}
\qquad
\text{B. }
\begin{array}{r}
9 \\
12\,\overline{)\,113\ 1/4} \\
108 \\ \hline
5\ 1/4
\end{array}
$$

The answer indicates that there are 9 feet with 5 1/4 inches left over. This should be written as 9′ - 5 1/4″.

CHANGING FEET TO INCHES

Example:

Change 7′-6 3/4″ to inches.

$$
\text{A. }
\begin{array}{r}
7\text{'-}6\ 3/4'' \\
\times\,12 \\ \hline
84
\end{array}
\qquad
\text{B. }
\begin{array}{r}
7\text{'-}6\ 3/4'' \\
\times\,12 \\ \hline
84 \\
+\,6\ 3/4 \\ \hline
\end{array}
$$

$$
\text{C. }
\begin{array}{r}
7\text{'-}6\ 3/4'' \\
\times\,12 \\ \hline
84 \\
+\,6\ 3/4 \\ \hline
90\ 3/4'',\ \text{answer}
\end{array}
$$

CHANGING FRACTIONS TO DECIMALS

Example:

Change 3/4 to a decimal fraction.

A. 3/4   4) 3.000

B. 4) 3.000   C. 4) 3.000
    2 8   2 8
    20   20
                          20
                          00

D. .75 = 3/4, answer

Example:

Change 11/16 to a decimal.

A. 11/16   16) 11.0000

B. 16) 11.0000   C. 16) 11.0000
    9.6   9.6
    1 40   1 40
                          1 28
                          120

D. 16) 11.0000   E. 16) 11.0000
    9.6   9.6
    1 40   1 40
    1 28   1 28
    120   120
    112   112
    80   80
                          80

11/16 = .6875, answer

It is a good idea for the student to memorize all of the decimal equivalents for the fractions used every day, for example, 1/16, 1/8, 1/4, 3/8, 1/2, 5/8, 3/4, and 7/8. If it is necessary to have the decimal equivalent of a sixteenth fraction, such as 11/16, the decimal for 1/16 (.0625) may be added to the next lower 1/8 fraction 5/8 (.625). This would give .0625 + .625 = .6875, which is the decimal equivalent for 11/16, Figure 7–10.

SUBTRACTION

Following are a few typical mathematic problems found in plumbing.

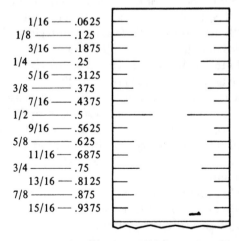

1/16 ── .0625
1/8 ──── .125
3/16 ── .1875
1/4 ──── .25
5/16 ── .3125
3/8 ──── .375
7/16 ── .4375
1/2 ──── .5
9/16 ── .5625
5/8 ──── .625
11/16 ── .6875
3/4 ──── .75
13/16 ── .8125
7/8 ──── .875
15/16 ── .9375

Figure 7–10

Example:

A pipeline has three fittings: two elbows and one tee. The overall measurement between elbows is 7'-6 1/4". The distance between the tee and one elbow is 3'-9". What is the distance from the tee to the other elbow? These are center-to-center measurements.

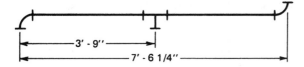

A. 7'-6 1/4"
   –3'-9"

Observe that this is a subtraction problem. 9 inches cannot be subtracted from 6 1/4 inches.

B. 6'-18 1/4"
   –3'- 9"

Borrow 12 inches from the 7 feet and make that 6 feet. 6'-18 1/4" is exactly the same distance as 7'- 6 1/4".

C. 6'-18 1/4"
   –3'- 9"
   3'- 9 1/4", answer

This is the distance from the tee to the other elbow.

Suppose that the problem has unlike fractions in it.

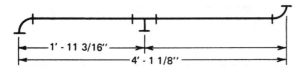

Example:

   4'- 1 1/8"
   –1'-11 3/16"

Three problem adjustments must be made before it can be solved, because:

❏ The fractions are unlike (1/8 versus 3/16).
❏ The lower fraction is bigger than the top.
❏ The number of inches in the inch column is bigger on the bottom than on the top (11 inches and 1 inch).

A. Remember that the top and the bottom may be multiplied or divided by the same number without changing the value of that fraction. There is a fraction of 1/8 and a fraction of 3/16. For fractions to be alike, the bottom of the fractions must have the same number. Change 1/8 into 2/16 by multiplying both the top and the bottom by 2. There are now 3/16 and 2/16 in the problem.

   4'- 1 2/16"
   –1'-11 3/16"

B. The lower fraction is bigger than the top fraction.

   4'- 1 2/16"
   –1'-11 3/16"

Every inch has 16/16 in it so borrow an inch worth of 16ths from the one inch.

   4'-1 2/16"   4'- 0 18/16"
                       –1'-11 3/16"

C. The number of inches in the inch column is bigger on the bottom than on the top.

Borrow a foot worth of inches (12) from the 4 feet making that 3 feet and increasing the number of inches in the inch column to 12.

   4'-0 18/16"   3'-12 18/16"
                        –1'-11 3/16"

We are now prepared to do the math. Remember that 3'-12 18/16" has exactly the same value as the original 4'-1 1/8"

3'-12 18/16"
–1'-11 3/16"
2'- 1 15/16", answer

ADDITION

The following problems will give the student practice in addition.

Example:

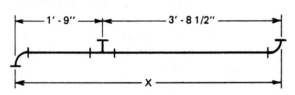

A. 1'- 9"   17 1/2" is the same as 1'-5 1/2" so add
  +3'- 8 1/2"   1 foot to the feet column and subtract
   4'- 17 1/2"   12 inches from the inch column.

B. 4'-17 1/2"
  +1'-12"
   5'- 5 1/2", answer

MULTIPLICATION

Multiplication and division have their own special problems when used with English measurements (feet and inches).

A common multiplier used by the plumber is 1.414. Suppose it is necessary to multiply 1'-11 1/4" × 1.414.

Example:

Change 1'-11 1/4" to decimal form.

A. It is usually easier to change the problem to inches.

1'-11 1/4" = 23 1/4 inches

B. Change 23 1/4 inches to a decimal.

1/4 = .25

Add the decimal to the whole number.
23" + .25 = 23.25"

C. Set up the multiplication problems.

$$
\begin{array}{r}
23.25 \\
\times\,1.414 \\
\hline
9300 \\
2325 \\
9300 \\
2325 \\
\hline
32.87550
\end{array}
$$

Round off answer to three decimal places.
32.8755 = 32.876, answer

CHANGING THE DECIMAL TO USABLE FORM

The answer to a multiplication or division problem involving a decimal constant is in decimal form. This must be changed to a usable fraction in order to use the answer with a 6-foot folding rule. The following instruction will help the student make the answers usable.

Example:

Consider a decimal answer of 25.099 inches.

A. Consider only the decimal fraction part of the problem for the present. The whole number part of the answer will be reserved for later.

25 .009

B. Multiply the decimal fraction by 16.

$$
\begin{array}{r}
.099 \\
\times\ 16 \\
\hline
594 \\
99\ \ \\
\hline
1.584
\end{array}
$$

C. The answer indicates that the sixteenth fraction should be 1.584/16. This, however, is not a proper fraction.

D. The plumber must make the decision: Is the answer going to be 1/16 or 2/16? If the number to the right of the decimal point is 5 or more, (1. 5 84) then the answer will be 2/16, or more correctly, 1/8 inch. If the number to the right of the decimal point was less than 5, then the answer would be 1/16.

E. Now, put the whole number reserved in step A and the fraction together. This makes 25 + 1/8 or 25-1/8 inches.

All fractions that the plumber requires can be calculated from the sixteenth fraction, that is, 2/16 = 1/8, 4/16 = 1/4, 6/16 = 3/8, 8/16 = 1/2, and so on.

SUMMARY

What you have studied are only the basic mathematical operations. *Mathematics for Plumbers and Pipefitters*, a book published by Delmar Publishers, takes these basic math skills and will amplify your abilities with formula handling, geometry, and basic trigonometry as they are applied to the pipe trades.

Much of the math we do is dictated by the instruments used to take the initial measurements. Errors in measurement creep in. The walls of the space we measure may not be square or parallel; the floors may be uneven. The more pieces there are in the final installed pipe system, the more chances there are for error. Take measurements with great care. Notice that in floor covering and other trades, the craftsperson routinely lays down indicator lines that are at right (90-degree) angles close to the center of the room. Then the craftsperson takes all measurements from that carefully constructed axis. This is so that walls that are not perfectly aligned will not cause errors in the craftsperson's work. In most plumbing and pipefitting jobs extreme accuracy is not necessary, but in some cases it is. Judgment must be exercised.

To minimize error and to avoid problems later:

❏ Take initial measurements as accurately as the measuring tools will allow.

❏ Check to make sure walls run square and floors are level.

❏ Do not guess, and write all measurements down carefully.

Making a sketch of the installation often helps to visualize the kinds of math needed to be used, especially if the sketch is drawn to scale. Remember that feet, inches, and fractional parts of an inch should be lined up in columns to add and subtract in the same manner as units, tens, and hundreds digits when operating on counting numbers (integers).

If your measuring tool shows inch numbers beyond the foot mark (for example, 13, 14, or 15 inches), on small jobs it is often convenient to take all measurements in inches and fractional parts of an inch even if the measurement comes to 71 inches or more. A large portion of errors occurs in the conversion process (feet to inches). Avoid the conversion when convenient, and you may avoid an error.

◆ ◆

TEST YOUR KNOWLEDGE

1. What kind of measurement is taken with the 6-foot folding rule, metric or English?

2. With which system are conversions easiest?

3. What is the smallest unit of measurement commonly found on the 6-foot folding rule?

4. Give one disadvantage of the steel measuring tape.

5. Where is the denominator of a fraction found?

6. What would be the mathematical name for the number above the dividing line or to the left of the slash mark?

7. Mathematical constants that the plumber uses are almost always in what form, fractional or decimal?

8. When we move a decimal point one position to the left, what becomes of the value that the number represents?

SECTION 2

THREADED
PIPE

◆ ◆ ◆ ◆ ◆ ◆ ◆ ◆ ◆ ◆ ◆ ◆ ◆ ◆ ◆ ◆ ◆ ◆

GAS AND
THREADED PIPING

KEY TERMS

BTU
BTUH
draft hood
gas
make up

OBJECTIVES

After studying this unit, the student should be able to:

❏ Obtain local information about piping materials.
❏ Calculate pipe sizes.
❏ Test and repair gas piping.
❏ Install gas piping correctly.
❏ Install domestic gas equipment safely.

The term gas refers to fuel in a gaseous form. The air that we breathe, for instance, is an a gaseous form. It is important that the student understand that the slang term "gas" for gasoline is not referring to the same type of thing. It is also important to understand that there are many different kinds of fuel gas. Some fuel gases are found in nature, like natural gas. Other fuel gases are manufactured from petroleum like MAPP, propane, or butane. These and other gases all have their own characteristics and installation requirements. When making an

75

installation, the gas supplier should be contacted to determine which pipe materials and methods are acceptable. Most appliances are equipped to run on natural gas, Figure 8–1. These must be adapted to use other fuel gases. Gas is an efficient and convenient fuel to use. One of its efficiencies is that the fuel can easily be delivered within the building right to the place where it is to be used by piping it there. The potential danger is that a leak in the delivery system could flood the building with a highly inflammable gas.

Figure 8–1 A gas range installed (*Courtesy of Whirlpool Corporation*)

SAFETY CONSIDERATIONS

When testing for leaks in an existing gas line, the proper procedure is to paint the pipe with a soap-and-water mixture. If there is a leak present, large bubbles will appear on the joint that is leaking. There are also commercial electronic leak detectors that will find leaks safely. Under no circumstances is an open flame an acceptable way to discover leaks. Because fuel gases are often heavier than air, concentrations of gas may accumulate near the floor. Therefore, the sense of smell is not reliable for detecting dangerous amounts of gas.

If the plumber suspects that a large amount of gas has leaked into the building:

❑ All people should be evacuated.
❑ Windows and outside doors should be opened.
❑ Allow no cigarettes or open flames.
❑ Do not turn on any lights or electrical equipment within the building.

The space flooded with gas may need to be ventilated from outside. Remember that electrical switches produce a tiny electrical arc when they are operated. This is also true of most electric motors.

Leaking pipe and fittings should be replaced with new materials. Liquids that may accumulate within a gas pipe system are to be treated with extreme care because they may be flammable. When opening up an existing system for repairs, the gas cock that controls the area being worked on should be closed and then padlocked. This will prevent unauthorized persons from turning the gas back on while the system is being worked on.

SIZING THE SYSTEM

When sizing gas pipe, consideration must be given to possible future additions to the system. Sizing means to determine the size of the mains and branches in the gas delivery system for the particular installation. Taking into account the BTU, or a unit of heat measurement known as the British thermal unit, per hour requirements for the individual appliances, the installer works from the gas source, which would be the meter or an existing gas cock, to the farthest appliance. As the branch lines to appliances join the main line, the size of the pipe is increased. It is essential to start with a sketch or plan. The lengths of the branches and mains must be taken into account. Because of the low pressures at which most gases are supplied, pressure losses caused by the friction of the gas against the side of the pipe are

important. Fortunately, most gas piping tables will provide a way to allow for this. Whenever possible the gas supplier's recommendations for type and size of pipe must be followed.

This is an outline of one sizing process:

1. Obtain the following:
   a. the heating value in BTU per cubic foot of the particular gas being used. (In our example, we'll use 1 cubic foot = 1,000 BTU.)
   b. the specific gravity of the gas.
   c. the supply pressure of the gas.
2. Draw up a sketch with lengths and BTUH (British thermal units per hour) requirements of the various appliances shown. See Figure 8–2.
3. Convert BTUH requirements to cfh (cubic feet per hour) requirements at the terminals.

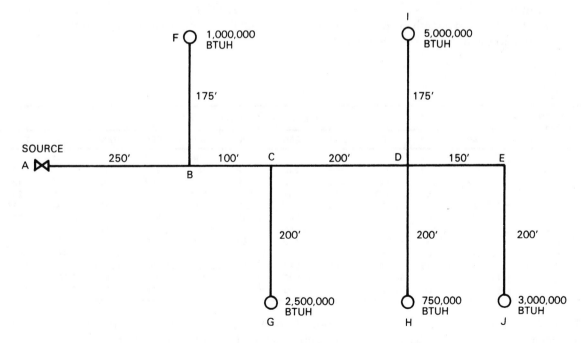

Figure 8–2

Terminal F will now equal 1,000 cfh
Terminal G will now equal 2,500 cfh
Terminal H will now equal 750 cfh
Terminal I will now equal 5,000 cfh
Terminal J will now equal 3,000 cfh

4. Calculate the length of the assembled pipe to the farthest appliance away from the source. From now on this will be the number used when calculating pipe sizes. This number remains the same even for short branches.

A to B is 250 feet
B to C is 100 feet
C to D is 200 feet
D to E is 150 feet
E to J is 200 feet

The total distance is 900 feet to the farthest terminal, J.

5. Start at the first length of pipe from the source to the first tee, Table 8–1 and Table 8–2. The chart must correspond to the information you gathered in step 1. Add up the cfh requirements for all of the appliances because this piece of pipe must supply enough gas for all of them. Using the length from step 4 and the cfh requirements, read off the pipe size from the chart. Working toward the farthest appliance and subtracting the cfh requirements for appliances that are passed along the way, all of the pieces and branches may be looked up on the chart.

NOTE:
On pipe sizes larger than 4 inches each fitting is given a footage equivalent. This is then added to the total pipe length. The purpose of this is to allow for the friction of the gas against the pipe walls. Friction loss, as it is called, is not significant in smaller pipe sizes.

Iron Pipe Size Schedule 40 (Inches)	Total Equivalent Length of Pipe in Feet										
	50	100	150	200	250	300	400	500	1000	1500	2000
1	244	173	141	122	109	99	86	77	54	44	38
1-1/4	537	380	310	268	240	219	189	169	119	97	84
1-1/2	832	588	480	416	372	339	294	263	185	151	131
2	1,680	1,188	970	840	751	685	594	531	375	306	265
2-1/2	2,754	1,952	1,591	1,379	1,232	1,123	974	869	617	504	436
3	5,018	3,549	2,896	2,509	2,244	2,047	1,774	1,587	1,121	915	793
4	10,510	7,410	6,020	5,170	4,640	4,480	3,660	3,340	2,360	1,910	1,660
5	19,110	13,480	10,960	9,410	8,440	8,150	6,660	6,070	4,290	3,480	3,020
6	31,140	21,960	17,860	15,320	13,760	13,280	10,860	9,890	7,000	5,670	4,920
8	63,310	44,740	36,380	31,220	28,020	27,040	22,120	20,150	14,250	11,550	10,030
10	113,020	79,720	64,830	55,630	49,940	48,180	39,420	35,920	25,400	20,580	17,870
12	177,450	125,180	101,790	87,350	78,400	75,650	61,900	56,400	39,890	32,320	28,060

Table 8–1 Pipe sizing table for pressure under 1 pound. Approximate capacity of pipes of different diameters and lengths in cubic feet per hour with pressure drop of 0.5 inch water column and 0.6 specific gravity. (*Courtesy of National Standard Plumbing Code*)

Specific Gravity	Multiplier
.35	1.31
.40	1.23
.45	1.16
.50	1.10
.55	1.04
.60	1.00
.65	.962
.70	.926
.75	.895
.80	.867
.85	.841
.90	.817
1.00	.775
1.10	.740
1.20	.707
1.30	.680
1.40	.655
1.50	.633
1.60	.612
1.70	.594
1.80	.577
1.90	.565
2.00	.547
2.10	.535

Table 8–2 Multipliers to be used with Table 8–1 when the specific gravity of the gas is other than 0.60. *(Courtesy of National Standard Plumbing Code)*

INSTALLATION OF GAS PIPING SYSTEMS

The most common pipe material used for gas piping is threaded steel pipe. It is acceptable to convey all present fuel gases. Threaded brass, copper, plastic, or aluminum pipe may be used also under certain circumstances. These are:

1. The gas supplier approves that these materials will not be corrosive with the gas supplied.
2. Aluminum pipe will not be used outside or underground.
3. Aluminum pipe must be factory coated where it may come in contact with plaster, cement, water, detergent, or sewage.
4. Plastic pipe may only be used outside and underground.

With the same provisions as above, copper, steel, and aluminum tubing may be used also. The most important caution is that the gas supplier must approve the pipe or tubing proposed. Brass, steel, and malleable or ductile iron fittings may be used with steel pipe. Tubing that is to be soldered or brazed must be joined with a solder that melts at a temperature above 1000° F. This means that most soft solders are not acceptable.

The gas that may be supplied to a given location is subject to change. So the greatest number of installations is made with threaded steel pipe, which is compatible with all fuel gases.

All pipe must be reamed before installation so that the pipe joint presents a smooth surface to the flow of gas. The following chart will give approximate lengths of the threaded portions of pipe.

Pipe Size	Length of Threaded Portion
1/2	3/4
3/4	3/4
1	7/8
1-1/4	1
1-1/2	1
2	1
2-1/2	1-1/2
3	1-1/2
4	1-5/8

Because some threads on the pipe are left showing after a fitting is installed, the above chart does not tell us how far the pipe goes into the fitting.

Every joint should be made up with two pipe wrenches. The term **make up** means to apply joint compound to the male threads and tighten the fitting onto the pipe. An approved pipe joint compound must be used.

When locating steel gas piping within the building, care must be used to place it so that machinery or vehicles are not likely to damage it. No unions should be used, except at the appliance connections. At the bottom of each vertical section of pipe a drip should be placed. Because gas can have moisture in it, water may accumulate at low points. To provide a way to keep this condensate out of the way, a tee with a short nipple and a cap is used in place of an elbow. This is called a drip, Figure 8–3. Because of the possibility of moisture, horizontal runs of pipe must be carefully installed so that any moisture will run toward these drips. Branches should be taken off from the top or side of the main rather than the bottom. Also, gas pipe must be installed so that freezing will not be a factor. Gas pipe should be supported at each floor level for vertical runs. The following chart explains the support spacing for horizontal runs.

Size of Pipe	Spacing of Supports
1/2	every 6 feet
3/4 or 1	every 8 feet
1-1/4 or larger	every 10 feet

Every outlet that is not connected to an appliance should be capped off (a pipe cap installed), even if a gas cock is in place.

After a system is installed and before any pipe is concealed by flooring or wall coverings, it must be tested. This may be accomplished by

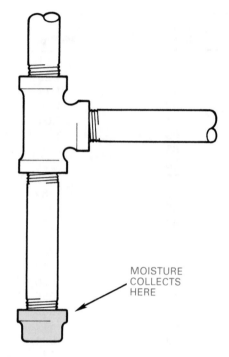

MOISTURE COLLECTS HERE

Figure 8–3   A drip

capping off the terminal ends and (a) using an electronic gas detector, (b) pressurizing the system with an air pump and painting each joint with a thick mixture of soap and water, or (c) placing a manometer at one terminal end and turning on the gas. If there is no real pressure drop in 3 minutes after the gas is turned off, the system is leak free. A manometer is a device for measuring very low pressures. Great care must be used in the testing process if the fuel gas itself is used, to prevent flooding an area of the building with fuel gas. The student will please take note that testing gas pipe with a lit match or a candle is *not* one of the options.

After testing, the system should be purged. This is the process of filling the system with the gas to be used:

1. Turn on the gas.
2. Open the union at each appliance until gas is detected, and then retighten.

As can be imagined, great care must be exercised to ensure that (a) the combustion chambers of the appliances are not filled with gas prior to lighting them, (b) plenty of ventilation is available while the purging is going on, and (c) no open flames are present. After purging, the plumber should light all pilots on the appliances.

INSTALLATION OF GAS EQUIPMENT

In order for combustion to take place, there must be air or oxygen present. In all appliances for domestic use (use in the home) and most appliances for commercial use, the air containing the oxygen is introduced at a point where the fuel gas exits the piping system. Because there is no oxygen or air within the pipe, all fire takes place outside. An exceedingly dangerous situation could exist if air would become mixed with the gas inside of the gas pipe. In cases where air is mixed with the gas before it exits into the combustion chamber, a back pressure regulator or other check valve must be placed in the line to prevent air of oxygen from traveling into the piping system. A gas appliance is also a consumer of air. A water heater in a small utility room must be provided with air to operate properly. A door that is cut with a space underneath it or a door with slots or louvers must be installed to provide this air. Gas appliances meant for domestic use that have a pilot flame to light the main burner automatically also have a safety device that shuts off the flow of gas if the pilot flame goes out. Water heaters and other vented appliances have a **draft hood.** This is a sheet metal device that is placed on the appli-

ance to which the vent or smoke pipe is connected. The purpose of the draft hood is to prevent a down draft within the building's chimney from blowing out the appliance's pilot flame. The student plumber must install appliances in a manner that does not nullify the effect of these safety devices.

Installation information must be checked for clearance requirements for the particular appliance being installed. While a minimum clearance of 6 inches from any combustible material may be acceptable for a water heater, other appliances may require different clearance. While it is possible to use a range designed for one kind of gas with another kind with some modification, it is important that the gas supplier be consulted for his/her recommendations. Some gas appliances do not require vent piping; others do. The installation sheet that comes with the appliance is an important source of information. Follow the instructions exactly. This may mean that pieces of equipment that require a vent will need to be located near a chimney. Other appliances like clothes dryers may need to have wall penetrations made and commercial vent terminals installed. The plumber must exercise great caution when installing fuel-burning equipment. The possibility of fire or asphyxiation must be carefully considered.

SUMMARY

In this unit, gas refers to a fuel in gaseous form (not a liquid). There are many different kinds of fuel gases. Some occur naturally (natural gas), while others are manufactured from petroleum (MAPP), or like acetylene, are made from minerals such as carbide. Gases all have their own characteristics, and the installer must be aware of which kind of fuel gas is to be used so that the equipment may be matched to it.

The two major safety issues with fuel gas are the possibilities of both fire and explosion. Undiscovered leaks caused by faulty installation, and leaks that develop because of the effects of time and physical damage, are to be considered in the planning and installation phases.

Careful testing with purging gases under pressure or electronic leak detectors is an absolute must. Because some gases are heavier than air and fill spaces from the floor upward, smell is *not* a reliable indicator of dangerous gas levels. Leaking fixtures, pipe, or fittings must be replaced or repaired with new materials. When opening the gas piping system, the plumber should be aware that any accumulated moisture within the pipe may be flammable.

Although gases may be conveyed in various piping materials, in most cases steel pipe and ductile iron fittings have the advantage of being highly resistant to physical damage. Cast-iron fittings are permeable to some fuel gases, that is, gases can leak through the pores of the material itself.

The installer must pay careful attention to the installation instructions that come with any appliance. Draft hoods must be carefully installed, and clearance distances between the shell of the appliance and flammable building materials are vitally important considerations.

TEST YOUR KNOWLEDGE

1. Incorrect or careless gas piping installations can result in two kinds of life-threatening hazards. What are they?

2. Gas often has a scent enhancer, referred to as skunk oil, added to it so that the gas would be detectable by smell. Why might that not always work?

3. The unit of measure known as BTU is also known as _____ _____ _____.

4. Gas pipe is most often made of _____.

5. What is the danger of water trapped within gas lines?

6. What could be one drawback of looking for gas leaks with a lighted match?

AMERICAN STANDARD PIPE THREAD

OBJECTIVES

After studying this unit, the student should be able to:

❏ Describe pipe threads.
❏ Describe the relationship of threads per inch to pipe size.
❏ Select and use tools for threaded pipe properly

In 1862, Robert Briggs formulated the nominal dimensions of pipe and threads. These were adopted in 1886 by the manufacturers of pipe and became known as the American Standard Pipe and Pipe Thread, Figure 9–1. These are the standard pipe sizes used today.

The standard thread used on pipe is V-shaped with an angle of 60 degrees (°), very slightly rounded at the top. It has a taper of 1/32 inch per inch of length. This taper provides a

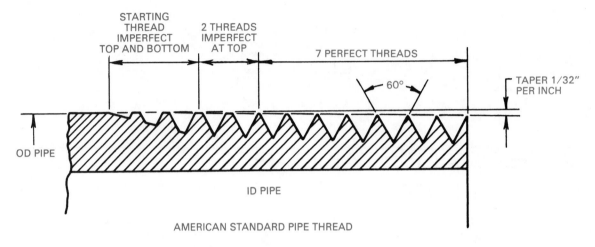

STARTING THREAD IMPERFECT TOP AND BOTTOM

2 THREADS IMPERFECT AT TOP

7 PERFECT THREADS

60°

TAPER 1/32" PER INCH

OD PIPE

ID PIPE

AMERICAN STANDARD PIPE THREAD

Figure 9–1

watertight joint. Examine the sketch of the pipe thread. Notice that the first seven threads are *perfect threads:* that is, they are sharp at the top and bottom. The next two or three are perfect at the bottom and imperfect at the top, and the rest are imperfect at the top and bottom. The latter are only starting threads and have no sealing value. If the perfect threads are marred or broken, leaks will occur.

The number of threads per inch for each pipe size is given in Figure 9–2.

Since pipe threads must fit perfectly into the thread of a fitting, particular attention must be paid to the location of the standard mark on the stock and dies when threading. If the dies are too close, the diameter of the pipe will be reduced and the thread will be loose. If they are too far apart, the thread will be shallow and will not screw far enough into the fitting. In either case, leaks will occur and the threads will have to be recut.

It is very important to use cutting oil when making threads. The oil should be the proper type for the job. Any good commercial-grade cutting oil may be used. Apply it to the dies and

Pipe Size	Threads per Inch
1/8"	27
1/4", 3/8"	18
1/2", 3/4"	14
1" to 2"	11 1/2
2 1/2" to 12"	8

Figure 9–2

pipe to reduce heat and friction that may cause torn threads and ruined pipe dies.

Before attempting to use any pipe dies, refer to the manufacturer's instruction sheet or ask for advice from your instructor. Improperly assembled pipe dies may not cut a proper thread.

Tools for threaded pipe fall into three categories: tools that cut, ream, and thread the pipe; tools that hold the pipe firmly for assembly or end preparation (end preparation refers to cutting, reaming, and threading pipe); and tools for the assembly of pipe and fittings.

END PREPARATION TOOLS

Hacksaw

An important tool for cutting pipe is the hacksaw. The hacksaw has the advantage of being light and able to cut pipe from one side. It does not require room all the way around the pipe as do some pipe cutters. The most important part of the hacksaw is the blade. Blades should be of high-quality steel and have between 18 and 24 teeth to the inch. The frame of the saw is usually of light steel and may or may not be adjustable for different blade lengths. Hacksaw blades come in 8-, 10-, and 12-inch blade lengths, Figure 9–3.

To use a hacksaw properly, start the cut carefully with one hand using the thumb of the other hand as a guide. If the thumb is held up high on the blade and very light strokes are taken, there is little chance of being cut by the blade. After the cut is started, the other hand can be used lightly on the frame. Remember that because of the angle of the teeth on the blade, it will only cut when it is pushed forward. Do not bear down when drawing the blade back.

Pipe Cutters

Pipe cutters may have from one to four cutting wheels, Figure 9–4. The more cutting wheels that a pipe cutter has, the less room it requires to cut the pipe. The most common pipe cutters have only one wheel. This provides room on the cutter for more rollers to keep the cutter aligned. However, this means that the cutter will have to be passed all the way around the pipe to cut it all the way through. In use, the pipe is placed in the center of the wheels and rollers. The T-handle is rotated until contact with the pipe is made. Then the cutter is rotated around the pipe while turning the T-handle about 1/4 turn for each revolution around the pipe. The cutter on a pipe machine is nothing more than a hand pipe cutter mounted on a machine.

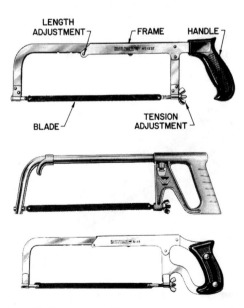

Figure 9–3 Standard hacksaws

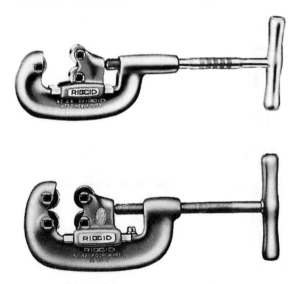

Figure 9–4 Pipe cutters. (*Courtesy of The Ridge Tool Company*)

Do not try to cut through pipe threads with the pipe cutter; this will ruin the tool. The hacksaw is the proper tool for cutting pipe threads.

Pipe Reamers

The **pipe reamer** is designed to remove the metal burrs that appear on the pipe after a pipe cutter has been used. The ratchet type is very common, Figure 9–5. The reamer is placed in the pipe opening with the palm of one hand placed against the back of the reamer. The reamer is then rotated clockwise with the other hand. A little experimenting on the student's part will quickly show how to use this tool. Do not remove more metal than necessary to obtain the full inside pipe diameter.

Pipe Threaders

Pipe threaders may be divided into hand threaders and power threaders. The threaded pipe sizes found in the average home rarely have inside diameters of more than 2 inches. Pipes up to this size can be threaded with either the fixed-die-type pipe threader, Figure 9–6, or the ratchet type, Figure 9–7. Both of these types are sometimes called **stock and dies**. The **stock** is the device that holds the thread-cutting part, which is called the **die**. Although these may frequently be termed hand threaders, they may also be used with a power drive or power vise, Figure 9–8, which threads automatically. A pipe-threading machine is usually used when large quantities of pipe are to be threaded. With this machine the end preparations can be done very quickly, Figure 9–9.

Figure 9–6 Fixed-die-type of pipe threader. *(Courtesy of The Ridge Tool Company)*

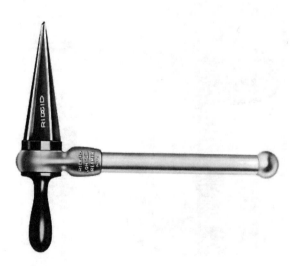

Figure 9–5 Pipe reamer. *(Courtesy of The Ridge Tool Company)*

Figure 9–7 Ratchet-type pipe threader. *(Courtesy of The Ridge Tool Company)*

Holding Tools

When threading or assembling pipe, a device that will hold the pipe firmly is very helpful. A pipe vise is the tool ordinarily used for this. The two most common pipe vises are the chain vise, Figure 9–10, and the yoke vise, Figure 9–11. Vises that may be bolted to work benches or fastened to vertical columns re also available. The common machinist-type bench vise is not suitable for pipe work because the flat jaws will not hold round pipe securely.

Figure 9–8 Power vise. *(Courtesy of The Ridge Tool Company)*

Figure 9–10 Chain vise. *(Courtesy of The Ridge Tool Company)*

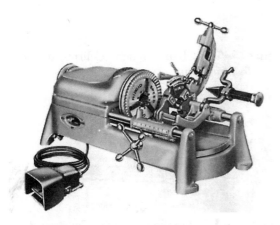

Figure 9–9 Pipe-threading machine. *(Courtesy of The Ridge Tool Company)*

Figure 9–11 Yoke vise. *(Courtesy of The Ridge Tool Company)*

Wrenches

Wrenches are used to attach fittings to threaded pipe. Since most pipe fittings are rounded, a special type of wrench is used for the fitting assembly. This is called a **pipe wrench**. It has teeth that are set at an angle in one direction. The teeth are designed to grip the rounded fitting firmly and give a ratchet effect so that the wrench only grips when pushed in one direction, Figure 9–12. Other kinds of pipe wrenches are the **offset pipe wrench**, Figure 9–13, for work in confined areas, and **chain wrenches** for work on larger pipe sizes in even smaller work areas, Figure 9–14. The chain wrench is sometimes referred to as **chain tongs**. Some fittings have flat sides. These may be assembled with the **monkey wrench**, Figure 9–15. The monkey wrench has no teeth and will not mar the finish on chrome-plated fittings. Another tool that is useful for chrome-plated pipe is the **strap wrench**, Figure 9–16. The strap wrench should be dusted occasionally with **rosin** (the dried resin from pine trees) to prevent slipping.

THREADED PIPE

In residential plumbing, threaded pipe is used mostly for gas piping. It may sometimes be used, however, for drainage piping and water service piping. Properly threaded and assembled steel pipe can contain pressures at the joints up to the bursting pressure of the pipe itself. End preparation of threaded pipe requires three steps: cutting, reaming, and threading

Cutting

Pipe may be cut with a hacksaw, a pipe cutter, or a power machine. When using a machine-type cutter, do not rotate the cutter handle too quick-

Figure 9–12 Pipe wrench. *(Courtesy of The Ridge Tool Company)*

Figure 9–13 Offset pipe wrench. *(Courtesy of The Ridge Tool Company)*

Figure 9–14 Chain wrench. *(Courtesy of The Ridge Tool Company)*

Figure 9–15 Monkey wrench. *(Courtesy of The Ridge Tool Company)*

Figure 9–16 Strap wrench. *(Courtesy of The Ridge Tool Company)*

Figure 9-17

ly. This causes a large burr to form on the inside of the pipe and excessively wears the cutter wheels. Oil should be used when cutting on the power machine.

When using a hand pipe cutter, the T-handle should be tightened about 1/4 turn for each revolution around the pipe, Figure 9–17. Do not force the wheel into the pipe too quickly. Do not attempt to cut pipe on threads. This will twist the cutter wheel and shorten the life of the tool. Oil is not necessary in this operation. One disadvantage of the single-wheel type of pipe cutter is that it requires more clearance around the pipe.

The hacksaw is often used to cut pipe that has been installed close to a wall or ceiling.

Great care should be used when cutting pipe to be threaded. An end that is not square will be difficult to thread properly, Figure 9–18.

Reaming

After the pipe is cut with a pipe cutter, it must be reamed. The purpose of reaming is to remove the internal burr created by the pipe cutter, Figure 9–19. If the burr were allowed to remain, the flow of liquid or gas in the pipe would be slowed down. It could also cause an obstruction by catching and holding solids from the flow. Use the reamer to remove only the amount of material necessary to restore the full internal pipe diameter.

Threading

Failure to cut clean sharp threads is usually because of dull dies, worn equipment, or poor quality cutting oil. More power is required to

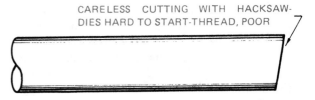

CARELESS CUTTING WITH HACKSAW-DIES HARD TO START-THREAD, POOR

Figure 9-18

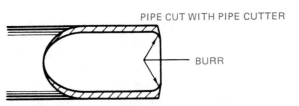

PIPE CUT WITH PIPE CUTTER

BURR

Figure 9–19

cut threads if the dies are dull. To secure water-tight joints, the threads must be cut smoothly, have the proper pitch or lead, and have 1/32-inch taper per inch of length. Threads are of the proper depth when a new fitting can be screwed on with 2 1/2 to 3 turns by hand.

Occasionally, the student will have difficulty starting a thread with the hand threader. This may be caused by dull or improperly adjusted dies. More often, however, it is caused by a bevel on the outside of the pipe end. If there is a bevel, a fresh cut should be made on the pipe. The bevel is caused by insufficient pressure being applied to start the die originally.

THREAD CUTTING

The *lip* is the cutting edge of the tool that must slant at a certain angle, differing with each metal. The angle for steel pipe is 15 degrees to 20 degrees, Figure 9–20. For wrought iron, brass, and copper, the angle should be 25 degrees. This type of die has its front edge on the centerline of the pipe.

The *lead* is the angle that is machined or ground on the first three threads of each die to enable the die to start on the pipe. It also distributes the cutting equally on all threads. The threads cut by the lead are shallow starting threads.

Chip space is the space in the die holder in front of the die that allows the chips to curl. Improper chip space would cause a packing of chips, tearing the threads.

Clearance is the distance between threads on the die and those on the pipe at the back of the die. Proper clearance will make threading easier and allow cleaner threads to be cut.

Thread cutting oil serves two purposes. It lubricates the dies and reduces the heat caused by friction. Oils with a high sulfur content are good because they keep heat to a minimum.

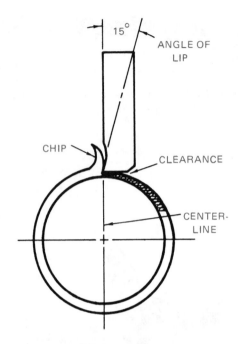

Figure 9–20

When dies are placed in stocks, it is very important that they be inserted in the numbered slot. Remove any chips that may lodge under or in back of the dies. These chips keep the dies from their proper position and result in torn threads.

Dies of the Armstrong pattern must have the same serial number. This is found on the bottom of the dies. Figure 9–21 shows the reverse side of a die.

Threads are cut standard length when the end of the pipe is even with the face of the die.

New plumbers should be careful not to use too much strength when screwing fittings or valves on pipe. This is poor practice because it wastes energy and the fitting is usually stretched in the process. A stretched fitting will leak and must be removed. Valves made of brass, a soft metal, are more apt to be stretched

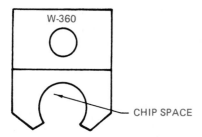

Figure 9-21 Reverse side of threading die

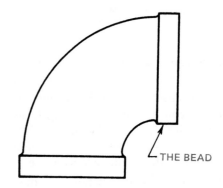

Figure 9-22

and twisted out of shape than are iron valves.

To screw a fitting on a pipe, place the pipe in the pipe vise, examine the thread, and apply a small amount of pipe done to the perfect threads of the pipe to fill any imperfections in the thread. **Pipe dope** is a compound for lubricating and sealing threaded pipe joints. Screw the fitting on clockwise by hand until it is tight, place a wrench on the bead of the fitting, Figure 9–22, and turn it one or two turns.

To tighten a valve on a pipe, use a smooth-jawed wrench on the hexagonal shoulder. The wrench *should be placed on the end into which the pipe is being screwed.* The pipe will prevent the valve from being crushed by the pressure of the wrench. If fittings are put on a pipe without a vise, place two wrenches closely together in opposite directions. The handles should be kept within 1/8 turn of each other, Figure 9–23.

A convenient way to use two pipe wrenches to assemble pipe follows:

1. Place the holding wrench on the floor with the jaws facing up, Figure 9–23.
2. If tightening, place the pipe near the fitting in the holding wrench, with the pipe extending to the right. Using the left hand on the tightening wrench, press down, turning pipe clockwise.
3. If loosening the fitting, the pipe should

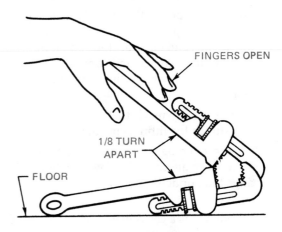

Figure 9-23

extend to the left and the wrench should be pressed down with the right hand, turning pipe counterclockwise.

Caution:
When pressing a wrench handle toward a solid object such as a floor or a wall, KEEP THE FINGERS OPEN. If this rule is not followed, injury could occur.

SOME HINTS FOR MAKING UP FITTINGS

❑ When given a choice, always press down on a tool handle. It is less tiring and safer.

❑ Always use two wrenches when a pipe vise is not convenient. In this way, fittings that have already been assembled will not get out of line; each joint will be properly tightened; and there will be less chance of a leak.

❑ When using two wrenches, keep the wrench handles 1/8 turn apart. More power can be applied and the assembly has less chance of being twisted out of line, Figure 9–23.

❑ Put pipe dope on the threads of the pipe only. Pipe dope will fill the imperfections in the threads and lubricate the joint. It may be a paste-like compound applied with a brush, or it may be in the form of a plastic tape.

❑ Do not use a wrench that is too large. This will stretch the fitting, creating a leak.

❑ Do not over tighten threaded pipe joints. More leaks are caused by overtightened fittings than undertightened ones.

Teflon tape is a convenient and easy way to apply pipe dope. The student should remember, however, that teflon products are also a good lubricant. This means that overtightening is a danger. And along with overtightening is the danger of stretched fittings. It will often be discovered that the joint will become loose when a fitting is *backed-off* even slightly. The term back-off means to loosen a joint for the purpose of *catching a thread* on a union or other connection perhaps. Catching a thread means to start one thread into another by turning. There are a number of good pipe dopes on the market. Hardening, electrical conductivity, vibration,

and pipe size should be taken into consideration when selecting a pipe dope.

The size pipe wrench to be used for several pipe sizes are:

8" wrench for 1/8" and 1/4" pipe
10" wrench for 3/8" and 1/2" pipe
14" wrench for 3/4" and 1" pipe
18" wrench for 1 1/4" and 1 1/2" pipe
24" wrench for 2" pipe

SUMMARY

The pipe thread has been around for quite a long while, but it was standardized in 1886 by pipe manufacturers from specifications laid down by Robert Briggs in 1862. Briggs stated that the pipe should be tapered at the end at 1/32 inch per inch slope, and then 60-degree, V-shaped threads should be provided on both the pipe and inside of the fitting so that the fitting and the pipe could be firmly drawn together with especially designed tools.

The taper provides a firm connection that becomes firmer as the parts of the joint are drawn together. If the dies are sharp, clean, and smooth, the parts will not leak even when pipe-joining compounds (pipe dopes) are not used. Old-time plumbers spoke of rust joints, which happened when the fitting and pipe were purposely drawn together without joining compound and allowed to rust together for a few days before use. Soldering acid was often applied to aid the rusting process.

The Briggs thread, now known as the American Standard Pipe Thread, is designed so that with a properly cut thread and sharp tools, the fitting should twist on 2 to 2 1/2 turns by hand and then be driven on another two complete revolutions with the correctly sized wrenches. The use of proper length wrenches is important. Wrenches that are too large provide too much leverage and may allow the pipe to be driven into the fitting too deeply, causing a leak when the end of the pipe pushes against the inside wall of an elbow thereby distorting the fitting. If the fitting is a tee or a coupling, excessive force may actually split the fitting's bead or hub.

Some special tools that are used with threaded pipe are pipe vises, pipe wrenches, pipe cutters, stocks and dies, and power threaders of various configurations. There is one non-specific tool: the hacksaw. The hacksaw blade used for cutting pipe should have a greater number of teeth per inch than one used on aluminum or other softer material. For steel pipe, the blade should have between 18 and 24 teeth per lineal inch. Always have a supply of new blades for the hacksaw. New blades and an even, smooth cutting pressure can actually make cutting pipe with a hacksaw easy. A little cutting oil makes the task even easier.

There are special tools for cutting pipe called, understandably, pipe cutters. By tightening the tee handle of the pipe cutter a quarter-turn for each revolution around the pipe, the job is done quite quickly and neatly. Watch for an excessive burr pressed into the inside diameter of the pipe by the cutting wheel of this tool. This burr must be removed.

Pipe vises are necessary to hold the pipe firmly so that the other pipe tools, such as threaders (stock and dies) and reamers, can be brought to bear with greater force.

Pipe wrenches have teeth machined into their gripping surfaces. The plumber always takes care to not place the wrench on the open end of a fitting. Malleable iron, which is a low-grade steel and which many fittings are made of, distorts easily when a pipe wrench is applied too strongly to the unsupported fitting end. When faced with no other recourse than to put a pipe wrench on an open fitting end, the knowledgeable plumber threads a small piece of pipe scrap or a nipple into the end to which the wrench will be applied. Chromed pipe requires strap wrenches to avoid marring the highly shiny finished surface. Fittings that are chrome-plated usually have hexagonal-shaped fitting beads so that smooth-jawed wrenches may be used to hold them.

Always put the joint sealing compound on the threads of the pipe, not on the internal (female) threads of the fitting, so that the sealant (pipe dope) will not be pushed into the inside of the fitting and left to harden into a flow-obstructing lump. Remember that the same pipe dope may not be the best thing for every piping job.

◆ ◆

TEST YOUR KNOWLEDGE

1. One example of a manufactured gas is _____.

2. Moisture collecting in the drip legs of a gas system may be dangerous because it may have become _____.

3. What is the most common kind of pipe material used for gas piping?

4. Can any kind of gas be used in any gas fixture or appliance?

5. What is the taper of the American Standard Pipe Thread?

6. Should or should not pipe cutters be used when it is necessary to cut through a pipe thread?

7. The tool used to remove the internal burr on a piece of pipe is called a _____ _____.

8. Why can a machinist's vise not be used to hold pipe firmly?

9. Which pipe wrench has no teeth?

10. Should pipe sealing compound be placed on the threads of the pipe or the threads of the fittings?

FITTING ALLOWANCE

◆ ◆ ◆ ◆ ◆ ◆ ◆ ◆ ◆ ◆ ◆ ◆ ◆ ◆ ◆ ◆

KEY TERMS

center-to-center	offset
cut list	run of an offset
end-to-center	set of an offset
end-to-end	the run of
face of the fitting	travel of an offset
laying length	

OBJECTIVES

After studying this unit, the student should be able to:

❑ Define fitting allowance.

❑ Obtain center-to-center measurements.

❑ Allow for fittings to obtain end-to-end measurements.

The object of fitting allowance is to obtain end-to-end measurements. End-to-end measurement is the actual length of the pipe to be cut. There are three steps to arriving at end-to-end measurements.

1. The space to be worked in or the object to be worked on is measured, giving special attention to the pipe connections and outlets. These measurements are recorded on a simple drawing in a notebook.

2. Calculations are then made to arrive at center-to-center and end-to-end and end-to-center measurements. This means the

measurement along the pipe including the fitting.

3. The end-to-center and center-to-center measurements are reduced to allow for the space that the fittings take up.

AN OVERVIEW

While it is possible to begin at an origin and cut and fit pipe or tubing one piece of pipe and one fitting at a time, it is the slowest, most labor intensive method, and it results in installations with a poor appearance. The craftsperson measures the job, calculates the distance from fitting to fitting, and then cuts the pipe and installs it. Occasionally, the working space has irregularly shaped walls and perhaps many obstructions. Or the job may be so large that doing all of the measurements before starting is unreasonable. On these occasions one can compromise and do the job in subassemblies. This means breaking the job down into areas, piping these up, and then joining the areas together at the end with "fill-in" pieces.

When allowing for fittings *always subtract* the amount the fitting takes up in the piping arrangement. If the student finds that he/she must add on for fittings so that the next measurement can be taken, then the inefficient one-piece-at-a-time method of fitting allowance is being used.

In a few rare cases the pipe does not slip into the fitting, i.e., with flanged fittings, butt-weld fittings, and miter work. But in the vast majority of the cases *the pipe or tubing slips or threads inside of the fitting.* When one learns how to allow for fittings with one kind of fitting, one knows how to allow for fittings with all kinds of fittings. See Figure 10–1.

Hint:
With threaded fittings, the particular threading die being used will have an effect on the allowance. Always check the fitting allowance with a piece of pipe threaded with the die which is going to be used.

When a job is started, it is a good idea to make a list of the fitting allowances for the various parts being used. If the fittings are new, of the same weight, and from the same manufacturer, the fitting allowances will not change.

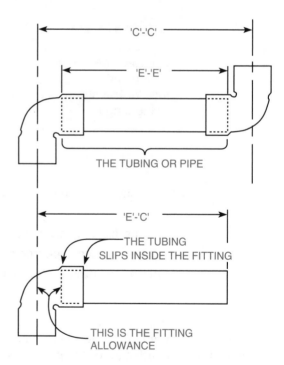

Figure 10–1

Some allowances will be the same as the pipe size. In other words 1/2″ Nominal Pipe Size, tees and ells will often have an allowance of 1/2″. This is true sometimes but not always. It is strongly recommended that the student check fitting allowances with the following method:

1. Select a short piece of pipe.
2. Measure the length of this piece of pipe and write it down.
3. Place a fitting on this piece of pipe. If this is threaded pipe, tighten the fitting with the same wrench and force that will be used on the job.
4. Place the assembly in a vertical position on a flat surface and measure from the surface to the center of the fitting.
5. Subtract the first measurement that was written down from this last measurement. This will be the fitting allowance. See Figure 10–2.

Tees and elbows of the same size and from the same manufacturer have the same fitting allowances throughout. Notice that the tee has fitting allowances at four points. See Figure 10-3.

The center-to-center and end-to-center measurements are referring to the center of the fitting and to the ends of the pipe. The student must be aware that the center of the fitting *is the point at which the center lines of the pipe would cross if they were continued on into the fitting.* This is often NOT the geometric center of the fitting. Refer to drawings in Figure 10–4. When we are *cutting in* fittings (installing tees and perhaps unions in an existing line of pipe), the laying length of the fitting needs to be known. This is the amount of pipe which must be cut out of the pipe line to allow the fitting to be installed. This amounts to two fitting allowances: the distances from the center to each end of the fitting added together.

While it is true that the plumber can usually make the assumption that if one end of the run of a tee has an allowance of `X' then the other end will have the same `X' allowance, sanitary tees and wyes are examples of fittings which have branches which intersect above or

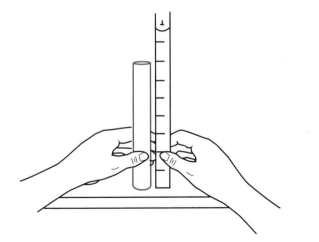
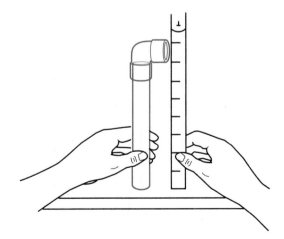

Figure 10–2

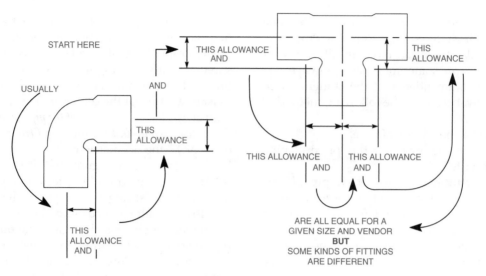

Figure 10–3

below the center of the run of the tee. These fittings do not obey that rule.

Notice where the centerlines intersect in the tee and the wye in Figure 10–4. Notice also that when there is a fitting with two different end preparations like the HUB × SLIP cast iron elbow in Figure 10–4, one side of the fitting has a different allowance than the other.

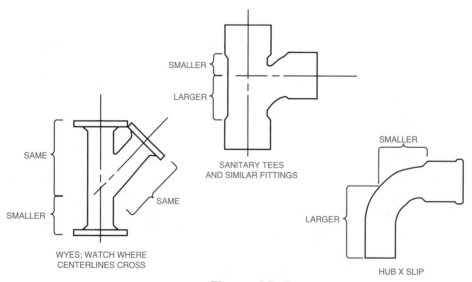

Figure 10–4

Hint:
The run of a tee or a wye is the centerline which is not on the branch. When you look into an opening of a tee and you can see straight through, you are looking down the run of the tee.

Primary Measurements

Figure 10–5 illustrates a piping job. The on-the-job supervisor tells the plumber to connect the two outlets on the tank to each other, keeping the pipe at least 12 inches away from the

tank for insulation. To do the job, the plumber must first take the measurements of the tank, Figure 10–6.

After taking the necessary dimensions of the tank, the plumber may fill in the rough piping dimensions, Figure 10–7.

When the sketch is developed to the point illustrated in Figure 10–7, the plumber is ready to figure the end-to-center (e-c) and center-to-center measurements (c-c), Figure 10–8. From Figure 10–8, the plumber makes a cut list, which contains the end-to-end (e-e) measurements for the four pieces of pipe in order to arrive at the end-to-end measurements by subtracting the distance that the fittings take up from the c-c and e-c measurements.

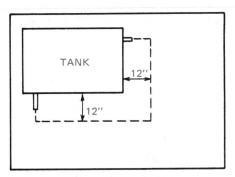

Figure 10–5

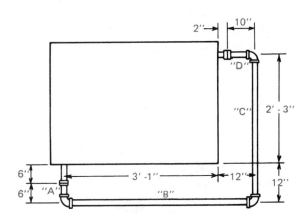

Figure 10–7

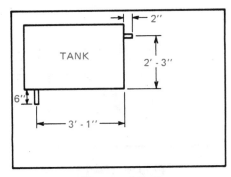

Figure 10–6

A = 6" - E - C
B = 4' - 1" C - C
C = 3' - 3" C - C
D = 10" E - C

Figure 10–8

Offsets

Center-to-center measurements for 45-degree offsets can present a problem. Often 45-degree elbows are used in offsets. An **offset** can set over a pipeline without changing its direction, Figure 10–9.

If the amount of offset is known, the distance between the centers of the 45-degree elbows (the travel) can be found by multiplying the offset by 1.414. The number 1.414 is a constant used for 45-degree angles.

It is helpful to picture any pipeline running at a 45-degree angle as running to opposite corners of a square, Figure 10–10.

All sides of a square are equal in length. Therefore, distance A-B, Figure 10–10 will always be the same as distance B-C. Distance A-C will be equal to 1.414 times A-B or B-C. It is also true that A-B and B-C will be equal to A-C divided by 1.414.

An offset is often described as having **set**, run, and travel, Figure 10–11.

The set and the run are always the same in a 45-degree offset.

Examples:

1. What would be the travel of a 45-degree off-set having a set of 26 inches?

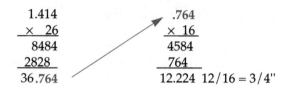

$$
\begin{array}{r}
1.414 \\
\times\ 26 \\
\hline
8484 \\
2828 \\
\hline
36.764
\end{array}
\qquad
\begin{array}{r}
.764 \\
\times\ 16 \\
\hline
4584 \\
764 \\
\hline
12.224\ \ 12/16 = 3/4"
\end{array}
$$

36 3/4 inches, answer

2. What would the run be for a 45-degree off-set having a travel of 28 1/4 inches?

$$1.414\ \overline{)\ 28.25} = \text{A-B or B-C}$$

$$
\begin{array}{r}
19.97 \\
1414\ \overline{)\ 28250.} \\
\underline{1414} \\
14110 \\
\underline{12726} \\
11140 \\
\underline{9898}
\end{array}
$$

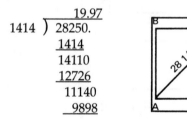

20 inches, answer

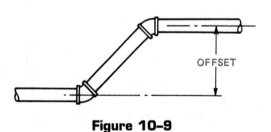

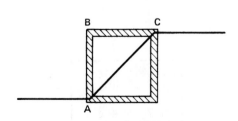

Figure 10–9

Figure 10–10

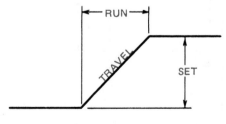

Figure 10–11

After the center-to-center or end-to-end measurement is obtained, an allowance must be made for the fittings. Figure 10–12 shows an end-to-center measurement with a 90-degree elbow.

Using the end-to-center measurement as shown in Figure 10–12, notice that the end of the pipe will not screw into the center of the fitting. Therefore, an allowance, "A," must be made for the fitting, and the pipe must be cut that much shorter.

Fitting measurements are not the same for all manufacturers, so it is very important that the allowance be obtained from the catalog of the manufacturer of that fitting.

How to Measure for the Fitting Allowance

The fitting allowance for small pipe jobs may be measured on the job. Otherwise, it is found in the manufacturer's catalog.

The measurements given are from the center to the **face of the fitting** as illustrated in the standard malleable elbow, Figure 10–13.

C is the center-to-face measurement.
T is the thread engagement, or the distance

the pipe screws into the fitting.

A is the fitting allowance, the space between the center of the fitting and the end of the pipe.

Using these identifying letters, the formula for determining the fitting allowance is A = C – T.

Example:

What is the fitting allowance for a 2" threaded tee?

C, the center-to-face measurement, is shown in the table for threaded tees as 2 7/8". T, the thread engagement, is shown in the table for threaded tees as 1/2".

$$\begin{array}{cccc} & A = & C - & T \\ \text{or} & 2\ 3/8 = & 2\ 7/8" - & 1/2" \text{ — the fitting} \\ & & & \text{allowance.} \end{array}$$

SUMMARY

The purpose of fitting allowance is to determine how long the length of the pipe should be. The length of the pipe is an end-to-end measurement. First the experienced plumber obtains the

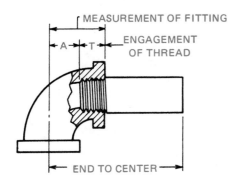

Figure 10–12

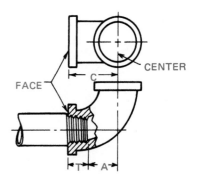

Figure 10–13

THREADED 90-DEGREE FITTING ANGLE

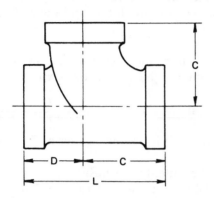

L is Face-to-Face   C is Center-to-Face
D is Center-to-Face   T is Thread-in

Nominal Pipe Size	L	C	D	T
1 1/2″	4 1/4″	2 1/2″	1 3/4″	1/2″
2″	5 1/8″	2 7/8″	2 1/4″	1/2″
2″ x 1 1/2″	4 1/2″	2 3/4″	1 3/4″	1/2″
2 1/2″ x 2″	5 7/16″	3 1/4″	2 3/16″	3/4″ x 1/2″

allowances for all of the fittings that are going to be used. This information may be obtained by consulting fitting manufacturing charts, or the plumber may measure the length of a short piece of pipe and then measure the end of the pipe to the center of the fitting with a fitting installed. The difference between the two measurements is the fitting allowance. The plumber writes these measurements in a pocket notebook and then, as long as the new fittings are of the same manufacturer remeasuring is no longer needed. If different threading dies are used, however, there may be a slight difference.

The plumber then measures the job and calculates all of the pipe lengths at once, then installs the system. This is the fastest and easiest way to fit pipe and results in the neatest installation. Of course, if the job is very large, it should be done in stages, and the same holds true if the measurements cannot be relied on.

If the walls and floors are exceptionally irregular, the plumber will proceed in smaller assemblies.

The most labor-intensive method of pipefitting is to measure then cut one piece of pipe at a time.

Remember that fitting allowance is *subtracted* from overall measurements.

◆ ◆

TEST YOUR KNOWLEDGE

1. Fitting allowance is always _____ from overall measurements.

2. Is it true that similar fittings from the same manufacturer will always take up the same amount of space in the installation?

3. Where the centerlines of the pipe would cross if the fitting were not there defines what kind of measurement?

4. The amount of space taken up by a fitting in a straight pipeline is the _____ _____ of the fitting.

5. Of set, run, and travel, which is the longest?

SECTION 3

WASTE SYSTEMS

UNIT 11

SEWAGE DISPOSAL

OBJECTIVES

After studying this unit, the student should be able to:

❏ Describe the essentials of waste disposal.
❏ Explain the biological and chemical action upon the sewage disposal systems.
❏ Understand the hardware used in a waste disposal system.

SEWAGE

Sewage is poisonous liquid waste that contains animal or vegetable matter. The poisonous, or toxic, elements of sewage must be disposed of before the used water can be returned to the environment. This is done by either a municipal or a private sewage disposal system. The term *sewerage* is used to describe the actual sewer system, disposal and treatment plants, and other elements of sewage disposal and treatment.

In order to make the sewage nontoxic, it is broken down into water, gases, and solids. In the ideal sewerage system, sewage can be made completely harmless. In systems designed for use in outer space, this is done so effectively that the water in the sewage is returned to the drinking water supply.

Sewage may be broken down by a number of methods. In general, some combination of the following methods will be used:

❏ Screening and filtering to remove solids.
❏ Action of bacteria, which does a major part of the work in the breaking-down process.
❏ Introduction of oxygen.
❏ Precipitation, or the settling out of solids.
❏ Application of ultraviolet light and chlorine compounds, which destroy bacteria.

MUNICIPAL SEWAGE DISPOSAL

The modern sewage disposal system may use all of these methods in combination. If the system is operating efficiently, the products of the operation will be:

❏ Gases, which may be used for heating purposes.

❏ High-quality water, which may be returned to the environment.
❏ A dry, rich, earth-like soil, which may be used for fertilizer.

However, most municipal systems are overloaded. This means that chlorine must be added to the waste water to kill the bacteria that are still present. This water may then be returned to rivers or streams. Although this water cannot be considered pure, it is not dangerous to human life. If the detoxification is incomplete, disposing of the waste products can be a serious problem.

PRIVATE SEWERAGE SYSTEMS

Private sewerage systems are constructed where city sewers are not available. If properly designed and constructed, they will give many years of satisfactory service.

The type of system used will be determined by the type of soil and size of the dwelling or dwellings. Before a building permit is issued, health authorities insist that a satisfactory percolation test be obtained. The percolation test demonstrates the speed at which the soil will absorb water. If the soil absorbs water very slowly, or not at all, a building permit will not be granted. A typical installation would have the main house drain emptying into a septic tank. The septic tank would then empty into a drainage field by means of a distribution box, Figure 11–1.

Septic Tank

In these days of environmental concern, it is rare indeed to find a place which does **NOT** require a permit from the local or state **Department of Health** before beginning the design or installa-

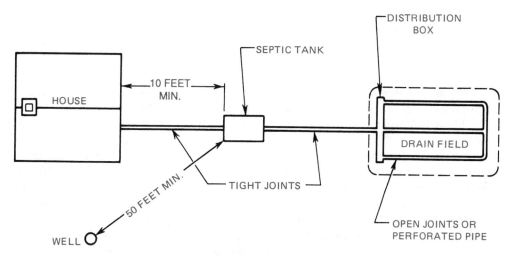

Figure 11-1

tion of an on-site sewage disposal system. Some of the diseases which are encouraged by faulty disposal systems are dysentery, typhoid fever, and hepatitis. Bacteria from septic systems have been found at surprising distances from their source. Environmental agencies and Boards of Health have been continually updating regulations concerning sewage disposal. Responsible tradespersons will always go through the permit process to ensure the safety of their customers.

Some localities and plumbing codes define plumbing as a system of drainage and water supply piping which extends to a point five feet outside of the building. In other words, the piping in the drainage system beyond five feet from the house is not actually plumbing from this viewpoint. It is not unusual for a different contractor than the plumber to be the one responsible for the installation and inspection of the private sewage disposal system. In these areas, it may well be that plumbers also do the heavy excavation and installation of the outside disposal system. In other areas most of the plumb-

ing contractors do the private disposal system as a matter of course. Whatever the case, *the plumber must understand the theory of human waste disposal.*

The city plumber may never have to deal with a private septic system. Keep in mind, however, that somewhere in the city there is a disposal terminal which makes use of all of the detoxifying agents and methods which will be discussed here. There is little difference between the principles of the private disposal system and the municipal system. One is much larger and uses almost every technique. The other is smaller and uses just the techniques necessary for that small disposal site.

PARTS OF THE SYSTEM

The pipe which exits the building should be of a high quality which has watertight joints. This pipe joins to the septic tank at a point at least ten feet from the building. One might consider the possibility of placing the whole disposal system far from the building. But there are some other considerations.

1. The septic tank usually uses the house main vent to allow gases from the decomposition of the wastes to escape high in the air. It would not be good to be too far from this vent.

2. When there are un-decomposed elements in the waste water there is more likely to be a blockage problem. Indeed, most blockages occur between the building and the septic tank. Why not keep this distance as short as is environmentally sound?

Septic and Aeration Tanks

The task of the first receptacle outside of the building, the septic tank, Figure 11–2, is to break down the sewage into sludge, effluent, scum, and gas. The only component of the sewage which does out of this tank is the effluent. Effluent is a liquid with a foul odor which may range from clear to cloudy in color. The sludge is a heavy organic material which falls to the bottom and slowly accumulates. This will slowly build over the years. *A septic tank which is never pumped out will eventually fill up with sludge and cease to function.* When this sludge reaches the inverts of the tank, sludge mixed with effluent will begin to be discharged out to the absorption or distribution field. (The student should be aware that the terms, *distribution field, absorption field,* and *drainage field* are used almost interchangeably.) This will soon block the openings in the pipe there and destroy the field's effectiveness. *A septic tank should be pumped out every three to five years.*

The scum is a hard crust which forms on the top of the liquid in the tank and seals out most of the oxygen. The gases formed by the action of anaerobic bacteria (meaning it exists without air) on the sewage are vented up to the atmosphere through either the building vent or a special vent for this purpose.

The septic tank is a holding tank. It allows the raw sewage to settle and separate, Figure 11–2. The sewage separates into scum at the top, effluent in the middle, and sludge on the bottom. Gases will be given off as the sewage is broken down by the anaerobic bacteria. The tank is provided with compartments and baffles to slow down and even out the flow of sewage from the inlet to the outlet. This allows the bacteria more time to accomplish their task. The inlet and the outlet of the tank have tees or pipe halves installed on them so that the incoming sewage will be released and admitted below the scum layer. These are called inverts. See Figure 11–2. Every tank of this sort must have a cleanout hole in the top or removable panels at the top so that periodic cleanouts and inspections may be accomplished. The cleanout or manhole should be positioned above an invert so that a rod may be pushed through the sludge to measure its depth without disturbing the scum layer.

An aeration tank, Figure 11–3, is much more efficient than a septic tank. This uses a

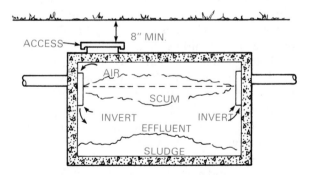

Figure 11–2 Concrete septic tank

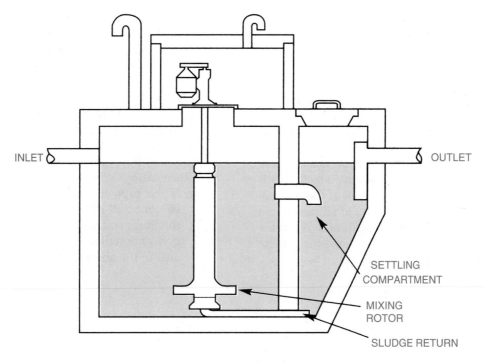

INLET

OUTLET

SETTLING
COMPARTMENT

MIXING
ROTOR

SLUDGE RETURN

Figure 11–3 An aeration tank

small electrically operated motor to stir the contents of the tank occasionally and to pump air below the surface. This has the advantage of allowing the more efficient aerobic bacteria, which needs air to exist, to break down the sewage. The mixing action allows the bacteria across to all of the solids over a longer period. The air which is pumped in contains oxygen which has a detoxifying property of its own. There is a small settling compartment on the outlet of the main chamber. Even in this highly efficient device, sludge will accumulate, but much more slowly than the septic tank. Local public health agencies have regulations regarding the types of systems which are allowable. Always check before planning a system.

Hint:
Inverts are technically the lowest point of the inside diameter of a horizontal pipe. They are often referred to as any pipe opening, however.

The aeration tank has one major disadvantage though, and that is moving parts. Moving parts occasionally stop moving and will need servicing or replacing.

After the septic or aeration tank, watertight pipe joints are assembled and joined to the next major system component, the distribution box.

The Distribution Box and the Distribution Field

The distribution field will be discussed first. This is a number of pipelines laid out in a parallel pattern which have perforations in them at 4 to 6 inch intervals. These perforations are always placed facing down. The pipes are laid out so that they are surrounded by crushed stone, Figure 11–4. The pitch or fall of these pipelines is very small, 1 inch in 50 feet. This is to slow down the flow so that all of the perforations will have an opportunity to allow effluent to seep out into the crushed stone bed. When the effluent from the distribution box enters these pipelines, it is important that:

1. Each line receive an equal amount.
   a. If the distribution box is installed or designed incorrectly, is not level, has outgoing lines placed, or is spaced incorrectly, one or two of the lines will receive the majority of the flow and a small area of the field will receive most of the effluent.
2. Each perforation in the individual branch lines gets an equal amount of effluent to seep out into the absorption area. This is much easier said than done, because:
   a. If the pipe is pitched too steeply the effluent will run quickly to the ends of the lines, and will always use the same perforation holes and the same small area of the absorption trench.

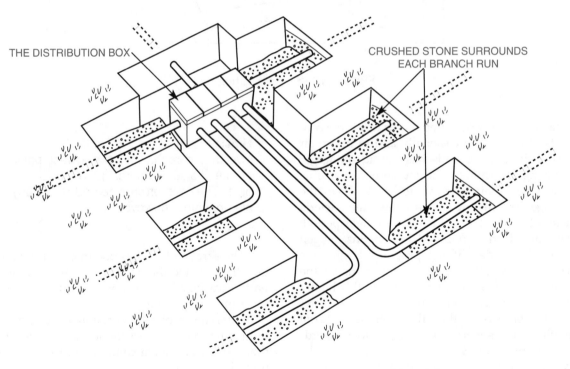

THE DISTRIBUTION BOX

CRUSHED STONE SURROUNDS EACH BRANCH RUN

Figure 11–4 A cutaway drawing of an absorption field

b. If it is not pitched enough, the effluent will flow so slowly that only the holes near the distribution box will be used, and again, only a small part of the absorption field will be utilized.

The distribution box may not even be a box. It may simply be a pipe with some tees in the middle and elbows on the end. The purpose of the distribution box is to effectively use all of the pipe in the distribution field. It must be level so that the incoming effluent flows equally to both ends. The branch lines cannot be in front of the inlet pipe or be unequally spaced.

The distribution box may also be a concrete box with cleanout panels on the top similar to a septic tank. See Figure 11–5. The larger box has the advantage of greater volume which allows incoming effluent to have time to slosh back and forth and then settle down before it divides itself equally between the distribution lines. It should be noted that the amount of flow which

can be expected here is directly related to the number and kind of fixtures in use in the building which the system serves. If one toilet is flushed releasing four gallons of water into the system, then four gallons of effluent will be discharged from the septic tank to the distribution box. It will not be the same four gallons, however. The septic tank operates in an always full condition. Therefore, one gallon in equals one gallon out.

This creates a problem. The continual but occasional admittance of small amounts of effluent will only utilize the same small area of the distribution lines and the absorption field, no matter how well laid out the system is. These frequently utilized holes in the pipe will be the first in the system to close up from tiny particles. Then the process will begin farther down the lines, only this time the system will be effectively smaller than it originally was. The area available for the seepage of effluent will progressively get smaller. Eventually the system will not be

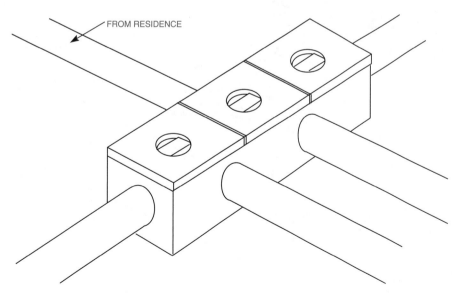

FROM RESIDENCE

Figure 11–5 A distribution box

able to handle peak-load discharges and it will back up (the sewage will cease to be absorbed by the system). Much of the absorption field may be in good shape. But this area may be too small now and a new field will have to be installed.

A distribution box with a bottom which is lower than the outlet inverts will provide an additional settling chamber. This will assure that the effluent is as clear as possible before it arrives at the distribution lines. This, of course, will establish another part of the system which will need periodic maintenance. The top will need to be removed and the solids taken out.

Some authorities maintain that a **serial distribution** system is better. They believe that filling one pipe with a lot of effluent will create a

higher flow through each perforation, which will help keep the holes clear. This pipe is arranged in such a way that one pipe must fill first. Then if there is too much for the one pipeline to handle, the excess will overflow into the next, and then the next, and so on. This style lends itself well to sloped terrain better than flat, Figures 11-6A and B. Actually, no distribution box is utilized with the serial distribution system. At the end of each run, on some serial distribution systems, there is a "drop inlet box" This box or large tile has a cleanout cover for accessibility. You will notice from the illustration that the pipe entering and leaving this access hole is 7" from the bottom, Figure 11–7. This will always have liquid effluent in it and settling out of the solids will eventually require cleaning out.

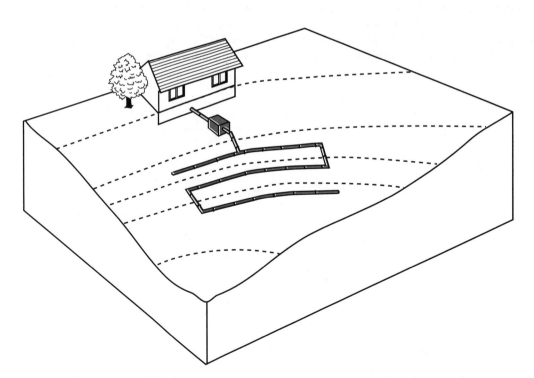

Figure 11–6A The serial distribution works best on sloped ground

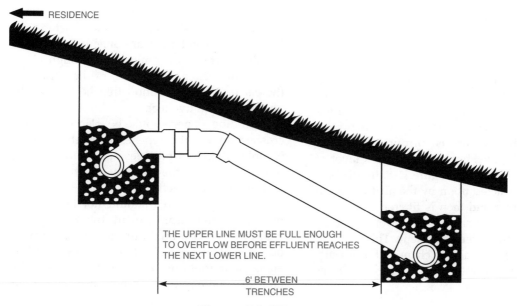

Figure 11–6B Serial distribution

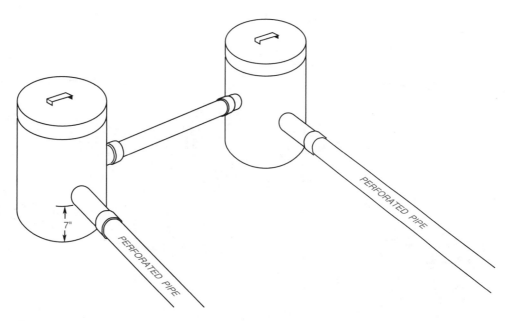

Figure 11–7 Serial distribution for commercial projects might use the "drop box" at the ends of distribution lines.

A Sand Filter Trench or a Sand Mound

These are used in places which have slow percolation. Using the sand filter method is actually sewage treatment rather than sewage disposal. One method has two layers of distribution pipelines, each pair of lines in their own individual trench. Each line is surrounded with its own layer of gravel but there is a thick layer of sand between the lines, Figure 11-8. In this way the effluent is acted upon by the air trapped in the gravel layer and then is filtered as it seeps down through the sand. The lower line then accumulates the filtered effluent. This line can exit into a surface drainage channel if necessary. The design of this system *must be left in the hands of a competent engineer.*

One caution should be remembered when running distribution lines: aerobic bacteria are far more efficient than are anaerobic bacteria. Aerobic bacteria will only have enough air to exist within three or four feet of the surface of the earth. The distribution trenches should be no deeper than five feet.

Building paper should be placed over the crushed stone in the trenches before the earth fill is replaced so that particles of soil do not filter down in between the much larger gravel stones and render them ineffective.

The disposal area should not be situated near the water supply, or in low areas which are apt to be saturated during long rainy periods. It is not a good idea to have trees near this area. Their roots not only tend to invade the area and sometimes the piping itself, but their

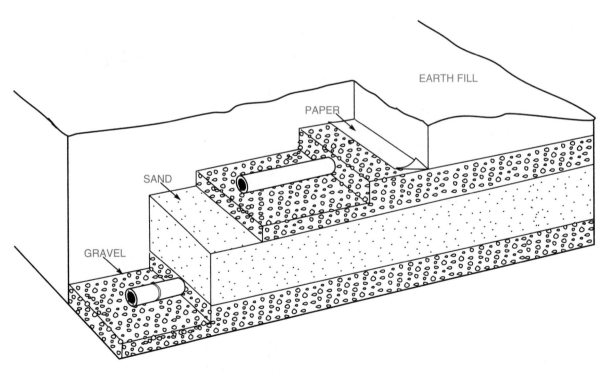

Figure 11–8 A sand filter trench

leaves catch rain and concentrate the runoff water in a circle which coincides with the limit of their branches.

Testing The Soil for Porosity: The Percolation Test

The soil must be tested before any kind of an on-site sewage disposal system is installed. Soil can be so slow in absorbing the effluent that it will collect on the surface and be a health hazard. It is also possible that the soil will absorb the effluent too quickly. This situation is potentially more dangerous than the first, if the subsurface water in the area is used for drinking and cooking. The percolation test determines the soil's ability to absorb water *even when the soil is water-saturated.*

To do a percolation test, select six or more evenly spaced sites throughout the area where the absorption field is planned. Clear away the sod at each point and dig a post hole, 4" to 12" in diameter, Figure 11–9. Remove all of the loose soil from the hole and pour a couple of inches of gravel in the bottom of each hole prepared. The sides of the hole should be scratched up so that the pores of the soil smear closed by the shovel or post-hole digger are opened. Now saturate the soil by filling the hole with water. The saturation of the soil may take up to 24 hours. It is important to not allow the water level to go below 12 inches in the hole. If the soil is sandy keep watering the holes for 4 hours. If the soil is loamy or clay-loam keep water in the holes overnight.

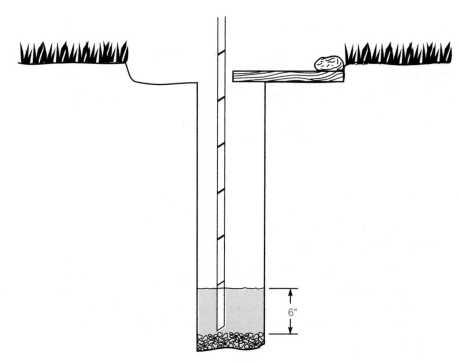

Figure 11–9 A percolation test hole

After 24 hours of soaking, fill each hole to 6 inches above the gravel and record the depth measurement and the time. As 30 minutes of time elapse for each hole, record the new depth measurement of the water. Now divide 30 (minutes) by the difference in inches above the gravel for the first and second measurements for each hole. This will give a minutes per inch drop rate for each hole. Average the rates together for all of the holes and the percolation test will be completed. Now consult Table 11–1 and Table 11–2.

If the soil is so sandy that it is difficult to keep water showing in the percolation hole, use the following procedure. After allowing the soil to "swell up" overnight, keep adding water every ten minutes to the 6" mark on your rule. During the last ten minutes of the hour record the distance that the water level has dropped. Divide this distance into the ten-minute period. Consider a problem like this:

Problem: The water level dropped 1 1/2" in the last ten minutes:

Step 1: divide 10 by 1 1/2".

Step 2: 10 / 1.5 = 6.66.

Solution: The hole has a percolation rate of 6.66 minutes per inch.

CARE OF THE SEPTIC SYSTEM

The people who make a living pumping out septic systems are often surprised when they are

Percolation Time*	Absorption Area Required** Septic Tank	Absorption Area Required** Aerobic Tank
(minutes)	(Sq Ft/Bedroom)	(Sq Ft/Bedroom)
15 or less	175	123***
16 to 30	250	210***
31 to 45	300	300
46 to 60	330	330

   *   *Time required for water to fall one inch in the hole.*
  **   *Divide figures in these columns by 1.5 to determine the length of trenches 18 inches wide or by 3.0 for the length of trenches 36 inches wide.*
 *** *These reduced areas will be authorized for experimental purposes only.*

Note: Most codes say something to the effect: "That if calculations result in an area of under 1000 sq. ft., then the area required will be 1000 sq. ft." This chart is included only for demonstration purposes. It is unlikely that the area in which you ply your trade will have exactly the same requirements. Requirements for septic systems change almost annually. Always consult local authorities.

TABLE 11-1 Sizing the absorption field from the percolation test.

Number of bedrooms	Minimum capacity in gallons	Approximate inside dimensions of rectangular tanks				
		width		length		depth
3 or less	1000	4'0"	X	8'4"	X	5'0"
4	1250	4'0"	X	10'5"	X	5'0"

For each additional bedroom add 100 gallons to the maximum capacity, or about 8 inches to the length indicated.

Note: Again the chart is only for demonstration purposes. Consult local authorities.

TABLE 11-2 Calculating the septic tank size for residences.

called in to pump out a tank which is **not blocked-up.** Few home owners pay attention to the maintenance of their sewage disposal system until it ceases to work properly. Septic tanks and all accessible parts of the disposal system need periodic cleaning out. If the system is allowed to operate until it backs up (fixtures in the building cease to drain out), *damage has been done.* The sludge on the bottom of the septic tank has built up to the point where sludge has been flowing down into the absorption field with each fixture use. This condition will go on for some time before the system finally ceases to function. The septic tank pumper, called a Honey Dipper at one time, may come and pump the tank out. But many holes in the perforated pipe in the absorption field have been blocked by sludge. As a result it may well be that the system only operates satisfactorily until the tank once again fills up with perhaps its 1000 gallons of capacity. When the internal level once again builds up to the point where additional water added results in the same amount flowing out of the outlet, this outflow may be met with no place to exit in a plugged-up absorption field.

Hint:
The septic tank operates in a full condition. If the top of the invert can be seen just above the liquid level, conditions are normal.

Chemicals and excessive amounts of laundry detergent will kill the bacteria operating in the septic tank. The bacteria living in the tank consume a high percentage of the effluent components. If there are no bacteria the sludge will build up very quickly. Some customers will dump gasoline, oil, insecticides, and other chemicals into the openings of their septic system. If their concentration is high, most, or all of the bacteria will be killed. During the period of time it takes the population of bacteria to recover, the sludge level will rise at a rapid rate. Products containing chlorine are especially dangerous. Chlorine is used to kill bacteria at water treatment plants and anywhere where disinfection (the destruction of bacteria) is required. Chlorine is also present in many laundry products. The overuse of these products will damage the system. *Despite advertisers' claims, there are no additives which will restore a septic tank or an absorption field to full usefulness or even help maintain an existing system.* These products are often yeast which foams up when wetted. Yeast will not break down sewage. It is a member of the plant family which is helpful when baking bread.

As a rule of thumb, septic tanks should be pumped out every 3–5 years to remove accumulated sludge. Some systems, depending on design and soil type, are more efficient than others. The pumper will be able to advise the homeowner on the best time interval between cleanouts after the system is pumped and the previous pump-out date is known.

It is perhaps misleading to say that septic tanks are merely pumped out. If this were the case, most of the sludge would be left in the bottom of the tank and the pumpout would have accomplished nothing. The solids in the bottom of the tank must be broken up with a long scoop-like device so that the solids become liquified once again and become pumpable. Some trucks have a "reverse flushing" mechanism. This is a reversing valve on the pump. When the pump has loaded some of the tank effluent into the truck, the operator reverses the direction of the flow. This backflow is supposed to help break up the sludge and make it pumpable.

The rules for good septic system maintenance are simple.

1. Don't allow chemicals to be admitted to the system.
2. Don't overuse laundry products.
3. Have the septic tank cleaned every 3–5 years.

SUMMARY

Untreated sewage is poisonous. The job of the private sewage treatment system is to detoxify the natural waste that any building with human or possibly animal inhabitants might generate. Some sources claim that the work of the plumber ends five or ten feet outside of the foundation walls of that building. At that point, excavation contractors take over the job of setting the septic tank, the distribution box and the disposal field. Some of these contractors specialize only in this kind of work.

But this is not true in all localities. Some plumbing contractors have the excavating equipment and do this work also. Plumbers are often called when a drainage system stops for any reason and may find the need to completely understand the source of the problem. For instance, vigorous use of a sewer rod to clear a blockage may break or knock off the baffle, which is located just inside the wall of the septic tank.

Depending on the complexity of the treatment system, the methods used to break down the sewage and help to render it harmless are:

1. allowing the normal activity of naturally occurring bacteria;
2. screening and filtering;
3. introduction of air; and
4. the application of chemicals and/or ultraviolet light. Large scale systems may use all of these methods.

Given a good percolation test, a simple septic system has a septic tank with a capacity of 1000 gallons or more located 5 to 10 feet outside of the building. This is connected to a distribution box, which in turn is connected to the distribution field. The septic tank holds a quantity of the soil discharge and allows solids to settle to the bottom, where they are broken down by anaerobic bacteria, leaving only an inert sludge on the bottom. The distribution box by its design distributes the effluent evenly to the branches of the distribution field. The distribution/absorption/tile field allows the effluent to trickle out into the soil where the detoxification is completed.

The sludge on the bottom of the septic tank remains there and builds up over the years. If this is not periodically removed, eventually the tank will fill with sludge and the system will fail. This failure will manifest as a sewage line blockage at the building's trap openings.

The percolation test determines the capacity of the soil to absorb sufficient effluent on a more or less continual basis.

◆ ◆

TEST YOUR KNOWLEDGE

1. Can sewage be rendered completely harmless?

2. Engineers at overloaded municipal sewage treatment plants may have to add _____ to the final discharge water.

3. A septic tank should be pumped out at approximately _____ __ _____ year intervals.

4. The serial distribution box system works on the theory that each horizontal drainage line should be _____ before the next lower one is started.

5. Is it normal for the septic tank to be filled or to be half empty?

6. What kind of bacteria live in the septic tank?

7. One disadvantage of an aeration tank is _____.

PUBLIC SEWER TIE-INS

KEY TERMS

building or house sewer	sewer saddle
combined sewers	slants
rain leaders	storm sewer
sanitary sewer	Y-connections

OBJECTIVES

After studying this unit, the student should be able to:

❏ Describe the function and construction of sewers.
❏ Define the sewer lateral.
❏ Discuss the installation of sewer laterals.

PUBLIC SEWERS

A public sewer is maintained and constructed by the municipality. It will usually, but not always, run under and parallel to the street. This pipeline is maintained and inspected with the aid of *manholes*. These are placed at strategic locations in the street surface itself. There are often two sewer lines under the street, the sanitary sewer and the storm sewer. The sanitary sewer carries human waste and the storm sewer carries runoff water.

The individual branches of the public or city sanitary sewer eventually join together and terminate at a *sewage disposal facility*. The branches of the storm sewer may terminate at any water runoff channel, stream, or body of water.

In cities and towns, the street corners often

have inlets to the storm sewer located at the curbs. The **rain leaders** from the roofs of buildings and homes may exit at the street curbs or directly into the city storm sewer. The plumbing student might pause and reflect on the result of tying a sewer lateral into the wrong kind of the city sewer line.

Most plumbing codes require that homes within a certain distance of a public sewer (200 feet in the case of the BOCA plumbing code) be connected to that sewer.

The exact depth and location of the sewer line and the location of existing sewer inlets may be obtained from local administration officials.

Sewers are constructed of plastic, cast iron, reinforced concrete, or vitrified clay pipe. Large city sewers may be built of bricks and mortar.

Combined sewers receive both storm water and sewage. Combined sewers are still used in some parts of the country, but federal and state governments no longer give grants for their construction.

Sewers receiving storm water are egg-shaped, Figure 12–1. This allows them to be kept clean with a minimum of water. Their discharge is carried to rivers, lakes, oceans, or an approved terminal. Where sewage is treated in a disposal plant, rainwater is not discharged into the sewer because it does not require treatment and overloads the plant.

When a new sewer is laid, outlets are inserted for each building lot. These outlets are called **slants** or **Y-connections**. The branch is inserted above the center of the sewer at a 45-degree angle, if possible, to keep the house sewer above water level. The branch pipe must never extend inside the public sewer because it would form an obstruction, Figure 12–1 and Figure 12–2.

There is some danger, when inserting a new sewer outlet into a public sewer, that the main pipe will be broken by the cutting-in process.

Often the plumbing inspector will require that he or she be there to observe this process.

SEWER CONNECTIONS

Slants, or sewer connections, are usually inserted in sewers when they are constructed, Figure 12–3. Occasionally, however, it may be necessary to add a slant to an existing sewer.

To insert a terra-cotta slant in a brick sewer, mark the location on the outside of the sewer above the center (up on a 45-degree angle if possible) to keep it above the water in the sewer. With a hammer and cold chisel, carefully cut a round hole slightly larger than the slant through the wall of the sewer.

Some codes require that the hole in the sewer wall be cut with a special core bit. Always check with the local municipal authority before cutting into a city sewer.

Measure the thickness of the sewer wall and cut a piece of terra-cotta pipe the same length from the back of the bell to the end (L). The slant should be flush with the inside of the sewer, Figure 12–4.

The bell of the slant is placed against the sewer wall to prevent it from slipping into the

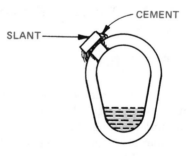

Figure 12–1 Storm sewer

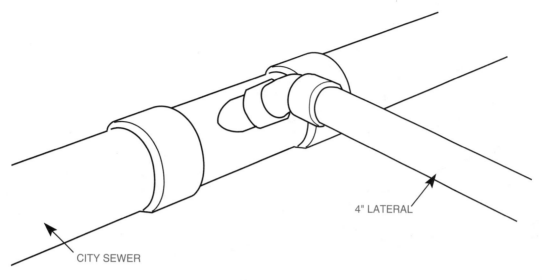

4" LATERAL

CITY SEWER

Figure 12–2

sewer if the cement joint does not hold. The cement should be mixed in small quantities consisting of one part cement to two parts clean sand. If mixed in large quantities, it becomes hard and unfit for use.

Cast-iron pipe connections are made similarly, except that a whole length may be used at the sewer. The joint of the cast-iron pipe prevents the pipe from slipping into the sewer.

A slant may be inserted in a terra-cotta sewer, but great care must be used when cutting the hole in the sewer.

It may be a requirement that a concrete block be poured around the sewer saddle to give it secure support. A sewer saddle is a fitting complete with straps to go around the main sewer, which is a requirement in some localities, for an added outlet (Figure 12–5).

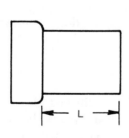

Figure 12–3

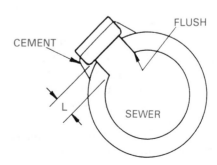

CEMENT

FLUSH

L

SEWER

Figure 12–4

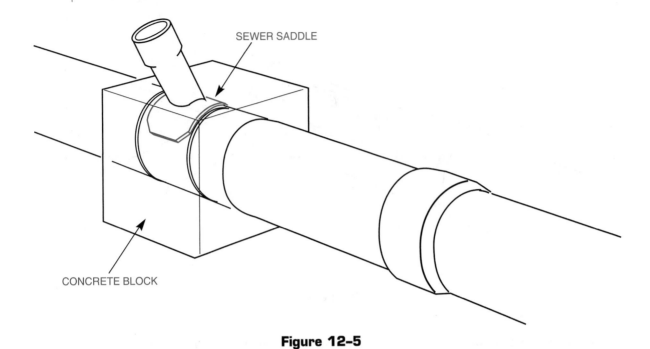

SEWER SADDLE

CONCRETE BLOCK

Figure 12–5

HOUSE SEWERS

The **building (house) sewer** is the part of the drainage system that extends from the end of the building drain and carries its discharge to a public or private sewer, an individual sewage disposal system, or some other approved point of disposal. Terra-cotta pipe is used for the construction of house sewers in some places, while cast-iron pipe is required in others. The house trap, Figure 12–6, is a subject of some controversy among authorities and is not permitted in some localities. Local codes may permit materials other than cast-iron soil pipe and terra-cotta pipe. Consult the local or state code for a list of approved materials. If terra-cotta pipe is used, the size should be one size larger than the house drain. The objection to terra-cotta pipe is that the joints sometimes break and tree roots enter and grow inside, eventually clogging the drain.

Trenches should be hollowed out under each bell when preparing for terra-cotta pipe. This permits the side of each length to lie flat on the bottom of the trench, giving it a solid bearing.

The hollow spots permit the making of a good joint on the bottom of the pipe. A ring of oakum should be packed into the joint and then cemented with a mixture of one part cement to two parts clean sand.

It is important that the insides of all lengths line up evenly. Otherwise, the joints would form an obstruction to the flow of sewage. After the joint is finished, a swab should be inserted and the excess cement should be removed. If this step is neglected, an obstruction is formed at each joint.

Terra-cotta or vitrified clay pipe will last indefinitely. Cement joints, however, do not. This weakness has been corrected in recent years by a new type of joint. This is the prefab-

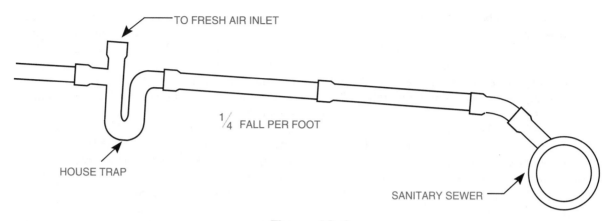

Figure 12–6

ricated joint. It is sometimes referred to as slip-seal. This type of joint remains watertight and strong and is much simpler to install.

Some pipe materials that may be permissible for house sewers are:

Acrylonitrile butadiene styrene (ABS plastic)
Asbestos cement
Bituminized fiber
Cast iron
Concrete
Copper tubing (type K or L)
Polyvinyl chloride (PVC plastic)
Terra-cotta clay

All of these pipe materials must be laid on a firm soil base. Care should be taken so that no more soil is removed from the trench than is necessary. If the trench is too deep, crushed stone must be added in 6-inch layers. Each layer should be tamped down before more stone is added.

Thermoplastic Sewer Laterals

Plastic pipe materials are very durable, and quite suitable for sewer pipe installation if the work is done carefully and with special consideration for the material being used. This material must be installed with these thoughts in mind:

1. PVC and most plastics for sewer work have thin walls and therefore a "springy" nature. Pipe which is not properly supported underneath along its whole length, may be pressed down by heavy traffic on the surface above and split. Earth fill which is heavily compacted over the pipe may compress the circular cross-section of the pipe into an oval and create a static stress in the pipe. When the sewer lateral is going to be plastic, compact the fill in the bottom of the trench to provide support along the whole length of the pipe. Soft, spongy soil will need the addition of compacted, crushed stone, followed by a finely divided material like sand to separate the pipe from the stone. When backfilling make sure there are no rocks in the fill until the pipe has been covered by at least six inches. Do not overdo the soil compacting above the pipe. Bury plastic sewer pipe

at least 12″ below the deepest expected frost line.

2. Plastic pipe materials are especially subject to abrasion damage. Pipe underground does move from vibration and, to a small degree, from changes in temperature. *The pipe should touch no rocks larger than 1/2″* and should be protected with a plastic sleeve at all pipe penetrations through masonry construction.

3. Plastic pipe has a high expansion rate. Even though the temperature underground is exceedingly constant throughout the seasons, it does change. Hot water from bathing and laundry facilities will also cause expansion in the pipe. Long runs of pipe should be snaked from side to side within the trench to allow for this.

4. If solvent/cement (glued) joints are being used, two men are necessary to join the pipe when it is 4″ or larger in diameter. The glue used on plastic pipe joints ``sets up'' too quickly for one person to spread the glue evenly on both pipe and fitting socket, and then have enough time to press the spigot end of the pipe all the way into the bottom of the fitting and give it a 1/4 turn.

SUMMARY

A public sewer is constructed and maintained by the municipality or one of its designated agents. When a public sewer comes within 200 feet of a home or property, the municipality requires that the homeowner use the public facility.

Often there are two sewer lines provided. One is a sanitary sewer to handle soiled water, and the other is a storm sewer whose purpose is to provide a safe channel for excess storm water.

To hook a rain leader to a city sewer will overload the disposal plant. To hook a sanitary waste line into a storm sewer will pollute the water of the creek, river, or runoff channel that the storm empties into. Storm water is never treated.

Outlets provided by the builders of sanitary sewers have sewer slants provided at each anticipated building lot. Occasionally, as the community matures, more buildings than originally anticipated are constructed. These require new sewer slants to be inserted in the old pipeline. Great care must be exercised to: (1) not break the original pipe, and (2) not allow the new sewer slant to protrude inside of the sewer's pipe wall, thereby providing a place for an obstruction to build up. One way to keep the pipe from protruding in too far is to cut the first piece of pipe (the sewer slant) so short that the hub will prevent excess penetration.

Houses have sewers too. A house sewer starts at the end of the main house drain, which extends 5 feet from the house all the way to the city sewer. The city sewer often runs under the street. It was the practice at one time to install a trap in the house sewer with a vent so that the gases from the city sewer could not pass into the house's system. This has fallen into disfavor and is not seen much except in old installations.

The correct operation of the house sewer depends greatly on the trench originally dug to receive it. If the trench is dug too deep, then flat, firm masonry must be embedded in the bottom of the trench to support the pipe. If the pipe is plastic, then there is danger of the pipe wearing on these foundations. In the case of plastic pipe, the better solution is to pour and tamp stone no larger than 1/2 inch in the bottom of the ditch to bring it up to proper grade. And, of course, the very best solution is to dig the trench carefully and to perfect grade from the beginning.

TEST YOUR KNOWLEDGE

1. To find the depth of a sewer line in a particular locality, you should _____.

2. Some municipalities use the same sewer for both the sanitary and the storm water drainage. What do you see wrong with that?

3. Why does the sewer slant go down into the city sewer at a 45 degree angle?

4. Plastic pipe must be protected from _____.

5. A storm sewer carries _____.

6. A combined sewer carries both _____ _____ and _____.

EXCAVATION

KEY TERM

backfilling

OBJECTIVES

After studying this unit, the student should be able to:

❏ Excavate trenches properly.
❏ Understand the safety procedures for excavation.
❏ Backfill a trench properly.

TRENCHING

Most trenching is done by machine. Trenching of shorter distances or inside the foundation walls of the house may be done by hand. The tools that are used most often are the round-nosed shovel, the pick, and the mattock. The pick and the mattock are used for breaking up harder soils before the soil is removed with the shovel. A trench that is carefully laid out and dug makes the job of pipe installation much easier.

Before trenching is begun, the proposed ditch should be laid out with a string. The string should be suspended on one side of the ditch. The string should also have the proper fall for the pipe that is to be installed. In this way, time-wasting mistakes may be avoided. The person who is excavating the ditch needs only to follow the string. If the depth of the trench is kept the same distance below the string throughout, the trench will have the proper fall. Avoid digging too deeply. The pipe should lay on unexcavated soil. This ensures that a minimum amount of settling will take place. The deeper a trench must be excavated, the wider it must be opened from side to side. For example, a ditch that is 9 or 10 feet deep may be 4 feet wide. Bell holes must be provided at each joint when bell-and-spigot piping is to be installed. By digging a bowl-shaped hole at each joint, the plumber allows hubbed pipe to lay flat in the trench rather than be supported only at the bells. The bell hole also provides room to work the tools all the way around the pipe. If the soil is swampy or soft, the plumber will have to tamp crushed stone or masonry into the bottom of the trench to support the pipe.

SAFETY

The plumbing student, perhaps more than any other trade, must be aware of the danger of working in holes and trenches. The most common fatal accident for this trade results from cafe-ins, or asphyxiation in manholes with poisoned air or with air containing insufficient oxygen. The student is referred to the Occupational Safety and Health Administration's (OSHA) free publication, "Selected Construction Regulations (SCOR) for the Homebuilding Industry (29 CFR 1926)." This publication is maintained by OSHA on the World Wide Web in a convenient downloadable format at: http://www.osha.gov.

Some kind of a safety assurance system must be provided to workers who are working in ditches over 5 feet in depth. Excavations made entirely in stable rock are considered safe. In excavations less than 5 feet in depth that have been examined by a competent person and declared hazard free, work may proceed without special shoring or special excavation techniques. The philosophy here is that a standing worker's head would be above the earth in the event of a cave-in. It must be pointed out that the plumber and pipefitter working in trenches spend much of their time bent over or on their knees. Earth may begin to fall without discernable sound. For this reason, there should always be a watcher standing near the trench to call out a warning. Be especially alert when bulldozers and other heavy equipment pass close to the trench. Water in a ditch weakens the sides as the earth may soak up the moisture and lose its structural strength. Therefore, it is dangerous to work in ditches that have any great amount of water in them. Ditches over 5 feet in depth must have sloping or benched sides, be provided with an approved shoring system, or be designed by a professional engineer. Remember that weather changes things. The safe ditch before a prolonged rainstorm may no longer be safe and must be reinspected.

A ditch left unattended must be barricaded and marked with safety lights.

Beware of entering tanks and manholes that have *not* been mechanically vented after opening. Often there is not enough oxygen in the atmosphere of these spaces to support human life.

BACKFILLING

Backfilling is the replacement of soil that was removed during excavation. All pipe trenches must be backfilled by hand until the pipe is completely covered. This gives the pipe protection from rocks that may fall in during machine backfilling. As the backfilling proceeds, the fill material should be tamped down from time to time.

TEST YOUR KNOWLEDGE

1. How deep should the trench be dug?

2. What is a bell hole?

3. Should all ditches, even one that is only 3 feet deep, be approved for work by a competent person?

4. At what depth must a ditch have some kind of safety system in place for the workers' protection, whether it be sloping, stepping, shoring, or some other approved method?

5. Why should the first backfilling be done by hand?

GRADE

OBJECTIVES

After studying this unit, the student should be able to:

❑ Understand how to determine the grade of drains.
❑ Apply this knowledge to a specific job.
❑ Understand the purpose of grading hot-water heating pipe, steam mains, and drainage line.
❑ Grade pipe to the proper level.

DETERMINING THE GRADE

The term grade refers to the slope of a pipe line. Most Bureau of Health Regulations require a 1/4 inch per foot minimum grade for main house drains. The student will note that this is the minimum grade permitted. Because of flow characteristics in different sizes of partially filled pipes, the optimum fall per foot is different for the various pipe sizes. With the proper grade, sewage should flow through the pipes at a velocity of 260 feet per minute. At this speed, pipes are self-scouring.

If the pipe is on too steep a grade, the water runs faster than the sewage. This leaves the sewage in the pipe, which causes additional

137

gases in the system. If the grade is not steep enough, the pipes may not be properly flushed, possibly causing stoppage. Small waste pipes must have a steeper grade than large ones in order to obtain the same results.

The formula for grading drains is shown on page 140. This formula can be applied to a typical trade situation.

Hot-water heating pipes must be on an upward grade about 2 inches to 3 inches per hundred feet to assist the circulation of water and to permit all the air that is liberated from the water to ascend to the air elimination points where it may be relieved.

Steam mains must be on a downward grade from a high point at the boiler to allow condensation to flow in the same direction as the steam. The grade of steam mains is 2 1/2 inches per hundred feet.

Drainpipes are on a grade toward the sewer at 1/4 inch to 1/2 inch per foot, depending upon the size of the pipe and the length of the drain so that the sewage will flow at a rate of 260 feet per minute. At this rate, the pipe is self-scouring and sewage is properly disposed of.

Grading is done with the use of a *level*. The level is placed on the pipe, and a block of wood corresponding in thickness to the grade required is placed under the low end of the level. When the air bubble is level (in the center of the glass), the pipe is on the proper grade.

The length of the level must be considered. If the level is 2 feet long and the grade is 1/4 inch per foot, the block under the level must be 1/2 inch thick, Figure 14–1. In running long lines of pipe within a building, the grade is figured by measuring from the floor or ceiling. It must first be determined if the floor or ceiling is level.

If it is impossible to place the level on top of the pipe, it may be held on the bottom of the pipe by placing the block at the opposite end, Figure 14–2.

To install a long line of pipe supported from the ceiling determine the proper grade, install the pipe hangers according to the line of the grade, and erect the pipe.

On longer runs of pipe, the use of the level can be inconvenient. And, of course, as this short measuring instrument is used again and again, errors will begin to accumulate. Therefore, plumbers will use the builder's or architect's level, Figure 14–3, on longer trenches. This is an optical instrument that is set up well to one side of the proposed ditch. This optical level has adjustments that allow it to be adjusted so that the telescope is perfectly level in whatever direction it is pointed. The telescope has crosshairs in it much like a rifle's telescope. As the telescope is moved about, all points upon which the crosshairs rest can be relied upon to be level with all other points upon which the crosshairs have touched. As can

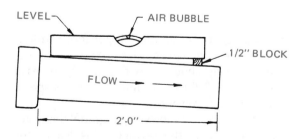

Figure 14–1

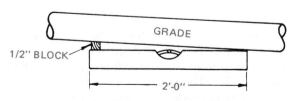

Figure 14–2

be imagined, this can be a very valuable tool.

With the use of a stadia rod, Figure 14–4, it becomes simple to measure down from this line of sight to the ground surface or, as the case may be, to the bottom of a trench. The stadia rod is essentially a measuring rule. If the lower numbers on the rule are placed down and the rod is held perfectly vertical, a measurement to the bottom of the ditch can be taken by looking through the telescope and reading off the measurement upon which the cross hairs are resting. The drop of the ditch between two points can be easily calculated by subtracting stadia rod measurements taken at both positions.

A laser beam is a small diameter beam of light of very high intensity. It has the advantage of not spreading out or becoming diffused like an ordinary flashlight beam, Figure 14–5. The laser level can project a small red dot of light on any surface many yards away. If the laser is

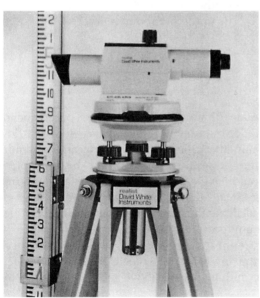

Figure 14–3 *(Courtesy of David White Instruments)*

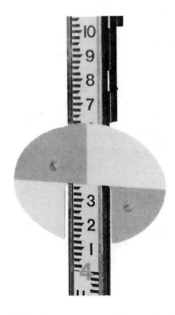

Figure 14–4 *(Courtesy of David White Instruments)*

Figure 14–5 *(Courtesy of David White Instruments)*

The Formula

The formula for grading drains is shown below. This formula can be applied to many trade situations.

$$F = \frac{L}{10d}$$

F = total fall or grade in feet
L = length in feet
d = diameter of the pipe

Example:

Find the grade per foot of a 2-inch sink waste 8 feet long to give a velocity of 260 feet per minute to the waste water.

1. 2″ diameter × 10 = 20
2. 8′ length of waste ÷ 20 = .4 of a foot fall
3. .4′ × 12″ = 4.8″ or 4 3/4″ total fall per 8′ length
4. The grade per foot would be 4.75″ ÷ 8′ or 5/8″ per foot (fall).

SUMMARY

Certain piping must be installed on an incline (grade) in order to utilize gravity to assist flow within the pipe. Different applications require different grades. Warm-water heating pipes should have a drop of 2 to 3 inches per hundred feet. Steam drains require about the same grade to facilitate the return of condensate to the boiler. Drainpipes are sloped so that the flow will be optimum for self-scouring. The optimum speed of the flow within the pipe is about 260 lineal feet per minute. Speeds above and below this by too

mounted level, this property of the laser can be used much as the builder's level is used. When installing larger pipe, the plumber can place the laser level in the trench and shine the dot of light on the pipe's destination. Then the pipe is installed so that the laser beam shines right up the center of the pipeline. In this way, each length of pipe is automatically installed straight and with the proper fall or grade built in.

Caution:
While this *cold beam* laser light is not dangerous to touch, do not look directly into the source while it is on.

great a margin will cause poor self-cleaning of the pipe's interior. Most plumbing codes accept 1/4 inch per foot fall as a good working average. New, smoother surfaced plastic and copper drainage pipe may make this application of different "falls" to different-sized pipes obsolete.

For short runs of pipe, the incline can be sufficiently controlled with a level and the application of spacers to allow the level to rest on the inclined pipe and yet remain level. Larger jobs require the use of the builder's or architect's optical level to establish a reliable reference line over a wide area. The building's level is positioned at any convenient place, and with the use of adjustment screws and leveling bubbles, the head of the instrument is set perfectly horizontal in any direction it may be pointed. Now any place seen through the crosshairs will be per-fectly level with any other place in the crosshairs. From these reference points, measurements can be taken to level or adjust grade in pipes.

The laser beam can be used to establish reference points in the same way, but in this case a red dot projected on any surface with the instrument will be at the same vertical height as any other red dot it projects. Another more interesting use of the laser is to place it in the trench and set it to project a beam with 1/4 inch per foot fall in it right up the center of the ditch. Now all the plumber must do is keep laying the straight lengths of pipe so that the red beam projects out of the center of the far end. In this way, perfect alignment and perfect fall are automatically achieved and the labor of leveling and aligning each length is much reduced.

◆ ◆

TEST YOUR KNOWLEDGE

1. The proper speed of waste water in drainage pipes is _____.

2. How thick of a block would you place under one end of a two-foot level so that the pipe it is used to install has 1/2 inch per foot fall in it?

3. The stadia rod is used to _____.

4. The builder's optical level establishes a horizontal plane from which _____ may be taken.

5. Why should pipe be lower on one end than the other?

6. Can the laser level be used to do the same job as the builder's level?

MAIN HOUSE TRAPS AND FRESH AIR INLETS

OBJECTIVES

After studying this unit, the student should be able to:

- ❏ Understand the function of the main house trap.
- ❏ List the types and locations of main house traps.
- ❏ Explain the flow of fresh air within the sanitary system.

MAIN HOUSE TRAP

The student may encounter a building trap. The fitting pattern used is that of a running trap, and this outdated feature is usually found in the main building sewer but may also be found outside of the house. Figure 15–1 shows a cast-iron running trap with two vent openings. In use, the house side of the trap has a fresh air inlet installed, and the city sewer side usually receives a trap screw

ferrule with a plug and is used for a cleanout opening. The fresh air inlet admits air to ventilate the house drainage system. These traps are made in sizes of 4″, 5″, 6″, 8″, 10″, 12″, and 15″. Inside buildings, both openings may be plugged and the fresh air inlet arranged as shown in Figure 15–2. Depending on where the trap is placed, it may be referred to as a yard trap or a curb trap. In earlier times, the city sewer was often constructed of brick and was made large enough to

Figure 15–1 Cast-iron running trap with two openings used for cleanouts

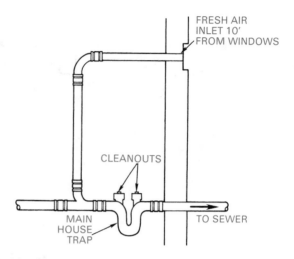

Figure 15–2

allow human access. This rough, often crumbling, and oversized interior was home to all types of vermin. The purpose of the fitting was to isolate the building's sanitary system from the city sewer with a water trap.

FRESH AIR INLET

Horizontal cast-iron and galvanized steel pipe may corrode from the inside and in time may actually become thinner on the top. Eventually, pinholes may develop, or in some cases, the top of the pipe may even break through. This is caused by sewer gases and moisture combining and hanging in droplets from the inside upper surface of the pipe. Small waste flows do not completely fill the pipe and so the droplets may remain for days. Eventually, the mild acid formed corrodes the surface from which the droplets hang. It was thought to be important to

maintain a continual flow of fresh air throughout the system. This would dry the droplets of remaining moisture. The plan for this air flow is illustrated in Figure 15–3.

The fresh air inlet is usually joined to the upstream (house) side of the main house trap. If there is little danger of the pipe becoming an obstruction, a simple mushroom-type cap is placed on top to keep dirt and debris from entering. On the other hand, if the house trap is located at the curb line, then the fresh air inlet may exit through the pavement. In this case, a vent box is used and it must be set flush.

Two types of vent boxes are shown in Figure 15–4 and Figure 15–5. In Figure 15–4 the openings are protected from debris by angular vanes on the inside cover. In Figure 15–5 a dirt pocket is directly under the louvered opening and the air must offset after it enters to reach the fresh air pipe itself.

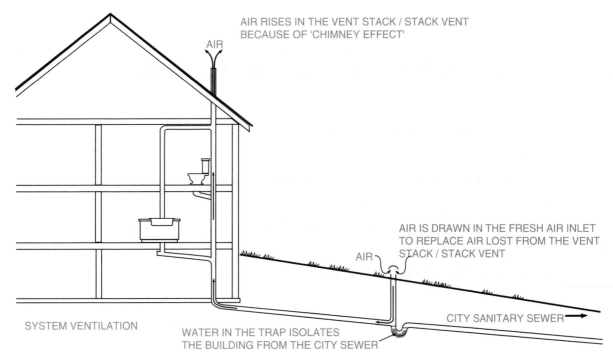

AIR RISES IN THE VENT STACK / STACK VENT
BECAUSE OF 'CHIMNEY EFFECT'

AIR

AIR IS DRAWN IN THE FRESH AIR INLET
TO REPLACE AIR LOST FROM THE VENT
STACK / STACK VENT

AIR

CITY SANITARY SEWER

SYSTEM VENTILATION

WATER IN THE TRAP ISOLATES
THE BUILDING FROM THE CITY SEWER

Figure 15–3

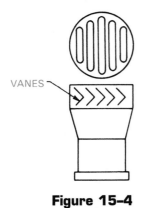

VANES

Figure 15–4

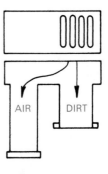

AIR   DIRT

Figure 15–5

TEST YOUR KNOWLEDGE

1. What is the purpose of the house trap?

2. Where would you expect to find a house trap in a system?

3. What is the fitting pattern used for a main house trap?

4. Where would you expect to find a sanitary fresh air inlet?

5. What was the purpose of the fresh air inlet?

◆ ◆ ◆ ◆ ◆ ◆ ◆ ◆ ◆ ◆ ◆ ◆ ◆ ◆ ◆ ◆ ◆ ◆

HOUSE DRAINS
AND SEWERS

KEY TERMS

back water valves
building or house drain
main house sewer

OBJECTIVES

After studying this unit, the student should be able to:

❑ State the location and purpose of house drains.
❑ State the location and purpose of house sewers.
❑ Describe the proper concerns about installation and mainte-
nance of each system.

HOUSE DRAINS AND HOUSE SEWERS

The building (house) drain is the lowest horizontal drainpipe *within* a building, Figure 16–1, although in some cases it may continue to a point 5 feet outside of the structure's walls. It receives discharge and waste from other branches and carries them to the building sewer. The building sewer begins where the building drain ends and continues to the city sewer or other approved point of termination. The main house drain (building [house] drain) is inside the house and the main house sewer is outside.

The size of the drain is determined by the number of fixtures to be drained, but other factors must be taken into consideration. Local or state codes outline the procedure for arriving at the proper size of drains and set minimum requirements regarding size and grade. House drains are usually on a grade of 1/4 inch per foot toward the sewer. Larger drains, however,

147

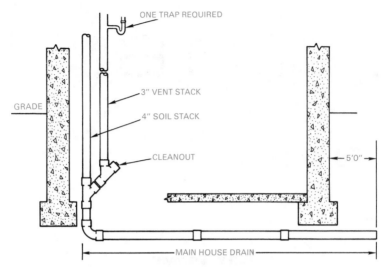

Figure 16–1

may have less grade. It is desirable to obtain a reasonable flow within the pipe, but not an excessive one or a speed that is too slow that allows solids to settle out. Waste and drainage systems are designed to carry their contents smoothly and at the proper rate of speed from the house stacks and branches to the point of disposal. All runs are therefore kept as straight as possible, and all turns in direction are achieved with long sweep fittings. Assembled fittings and pipe should show a smooth interior and present no projections to catch and hold undissolved materials. The minimum size for house drains is 3 inches.

HOUSE (BUILDING) DRAINS

If the house drain is above the basement floor, it may be constructed of service-weight cast-iron soil pipe and fittings, galvanized steel pipe with recessed fittings, ABS and PVC plastic pipe, brass pipe, or copper pipe and tubing. When the house drain is below the basement floor or in large buildings, the material must conform, in type and weight, with local code standards. The materials for below ground installation are generally the same, but the weights may be heavier.

HOUSE (BUILDING) SEWERS

The house (building) sewer can be constructed of any of the materials used for the house drain, in addition to asbestos cement pipe, concrete pipe, or vitrified clay pipe. Note that the weights of the materials used may, and probably will, be of greater weight and strength.

The foundation walls which the house drain passes through must be made watertight to keep ground water out of the basement. In low sections, near rivers, and other water channels subject to flooding, backwater valves (a kind of a check valve) are placed in house sewers. This prevents city sewer waste from backing up into the building's drainage system during extremely high tides or when storm conditions prevail.

When house drains are located within 2 feet of the floor, they are supported on brick or concrete piers. All brick or concrete piers

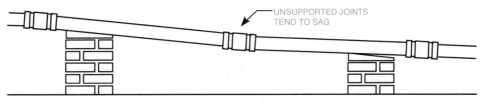

Figure 16–2

should be at least 8 inches square and built on solid earth or on the basement floor. In cold weather, piers should not be built on frozen ground. This would cause settling of the pipe when the ground thaws. Unsupported joints tend to sag, Figure 16–2. Pipes running horizontally at any height should be supported at every joint for cast-iron soil pipe, at 10-foot intervals for copper pipe, at 12-foot intervals for steel pipe, and 4-foot intervals for rigid plastic pipe. The most important place for a support on a house drain is directly under the stack. A stack in a small dwelling may weigh 300 pounds, whereas those in large buildings may weigh much more. The proper way to support a stack is shown in Figure 16–3. If the drain runs close to the floor, a stack base fitting might be a good choice. If the house drain is more than 2 feet above the basement floor,

hangers may be a better choice.

SUMMARY

The horizontal drains that carry the sewage from a building are called "main." These, the main house drain and the main house sewer, are treated in a special fashion by most plumbing codes. The materials used on one may be different from the other. In order to facilitate the smooth and efficient flow within the pipelines, authorities may strictly limit the angle that the line of pipe may turn with one fitting. Because this pipe is usually hidden, cleanouts may be required at specific locations such as at the base of the main house stack, where the main house drain exits the building, or every 50 running feet in the building sewer line.

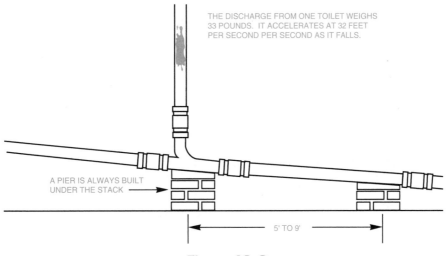

Figure 16–3

♦ ♦

TEST YOUR KNOWLEDGE

1. In what room of a house would you expect to see the main house drain?

2. The main house sewer runs between the main house drain and the _____?

3. What determines the size of the main house drainpipe?

4. What is the usual code mandated rate of fall for the main house drain?

5. What may have to be installed in main drains when the buildings they serve are located in low-lying areas or where canals or rivers might flood?

6. How close must you space the hangers or piers on a horizontal run of 4″ cast-iron soil pipe?

◆ ◆ ◆ ◆ ◆ ◆ ◆ ◆ ◆ ◆ ◆ ◆ ◆ ◆ ◆ ◆

ANGLES OF BRANCHES

KEY TERMS

combination wye and eighth bend
elbow or ell
return elbow
sanitary tee

OBJECTIVES

After studying this unit, the student should be able to:

❑ State how the angles of branches affect the flow in sewage lines.

❑ Describe how a branch enters a main drain.

Few branch wastes and/or main drains flow continuously. It is the nature of sanitary systems that flow be intermittent. When a fixture is used, the drain line may be full or nearly full for a time, then the branch or drain is empty again. When the plug is pulled on the fixture or the water closet (toilet) is flushed, water rushes down the pipe. The plug of water may accelerate to speeds approaching 125 miles per hour. Water has considerable weight (8.338 lb. per gallon). If the two lines are joined in a tee, Figure 17–1, there would be a tremendous hammering effect when this slug of water hits the bottom. When the pipe and fittings are made of cast iron or galvanized steel, rust will eventually accumulate at that point. In any pipe, especially kitchen branch wastes and laundry wastes, grease and soap crud may build up in time at the joining point if sufficient flow is not maintained.

When the flow from a branch joins with the flow of a horizontal main drain, the two streams must come together without undue splashing, backflow, and resulting loss of impetus. From a distance, the joining angle may appear to be at 90 degrees because the branch forms an upside-down tee with the main drain. Closer inspection, however, shows the fitting that joins the two turns so that the flow from the branch is traveling nearly in the same direction as the flow in the main drain before the two currents join together, Figure 17–2.

On vertical pipes, offsets or changes in direction are made with 45-degree bends so that rust scale, foliage, birds' nesting materials, and other accumulations fall to the bottom to be carried away with the normal day-to-day usage of the system.

BRANCHES IN SOIL AND WASTE STACKS

The best fitting to add a branch to a soil or waste stack is a sanitary tee, Figure 17–3. The necessity of directing the flow is to a large degree taken care of by gravity in a stack, which is, if course, a vertical pipe. If a fitting with a long downward curved branch were used, this would place the outlet of the pipe needlessly low with the risk of the water coming down the pipe filling the pipe completely at point. It is important that air be allowed to pass over the surface of the water in the pipe to prevent a siphonic action from taking place.

ANGLE MEASUREMENT

Angle is defined as the amount of space between two lines that are connected at one end. The point at which the lines meet is the *vertex*, Figure 17–4. The length of the lines makes no difference in the size of the angle.

There are 360 degrees in a circle; therefore, one degree equals 1/360 of a circle. Angles are measured by using a *protractor*, Figure 17–5. A protractor is marked in degrees around a curved edge.

To measure an angle, place the straight edge of the protractor along one side of the angle with the center mark of the protractor on the vertex. Read the number of degrees at the point where the opposite side of the angle crosses the

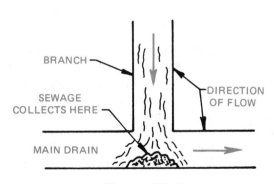

Figure 17–1

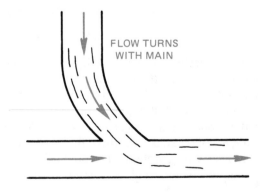

Figure 17–2

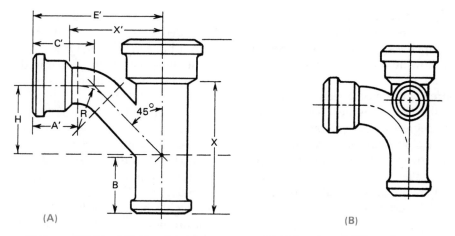

Figure 17–3 (A) Combination wye and 1/8 bend and (B) sanitary tee

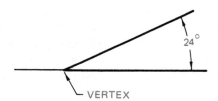

Figure 17–4

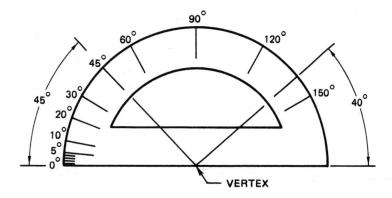

Figure 17–5 Protractor

curved side of the protractor.

An elbow (ell) is a fitting that changes the direction of pipelines. Elbows are generally used in the following angle patterns: 22 1/2°, 45°, 90°, and 180°.

To measure the angle of an elbow, place the protractor on the elbow, Figure 17–6, and read the number of degrees that the elbow turns.

In drawing, 180 degrees is considered a straight line. However, in plumbing, a 180-degree elbow is called a return elbow because it returns the pipe to the same direction from which it started, Figure 17–7.

SUMMARY

Fittings for drainage, waste, and vent (DWV) piping must be selected with care. The prevailing objective in this kind of design is to control and maintain the flow within the system. Occasionally, the purpose may be to *slow* the speed with offsets as in the stacks of tall commercial buildings. More commonly, the purpose is to maintain an orderly and smooth flow, which sweeps the internal surfaces clean without hammering.

The outlet of a drainage sanitary tee (threaded and recessed fitting) has a branch with a built-in 1/4 inch per foot fall besides a moderate sweep at the throat to direct the flow down the stack or drain. It is possible to install this tee upside down.

Examination of the combination wye and eighth bend reveals that the branch has two 45-degree turns in it, hence the name wye and eighth bend. The ordinary wye has a branch that diverges from the run of the fitting by an angle of 45 degrees. The name for a combination wye and eighth bend, which is applied to materials that are not made of cast iron is the *long turn `TY'* (occasionally called simply a *lateral*).

It is good for the branch to enter in a gradual fashion, but the vertical distance taken up by the branch has to be limited when a branch must be run between a ceiling and the floor above. Depending on the length of the branch, the plumber may need it to begin as close to the ceiling as is possible in order that it will arrive at its destination stack before the rise of 1/4" per foot brings it in conflict with the floor above. Compare Figures 17–3(A) and (B): Which one could be lower in the stack without the throat of the branch projecting through the finished ceiling?

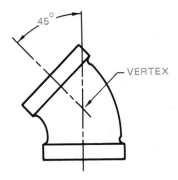

Figure 17–6 Fitting angle

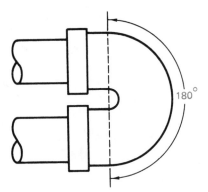

Figure 17–7 Return elbow

TEST YOUR KNOWLEDGE

1. Water weighs about _____ pounds per gallon.

2. The angle of entry into the main flow from a branch should never be more than how many degrees?

3. Rust may eventually be produced by _____ and _____.

4. The better choice for a branch fitting in a stack on the second floor probably would be the _____.

5. What tool is used to measure angles?

6. What fitting has a 1/4 inch fall per foot built in?

◆ ◆ ◆ ◆ ◆ ◆ ◆ ◆ ◆ ◆ ◆ ◆ ◆ ◆ ◆ ◆ ◆ ◆

SIZING THE DRAINAGE SYSTEM

KEY TERMS

drainage fixture unit
maximum fixture-unit load

OBJECTIVES

After studying this unit, the student should be able to:

❏ Explain what a fixture unit is.
❏ Describe how fixture units apply to various plumbing fixtures.

FIXTURE UNITS

After exhaustive research, the American Bureau of Standards has developed the drainage fixture-unit (d.f.u.) method to determine drainage system sizing. This system is based on the gallon per minute discharge of the smallest common fixture, the lavatory, and all other fixtures are compared to that. In other words, if a fixture discharged one and one-half times as much water in a minute as the lavatory, that fixture would be given a rating of 1.5 d.f.u. The quantity of water drained from a lavatory is 7 1/2 gallons per minute. A drinking fountain drains only half as much water as a lavatory; therefore it has a rating of 1/2 d.f.u. The question might be asked: "Why did they not use the drinking fountain for the basic unit?" The answer is that it is not a common fixture found in a home. It also should be kept in mind that these unit values are rough averages that have been smoothed to make sizing charts understandable and easy-to-use.

Table 18–1 shows the number of fixture units

157

Drainage Fixture Unit Values for Various Plumbing Fixtures

Type of fixture or group of fixtures	Drainage fixture unit value (dfu)
Automatic clothes washer (2" standpipe)	3
Bathroom group consisting of a water closet, lavatory and bathtub or shower compartment:	
public	8
nonpublic	6
Bathtub[a] (with or without overhead shower)	2
Bidet	3
Combination sink and tray with food waste grinder	2
Combination sink and tray with one 1 1/2" trap	2
Combination sink and tray with separate 1 1/2" traps	3
Dental unit or cuspidor	1
Dental lavatory	1
Drinking fountain	1/2
Dishwasher, commercial	2
Dishwasher, domestic	2
Food waste grinder	—
Floor drains with 2" waste	2
Kitchen sink, domestic, with one 1 1/2" waste	2
Kitchen sink, domestic, with food waste grinder	2
Lavatory with 1 1/4" waste	1
Lavatory (barber shop, beauty parlor)	2
Laundry tray (1 or 2 compartments)	2
Shower, domestic	2
Shower (group) per head[a]	2
Sinks:	
Surgeon's	3
Flushing rim (with valve)	6
Service (trap standard)	3
Service (P trap)	2
Pot, scullery, etc.	4
Commercial with food waste grinder	
Urinal, pedestal, syphon jet blowout	6
Urinal, wall lip	4
Urinal, stall, washout	4
Washing machine (commercial)	3
Wash sink (circular or multiple)	
each set of faucets	2
Water closet, nonpublic	4
Water closet, public	6
Unlisted fixture drain or trap size:	
1 1/4" or less	1
1 1/2"	2
2"	3
2 1/2"	4
3"	5
4"	6

Note a: A shower head over a bathtub does not increase the fixture value.

TABLE 18–1 Drainage fixture-unit values for various plumbing fixtures. (*Courtesy of National Standard Plumbing Code*)

for a variety of fixtures and the sizes of traps and waste or soil pipes required. For fixtures that flow continuously, such as air-conditioning equipment and sewage ejectors, allow 2 fixture units for each gallon per minute of flow.

The maximum fixture-unit load is the greatest number of fixture units that may be connected to a given size of a house drain, horizontal branch, vertical soil, or waste stack. Tables 18–2 and 18–3 give maximum fixture-unit loads.

PIPE SIZING

When determining the pipe sizes to install on a particular job site, the planner usually works back from the farthest fixture to a point where the main house drain exits the building. The widely accepted fixture unit method of calculating drainage, waste, and venting pipe sizes is easy and accurate. The student must be aware, however, that the locally used plumbing code must be consulted at each step along the way. It sometimes matters whether the building is commercial or residential, is a single-family dwelling, or the pipe is located on the top floor of the building. Drainage piping may have to be oversized to allow the passage of air over the passage of water.

SOME CONSIDERATIONS AND DEFINITIONS

There are two major considerations to keep in mind when sizing a drainage system. These are:

1. The drain or waste size of a drain from any fixture appliance *is never smaller than the trap size of that fixture.* There is one exception to this rule. A 4" drain from a water closet may enter a 3" stack.

2. The pipeline size going in the direction of the city sewer or private disposal system *never gets smaller in size,* with the above-mentioned exception.

These considerations will make sizing the drainage system much easier. The venting system is sized by its own rules and is not considered here.

Some Helpful Definitions

Automatic clothes washer (2" standpipe): If the clothes washer drains into a laundry tray, the value for that laundry tray is used. Nothing is added for the clothes washer. If there is no laundry tray and the clothes washer simply drains into a vertical 2" pipe (this is the minimum size for this kind of application), the pipe must have the capacity to carry the water away as fast as the clothes washer pump supplies it. The drainage fixture unit's value must be larger than when the clothes washer empties into the laundry tray.

Bidet: Pronounced "bid-ay"; a fixture with somewhat the appearance of a water closet, without the tank, for washing the genitals and posterior parts.

Branch interval: Branch lines enter a stack at or near each floor level in a building. This is called a branch interval. A building of offices or apartments will have a branch interval for each floor in the building. Some buildings, parking garages, warehouses, etc., may not have a branch interval for each floor in the building if they do not have waste lines connecting to the stack at all floors.

Domestic: Pertaining to a place where people live and used by people in ordinary living processes. Oppose this term with "commercial."

Drain: Any pipe which carries waste water.

Flushometer valve: A valve which replaces the tank of water on toilets and some urinals. It is

Maximum Number of Fixture Units that may be Connected to:

Diameter of Pipe	Any Horizontal[1] Branch Interval	One Stack of Three Branch Intervals or Less	Stacks with More Than Three Branch Intervals	
			Total for Stack	Total at One Branch Interval
Inches				
1½	3	4	8	2
2	6	10	24	6
2½	12	20	42	9
3	20[2]	48[2]	72[2]	20[2]
4	160	240	500	90
5	360	540	1,100	200
6	620	960	1,900	350
8	1,400	2,200	3,600	600
10	2,500	3,800	5,600	1,000
12	3,900	6,000	8,400	1,500
15	7,000			

[1] Does not include branches of the building drain.
[2] Not more than 2 water closets or bathroom groups within each branch interval nor more than 6 water closets or bathroom groups on the stack.

TABLE 18–2 Horizontal fixture branches and stacks. *(Courtesy of National Standard Plumbing Code)*

Diameter of Pipe	Maximum Number of Fixture Units That May Be Connected to Any Portion of the Building Drain or the Building Sewer Including Branches of the Building Drain.[1]			
	Fall Per Foot			
	1/16-Inch	1/8-Inch	1/4-Inch	1/2-Inch
Inches				
2			21	26
2½			24	31
3		36*	42*	50*
4		180	216	250
5		390	480	575
6		700	840	1,000
8	1,400	1,600	1,920	2,300
10	2,500	2,900	3,500	4,200
12	2,900	4,600	5,600	6,700
15	7,000	8,300	10,000	12,000

*Not over two water closets or two bathroom groups.
1. On site sewers that serve more than one building may be sized according to the current standards and specifications of the Administrative Authority for public sewers.

TABLE 18–3 Building drains and sewers. *((Courtesy of National Standard Plumbing Code)*

supplied with water from a large supply pipe. When the valve lever is depressed, a large quantity of water, under pressure, accomplishes the flushing action in the fixture. This is mostly used in commercial applications. The phrase "water closet—valve operated" refers to this.

House drain: The lowest horizontal drain which carries waterborne waste out of the building.

Lavatory: The small sink or basin found in most bathrooms.

Showers (groups) per head: An open shower area where many people shower at the same time. The drainage fixture unit value is added for each shower head.

Soil pipe: Pipe which carries waterborne human waste.

Stack: A vertical section of the drainage system.

Waste pipe: A pipe which carries any waterborne waste material, including human waste. Many plumbers consider the term **"waste,"** as it applies to plumbing, to refer to waterborne waste *other* than human waste. The word **"soil"** is reserved for human waste.

Water closet: A toilet.

DOING THE CALCULATIONS

Table 18–2 refers to sizing stacks and Table 18–3 is for sizing primarily horizontal drains. Table 18–2 is a multiuse chart. The second column sizes the branches feeding into the stack. The first column on the left contains the required pipe sizes for the fixture unit calculations of all columns to the right of it.

Columns 2 and 3 on Table 18–2 are combined for buildings with three floors or less. Columns 4 and 5 are used together for larger buildings.

In Table 18–3, notice that if the fall for a given size pipe is increased, its drainage capacity is increased. The blank entries in this graph are important. They mean that the size pipe for that fall and that number of fixture units *may not be used*. For a fall of 1/16 inch per foot, the smallest horizontal pipe that can be used is 8 inches in diameter or 8" N.P.S. (nominal pipe size).

Figure 18–1 shows a section of a building that will have the indicated rooms with fixtures added. In this drawing a venting pipe has not been included for clarity.

Step 1. How many branch intervals are there? Three is the correct answer. The cellar floor level doesn't count as a branch interval. This means that values will be taken from the first three columns of Table 18–2.

Step 2. Consider trap sizes and fixtures for this job in Table 18–4.

Step 3. Locate all of the bathroom groups that qualify for a reduced d.f.u. (drainage fixture unit) allowance. See Table 18–1. Since most domestic-type bathrooms are used by one person at a time, and that one person is not likely to run water in all of the fixtures at once, a reduced d.f.u. is allotted to these situations. This is for the purpose of using Tables 18–2 and 18–3 where they specify bathroom groups. The individual drain lines within these grouped fixtures must still conform to the tables.

Step 4. Determine the drain sizes leading from each fixture. A 3-inch drain line is sufficient for a water closet but a 4-inch line is often run, (we will use 3"). Drain lines from lavatories are usually 1 1/2" even though the trap size is 1 1/4" (we will use 1 1/2").

In Figure 18–1, some of the d.f.u. are left blank and some of the minimum pipe sizes (i.p.s.) are left blank. Can you fill in all of the missing values using the tables in the unit?

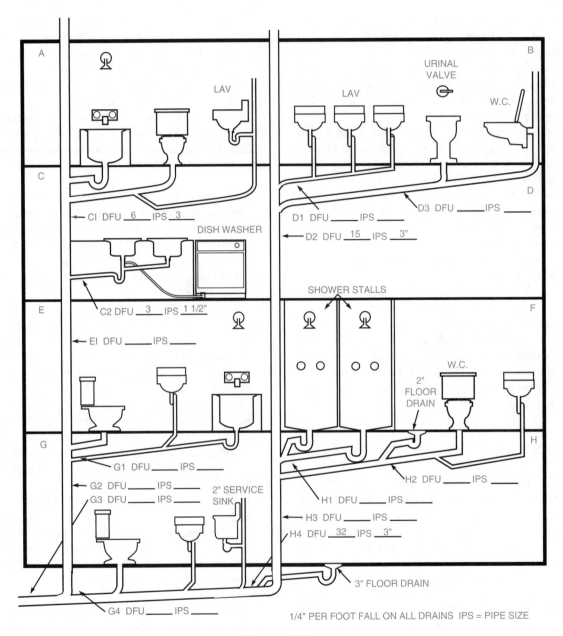

Figure 18–1

Fixture	Trap Size	What Space
Water Closet, tank type	3"	A,E,F,G
Water Closet, flush valve	3"	B
Lavatory	1 1/4"	A,B,E,F,G
Bathtub/shower	1 1/2"	A,E
Kitchen sink/dishwasher	1 1/2"	C
Stall Shower	2"	F
Service sink P trap	2"	G
Floor Drain 2"	2"	F
Floor Drain 3"	3"	H

TABLE 18–4

SUMMARY

The drainage fixture-unit method of sizing drainage systems was developed using the expected discharge of the common bathroom lavatory for a basic value. A 1 1/4 inch pipe handles the drainage from that fixture.

Picture the drainage system as a small tree, with the main sewer being the trunk. As the branches diverge, they become smaller and smaller until at the tips they are twig thin. Then traveling back toward the trunk we can see that as branch twigs and an occasional branchlet are added, the size of the main branch gets larger and larger. The part where the branch enters the trunk must be large enough to accommodate all of the twigs and branchlets that entered the branch. Thus the trunk must be large enough to accommodate all of the branches that enter into it.

This is the way the drainage system is sized. The designer starts out at the very tips of the system and keeps increasing the size of the branch drain or waste back to the main drain (the trunk). The smallest twig is the drain that accommodates the lavatory (1 1/4").

◆ ◆

TEST YOUR KNOWLEDGE

1. What plumbing fixture is the fixture unit based on?

2. How many gallons per minute drainage is a fixture unit equal to?

3. What factors may cause the sizes given on the charts to be increased?

4. Is each floor a branch interval?

5. What does "domestic" mean?

◆ ◆ ◆ ◆ ◆ ◆ ◆ ◆ ◆ ◆ ◆ ◆ ◆ ◆ ◆ ◆ ◆

SOIL STACKS AND STACK INLET FITTINGS

KEY TERMS

roof flanges
soil stack
stack

OBJECTIVES

After studying this unit, the student should be able to:

❏ Explain how stacks are installed and how water flows in them.

❏ Determine the capacity of fittings in gallons per minute.

STACKS

A stack is a vertical drainpipe. A soil stack is a vertical drainpipe that receives the discharge from toilets. Other fixtures may also drain into it.

Soil stacks extend full size throughout their entire length. They extend above a roof not less than 6 inches and not more than 24 inches without support. A stack running within 10 feet of any door, window, or air shaft cannot end less than 2 feet above that door, window, or air shaft.

Soil stacks are constructed of cast-iron soil pipe, copper tubing, galvanized steel pipe, ABS plastic pipe, or PVC plastic pipe. Depending on the individual local code, which may be applicable and superseding, the soil pipe should be of at least service weight, the copper tubing may be of DWV weight, and the steel and the plastic pipe materials should be at least schedule #40. All of the weights or strengths mentioned could be described as standard weight with the exception of the copper tubing. (Copper tubing will be discussed in detail in Unit 44.)

To prevent the gradual accumulation of debris, offsets on vent stacks are made at angles no greater than 45 degrees, that is, a 90-degree elbow would be unacceptable for vent stacks. The stacks enter the main drain at an angle of 45 degrees. If the stack is not directly over the drain, a long sweep bend with a cleanout may be used.

Stacks are supported at the base by piers or hangers. In large buildings, additional supports are placed at each floor by means of beam clamps, Figure 19–1, or riser clamps, Figure 19–2.

In small buildings, soil stacks are placed inside partitions or in a pipe chase (a channel in a wall specifically meant to conceal pipes). The wall must have at least a 7-inch space inside to accommodate a 4-inch pipe hub, Figure 19–3. Pipe chases are also used in very large buildings to provide easy access if repairs are needed, Figure 19–4.

Roof flanges are placed around stacks where they pass through the roof to prevent leakage. In cold climates, small stacks (less than 3 inches) must be increased in size where they pass through roofs to prevent them from being

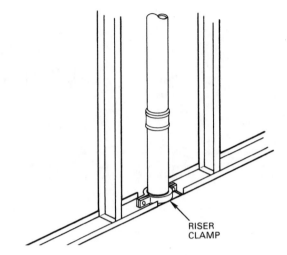

Figure 19-2

closed by frost. Stacks should never be placed outside of buildings. Such stacks are unsightly and may freeze closed with frost.

The size stack required for various combinations of fixture units and toilets is shown in Figure 19–4.

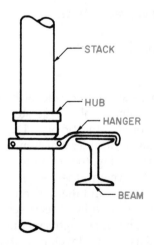

Figure 19-1 Supporting a stack

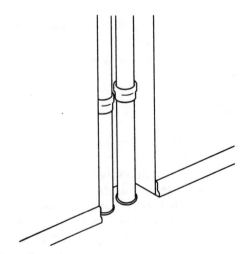

Figure 19-3

Size in In.	Fixture Units	Toilets	Lengths in Feet
1 1/2	4		50
2	10		75
2 1/2	20		100
3	48		150
4	240	33	300
5	540	80	500
6	960	120	unlimited
8	2200	225	unlimited
10	3800	400	unlimited

Figure 19–4 Stack sizes

STACK INLET FITTINGS

It was the custom in the past to place combination wye and eighth bend fittings into the stack to receive the branch flow wherever the large vertical dimensions of this fitting would fit. This was done because the *combo*, as it was called, would change the direction of the flow more gradually, Figure 19–5. It would direct the flow downward with less of a piling-up effect of the water. Modern housing often has limited space between ceiling and floor above and, therefore, the sanitary tee is used most often. Moreover, current practice is to make the sanitary tee the fitting of choice for soil and waste stacks. Because the branch of this fitting is shorter and exits into the stack higher, it is less likely to get choked with water thereby cutting off the flow of air along the top of the branch pipe.

The formula for determining capacity of fittings is:

Capacity = Kfactor × Pipe Size × Pipe Size.

❏ Capacity is stated in gallons per minute.
❏ The Kfactor for 45-degree turns is 22.5.
❏ The Kfactor for 90-degree turns is 11.25.

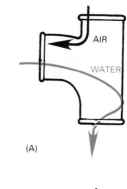

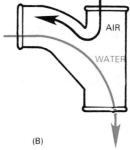

Figure 19–5 (A) Sanitary tee: air enters easily, water rebounds from far pipe wall. (B) Combination Y and 1/8 bend: air has more difficulty entering trap arm, water exits easily.

Because a combination wye and eighth bend is actually two 45-degree turns, the Kfactor would be 22.5. Remember: a combination is also a long turn TY. A sanitary tee turns the flow as abruptly as a 90-degree elbow; therefore its Kfactor would be 11.25.

The capacity of a 3-inch combination wye and eighth bend is therefore 22.5 × 3 × 3, which equals 222.5 gallons per minute. The capacity of a 4-inch sanitary tee is 11.25 × 4 × 4, which equals 180 gallons per minute.

The base fitting should be a 45-degree sweep or a long sweep bend. These cause less friction. Back pressure may result if this point is neglected.

◆ ◆

TEST YOUR KNOWLEDGE

1. Is a stack always vertical?

2. What is a roof flange for?

3. How far should a stack terminate above a window?

4. Which recommended pipe material is not standard weight?

5. What would a channel built into a wall and meant to conceal pipe be called?

6. A 3″ combination wye and eighth bend will carry 22.5 gallons per minute. How many gallons would a comparable sanitary tee carry?

UNIT 20

WASTE AND VENT STACKS

KEY TERMS

back pressure   soil stack
branch soil pipe   stack vents
branch waste pipe   trap arm
developed length   vent stacks
flood rim   waste stack

OBJECTIVES

After studying this unit, the student should be able to:

❏ Describe the different types of waste stacks and their use.
❏ Describe the purpose and use of vent stacks.

WASTE STACKS

A waste stack is a vertical drainpipe that receives the discharge of small fixtures. *Small fixtures* can be any plumbing fixture except toilets. Stacks may be constructed with the same pipe and fittings as other drainpipes and are subject to the same regulations as other stacks. The number of fixtures and fixture units permit-

ted on waste stacks may be found in the local plumbing code.

BRANCH SOIL AND WASTE PIPES

A branch waste pipe is a horizontal or vertical pipe that receives the discharge of small fixtures

169

and conveys it to the stack. If such a pipe receives the discharge of toilets and urinals, it is called a branch soil pipe. The size of this pipe is determined by the fixture-unit method described later.

A branch for a single fixture is shown in Figure 20–1. The distance from the trap to the vent is called the trap arm. The maximum length of the trap arm can be determined by using the chart in Figure 20–2.

Branches should have a pitch of 1/4 inch or more per foot and enter the stack or main drain at a 45-degree angle. A cleanout is placed on the end of the branch. If more than one fixture is on the line, it must be vented (have a special vent attached).

The length of the branch waste pipe is limited to 2 1/2 to 6 feet developed length. The developed length is the distance measured along the centerline of the pipe, from the trap to

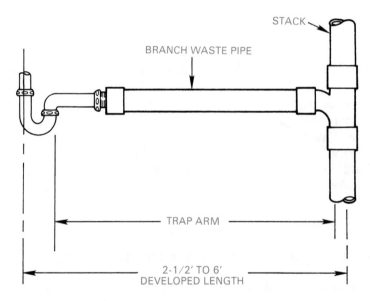

STACK

BRANCH WASTE PIPE

TRAP ARM

2-1/2' TO 6'
DEVELOPED LENGTH

Figure 20–1 Branch waste pipe

Size of Fixture Drain Inches	Distance — Trap to Vent
1¼	2 ft. 6 in.
1½	3 ft. 6 in.
2	5 ft.
3	6 ft.
4	10 ft.

Figure 20–2 Maximum length of trap arm

the vent. Some local codes may vary in their allowable developed length.

BRANCH CONNECTIONS TO STACKS

A sewage system is designed to remove sewage from a building quickly and without breaking it up. This directs the sewage in the direction of the flow in the main drain. Single or double wyes or combination wye and eighth bends may be used for this purpose. Fittings in the main drain used as branches should always

pitch up to allow the branch at least 1/4 inch per foot pitch, Figure 20–3.

On stacks, sanitary tees and short-turn tees may be used for branches. These are not the most effective but are permitted because of the limited space between floors and ceilings. Short-turn tees may retard the flow directly under the branch in stacks and cause slugs to form. A water slug completely blocks the pipe, preventing the passage of air.

Branch soil or waste pipes may be made of cast-iron, brass, wrought iron, galvanized steel, copper, ABS and PVC plastic, or lead. If lead is used, it is laid on a board to prevent sagging.

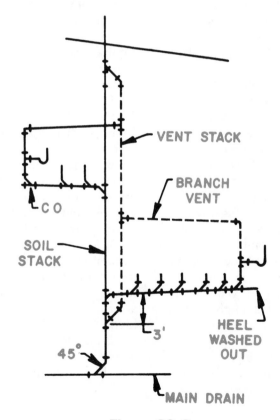

Figure 20–3

VENT STACKS

A vent pipe has two purposes. First, it provides a continuous change of fresh air within a drainage system. Second, it prevents siphoning and back pressure. Back pressure is caused by descending slugs of water. Because a slug completely fills the vent stack, it acts like a piston in a cylinder, pushing air ahead of itself. If there is no place for this excess pressure to escape, it may even cause bubbles of air to rise through traps in the system. Every fixture is provided with a water trap to prevent unhealthy gases from entering living spaces. The result is usually noisy but harmless. Water bubbles up in the system's traps but then it settles back into place.

Siphonage is more serious. This is again caused by a slug of water but moving down a trap arm this time and completely filling it like a piston. The air that the slug draws along behind it must be replaced by the atmosphere. If there is no venting attached near the trap, then the replacement air must be drawn through the water in the trap. This will carry some of the trap water with it and may destroy the trap's seal.

A vent stack is connected to the soil or waste stack with a wye fitting at or below the lowest branch, Figure 20–3. Connecting the bottom of the vent stack into the soil stack in this way permits rust scale to be washed away. It also relieves back pressure at this point.

A vent stack extends at least 6 inches above the flood rim of the highest fixture where it is reconnected to the stack vent. If the vent is more than 25 feet away, it must pass through the roof separately. In cold climates, small stacks must be increased to at least 3 inches before passing through the roof to prevent freezing.

The size of vent stacks and branches, permitted fixture units, and the developed length in feet are shown in Table 20–1.

SUMMARY

Stacks are pipes in a vertical position. A waste stack receives the discharge of smaller fixtures but not the discharge from water closets or urinals. If a stack receives the discharge from water closets (toilets) or urinals then it is a soil stack, no matter what else enters the stack below the entry point of the water closets and urinals. Stacks that supply only air are vent stacks. The student may take note that even soil and waste stacks have pipe at their highest points that carry nothing but air. And the student might well ask, "Aren't those parts at least vent stacks?" The answer is no. Those are stack vents (vents that vent stacks). Vent stacks are continuous vertical runs dedicated to nothing but supplying air. There is one exception: codes often require that at least one fixture empty low into a vent stack to "wash out the heel" of it, in other words, clear away accumulated debris.

From the foregoing information, the student should already be able to describe the difference between a branch waste pipe and a branch soil pipe.

Branched fittings that are installed in horizontal wastes and drains should have their branches rotated upward at a 45-degree angle, if possible, but never less than that angle that would provide a fall of 1/4 inch per foot.

If a vent stack is within 25 feet of another stack, it may be joined to that stack before exiting through the roof. The precaution here is that it must join the other stack at a vertical elevation at least 6 inches higher than the flood rim of the highest fixture (usually a lavatory). The flood rim is the surface of the fixture over which water would pour if the fixture's drain was plugged and the water supply was left on.

Size of soil or waste stack	Fixture Units Connected	Diameter of Vent Required (Inches)								
		1-1/4	1-1/2	2	2-1/2	3	4	5	6	8
		Maximum Length of Vent (Feet)								
Inches										
1½	8	50	150							
1½	10	30	100							
2	12	30	75	200						
2	20	26	50	150						
2½	42		30	100	300					
3	10		30	100	100	600				
3	30			60	200	500				
3	60			50	80	400				
4	100			35	100	260	1000			
4	200			30	90	250	900			
4	500			20	70	180	700			
5	200				35	80	350	1000		
5	500				30	70	300	900		
5	1100				20	50	200	700		
6	350				25	50	200	400	1300	
6	620				15	30	125	300	1100	
6	960					24	100	250	1000	
6	1900					20	70	200	700	
8	600						50	150	500	1300
8	1400						40	100	400	1200
8	2200						30	80	350	1100
8	3600						25	60	250	800
10	1000							75	125	1000
10	2500							50	100	500
10	3800							30	80	350
10	5600							25	60	250

TABLE 20-1 Size and length of vents

◆ ◆

TEST YOUR KNOWLEDGE

1. What is worse, siphonage or back pressure?

2. What kind of discharge would a branch waste receive?

3. How far away from its stack could you place a 2-inch unvented trap?

4. What is the stack pipe above the highest fixture called?

5. Why must vent stacks be offset using fittings with an angle no greater than 45-degrees?

♦ ♦ ♦ ♦ ♦ ♦ ♦ ♦ ♦ ♦ ♦ ♦ ♦ ♦ ♦ ♦ ♦ ♦

LOOP, CIRCUIT, CONTINUOUS, AND WET VENTS

KEY TERMS

back vent
circuit vent
continuous vent
loop vent
relief vents
seal of a trap
wet vent
yoke vents

OBJECTIVES

After studying this unit, the student should be able to:

❑ Explain the importance of continuous vents.
❑ Describe the uses of wet vents.
❑ Describe loop and circuit vents and how they are installed.

CONTINUOUS VENTS

In times past, it was common to see a trap that came up out of the floor and joined to the bottom of the sink or basin. This was an "S" trap. These traps were prone to losing their seals sim-ply because the plug of water going down would draw some of the water out of the trap behind it. These are now illegal for new installations. In current usage, the plumber runs a stack inside of the wall behind the fixture and the trap is inserted into a tee in this stack ("P" trap,

175

Figure 21–1). The tee in the wall must be less than the diameter of the pipe below the fixture trap outlet, or siphonage may occur. The portion of the stack that continues upward and eventually joins back into the main stack above the highest fixture is known as a continuous or back vent. It may also join with another vent stack or branch vent.

This vent and all vents may be constructed of brass, copper, cast-iron, galvanized steel, or plastic piping materials.

The size of the vent pipe should be no less than half the diameter of the drain that it supplies air to and under no circumstances less than 1 1/4 inches in inside diameter.

If the fixture is at some distance from the vertical waste pipe, the branch waste pipe should be extended to a point nearer the fixture, Figure 21–2. If the vent is run as shown in Figure 21–3, the waste backs up in the vent pipe, soap or grease is deposited, and the vent is eventually closed. The water level in the waste and vent lines assumes the hydraulic gradient level and rushes up the vent pipe when water is released from the fixture.

Another reason for keeping the horizontal pipe short is that air will not circulate in a branch pipe that is more than 1 or 2 feet. Moisture collects on the top of the pipe when it is not ventilated. This moisture absorbs carbonic acid gases and attacks pipe material along the top.

Figure 21–4 shows how moisture clings to and destroys the pipe. A vent pipe placed at A would remove the gases and prevent destructive action on the pipe.

Fixtures on three or more floors should be connected as shown in Figure 21–5. It is not necessary to vent the fixtures on the top floor since there are no fixtures above them.

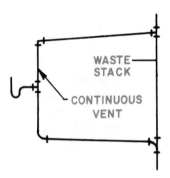

Figure 21-2 (Correct)

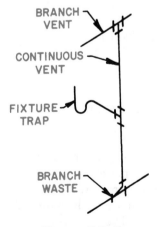

Figure 21-1

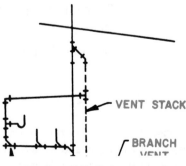

Figure 21-3 (Incorrect)

WET VENT

Wet vents can only exist where there is another floor above the one in question. Imagine a two-story dwelling and you are on the first floor. There is a toilet on this level and another toilet directly above and only one stack. The pipe between the two fixtures serves as a drain for the one above and as a wet vent for the fixture below. When the toilet above is being flushed, the toilet below does not have complete use of the stack for air supply and perhaps no air supply for a brief period. Because wet-venting is forbidden under some circumstances, the solution is to run two stacks, a vent stack and a soil stack. All drains go into the soil stack and all vents go into the vent stacks. See Figure 21–5 and Figure 21–6.

The upper fixtures, which drain into the waste stack, do not require an additional vent as long as trap arms are not longer than the maximum length allowed. The lower fixtures drain into the bottom of the vent stack. They also do not require an additional vent.

In all stacks it is best to use short-turn, single, or double-drainage TYs. If long-turn TYs are used where the branch is long (5 to 8 feet), the waste pipe at the junction of the stack may be below the weir of the trap, thereby causing siphonage. The pipe size depends upon the number of fixtures.

A wet vent must be a minimum of 1 1/2

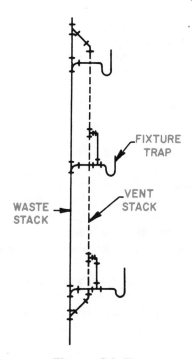

Figure 21–5

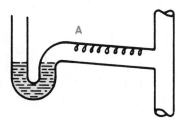

Figure 21–4

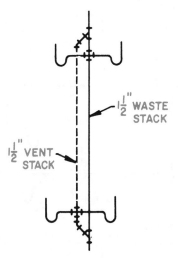

Figure 21–6

inches if one lavatory drains into it. A 2-inch wet vent may handle up to four fixture units. In both cases, the horizontal drain line should be 2 inches. On the lower floors of a multistory building, a lavatory drain may be used as a vent if the drain and vent are at least 2 inches.

LOOP AND CIRCUIT VENTS

Loop and circuit vents are used on a line of fixtures. A loop vent is used in single-story houses or on the top floor of a multistory building. In the loop vent, the vent branch goes back into the stack vent at a point above all drain inlets, Figure 21–7.

The horizontal branch vent pipe on the loop vent must pitch back to the traps. This prevents condensation from collecting at any point. Condensation also collects in drops along the top of horizontal vent pipes. The branch drain must continue full size to the last fixture connection.

A circuit vent is shown in Figures 21–8 and 21–9. In each case, it is limited in length so that the total pitch of the drain does not exceed 18 inches. The number of fixtures is limited by the size of the pipe.

The vertical vent at the end of the drain must be washed out by having a fixture connected at that point. A cleanout plug must be placed at the end of the drain.

Relief vents are sometimes installed to relieve air pressure on branch waste near the stack. See Figure 21–9. In tall buildings relief

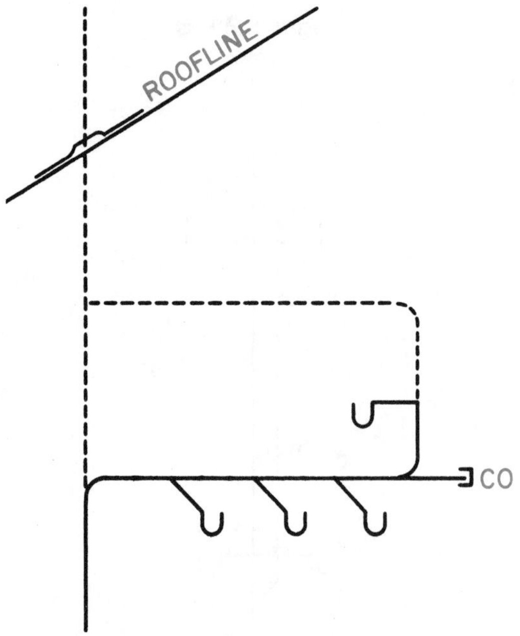

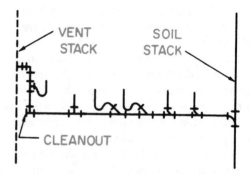

Figure 21–8 Circuit vent

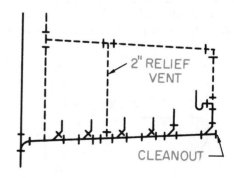

Figure 21–9 Circuit vent; note relief vent

Figure 21–7 Loop vent

vents, also called yoke vents, are installed at 10-story intervals from the top floor to relieve air pressure, Figure 21–10.

SUMMARY

The seal of a trap, which is the depth of the water held by the trap, is easily upset. Less than a half pound square inch of pressure will cause it. The purpose of venting and the reason for its complexity is to ensure that the trap seals in a building are not reduced, allowing noxious gases to enter the building. There is no such thing as a zero trap arm length; the stack must

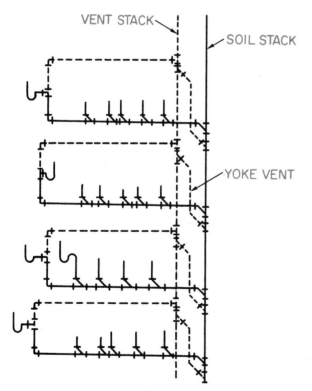

Figure 21–10 Yoke vent

run in the wall behind the fixture. Nevertheless, the length of the trap arm must be kept as short as possible to accommodate a small space above the water in the trap arm for air to move opposite the water flow, and to provide the means to break any siphon that might attempt to develop at the trap. In Europe and in some larger buildings in the United States, a system of venting called *Sovent* is being used. In this system, the trap arm is much larger in diameter than the trap outlet. This way, the discharge from the trap can never completely fill the trap arm (within limits) and cut off the siphon-breaking air over the flow of water.

Another consideration is the buildup of soap scum in a venting line. This is caused by the momentary rise of soapy water into the vent line each time a fixture drains. As the level of the water subsides, a thin film may be left inside the pipe. In time, this buildup can actually close the vent opening, see Figure 21–2 and Figure 21–3.

In stadiums, theaters, and so forth, we often see fixtures in banks or rows. This situation calls for some of the loop and circuit venting techniques. Observe in Figures, 21–7 through 21–10 that this presents new problems. Some fixtures by virtue of their positioning are actually being vented by a pipe that serves as a drain for another fixture. This situation is helped by the frequent installation of relief vents, see Figure 21–9. Also observe that the loop and circuit vents always have one fixture positioned on the stack at the end of the loop. This is to ensure that the heel of that short stack will be washed out. In practice the vertical stack is often run up between the farthest two fixtures in the bank to facilitate this.

A yoke, or relief vent joins two stacks in a way that equalizes pressure between the stacks and prevents the vent stack from receiving any of the discharge from the soil or waste stack to which it is joined. A yoke vent should occur once in every ten branch intervals.

◆ ◆

TEST YOUR KNOWLEDGE

1. What is the minimum vent size?

2. After first considering the minimum allowable vent size, what would a vent size be in relationship to the size of the drain it serves?

3. What is meant by hydraulic gradient?

4. What is a wet vent?

5. What is the difference between a loop and a circuit vent?

6. What does the abbreviation C.O. stand for?

7. A pipe that vents a fixture and then returns to the fixtures stack above the highest fixture on that stack is called _____.

8. On large buildings where soil and vent stacks are paired what should be installed at every tenth branch interval?

CLEANOUTS

KEY TERM

cleanout fitting
downstream

OBJECTIVES

After studying this unit, the student should be able to:

❏ Describe the purpose of cleanouts.
❏ List the various locations and sizes of cleanouts.

There are several causes for stoppage in drainpipes. Provision must be made for access to the pipe in order to clean out such obstructions. The cleanout fitting is made for this purpose.

Pipes are most likely to clog at places where the pipe changes direction. Cleanouts are installed at changes in direction of the building drain that are greater than 45 degrees. Horizontal pipes are equipped with cleanouts spaced not more than 50 feet apart for piping 4 inches or less in diameter and not more than 100 feet apart for larger piping. For underground piping that is over 10 inches in diameter, however, manholes with coverings are installed at each 90-degree change in direction and at maxi-

mum intervals of 150 feet.

All codes require cleanouts at the bases of all stacks and rain leaders.

There is a broad range of pipe and fitting materials that can be used for building drains and drain and waste branches. The code requirements apply to every kind of suitable pipe and fittings. The material illustrated in this unit is mostly cast-iron hubbed pipe. The student should easily be able to extrapolate for any other material from the materials shown.

The type of cleanout to be used must be the iron body trap screw type for cast-iron laid pipe illustrated in Figure 22–1 (A) and (B). This cleanout is made in sizes from 2 inches to 8 inch-

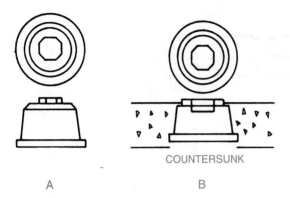

COUNTERSUNK

A   B

Figure 22-1

es. It is available in two types, the standard plug, Figure 22–1A, and the countersunk plug, Figure 22–1B. In places where the cleanout is flush with the floor, the countersunk plug type is used.

The iron body of a cleanout is caulked into the soil pipe. The brass plug may be removed with a wrench to provide access to the inside of the pipe if it was properly lubricated at the time of installation.

The locations of cleanouts are shown in Figures 22–2 through 22–5. When drains are underground, the cleanout is extended up to the floor line with pipe, and the hub for the cleanout

plug is cemented into the floor. If this occurs in a walkway or other area likely to receive people traffic, there should be no part extending above the surface of the floor; this would create a tripping hazard. If conditions permit, however, the top of the cleanout hub should be a little above the floor line so that the cleanout is not likely to be used for a floor drain.

Building maintenance personnel will often remove the cleanout plug and allow floor wash water to run down the cleanout inlet. The cleanout is not equipped with a trap, and if left open, may permit dangerous sewer gases to enter the building. Masonry, tools, cans, and other unsuitable debris may enter the drainage system and cause blockages.

A good way of supplying a cleanout which is less easily used for a floor drain is to use the underfloor type cleanout, demonstrated in Figure 22–5.

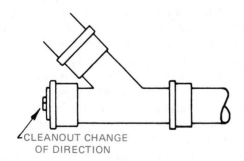

CLEANOUT CHANGE
OF DIRECTION

Figure 22-2

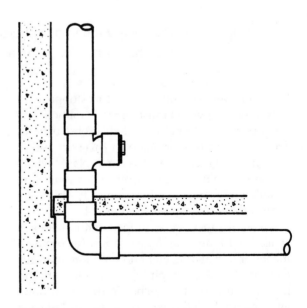

Figure 22-3 Cleanout at base of stack

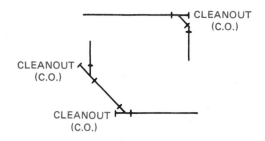

Figure 22-4

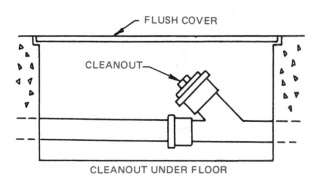

CLEANOUT UNDER FLOOR

Figure 22-5

SUMMARY

What goes into the drains of a plumbing system is a factor that cannot be completely controlled; this makes the judicious placement of cleanout fittings an important part of the system. Horizontal pipes are more prone to blockage than are vertical ones, but stacks must also have

cleanouts under certain circumstances. The first point of entry for clearing obstructions on a stack is on the roof at the open stack vent. If there are no horizontal offsets on the stack to stop the passage of a sewer rod, this will be sufficient down to the base of the stack where a cleanout is required anyway for access to the head of the main drain.

Offsets

An offset occurs when the line of pipe must move over to clear obstacles or to slow the acceleration of the pipe's contents. Take a bendable drinking straw and put two bends in it. Adjust the bends so that you can stand the straw on its end and the bottom section and the top section are perfectly vertical. That is an example of a simple offset.

On soil and waste stacks where there are horizontal offsets, a cleanout should be placed at the upstream end of the offset.

On building drains, every significant change in direction should have an accessible cleanout fitting. This fitting should be placed in a manner that allows sewer clearing equipment to be inserted in a manner that will allow **downstream** blockages to be cleared away. There are specified distances on drains within which cleanouts must be placed, whether or not there is a change in the line's direction.

◆ ◆

TEST YOUR KNOWLEDGE

1. What type of fitting must the plumber provide to allow sewer and drain cleaning equipment to be inserted?

2. On a perfectly straight main house drain 62 feet long, how many cleanouts should be installed?

3. Suppose the main house drain, which is 62 feet long is going to be covered by a concrete floor. What is the plumber going to do then?

4. How would you place a cleanout in the main house sewer that is 100 feet long and is 8 feet underground?

5. Suppose you are using a garage to repair cars, and you need some place to get rid of floor sweepings and oil changes. Is it all right to pull the plug from a flush-mounted cleanout in the floor?

♦ ♦ ♦ ♦ ♦ ♦ ♦ ♦ ♦ ♦ ♦ ♦ ♦ ♦ ♦ ♦ ♦ ♦

TRAPS

KEY TERMS

deep seal trap
seal of the trap
self-syphon
trap

OBJECTIVES

After studying this unit, the student should be able to:

❑ Discuss the uses of drainage traps.
❑ Identify the sizes and types of drainage traps.

DRAINAGE TRAPS

A trap is a fitting or an assembly of fittings that holds a small amount of water, Figure 23–1. This trapped water then presents a *seal* to the gases on the downstream side of the trap. This prevents those gases from entering the living spaces within the building. Traps have no moving parts and, therefore, will not wear out in a mechanical sense. However, it is vital that the pressures on both sides of the trap be equal to prevent the entry of sewer gas.

Cast-iron galvanized drainage traps are often used in large buildings where pipes are exposed on ceilings and below fixtures. They are used most often on urinals and floor drains in public rest rooms. However, they may also be used on any drainpipe above ground or inside buildings.

Drainage traps range in sizes from 1 1/4 to 10 inches and come in several patterns. Some have cleanouts and others have vent connections. Figure 23–2 shows four types of drainage traps.

185

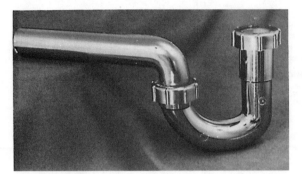

Figure 23–1 *(Courtesy of Genova Products Inc.)*

Carefully check the state and local codes concerning the use of these traps as some of them may be specifically forbidden for new construction. Full S traps and 3/4 S traps, popular when the crown method of venting was permitted, are easily siphoned.

All traps are based on the S pattern. There are four types of S traps: the *full S*, the 1/2 S ("P"), the *3/4 S*, and the *running trap*, Figures 23–3 to 23–6.

You may hear the *seal* officially described as the distance between the top of the dip and the crown weir. Figure 23–6 demonstrates the location of these old trap terms. No venting is permitted at the crown of a trap. The same figure shows the distance that vents must be kept away from the crown.

The **seal of the trap** is that vertical distance that gas must travel down through the water in the trap, before it can bubble up into the living space. When this happens, we say the *seal of the trap is broken.*

The S trap is no longer permitted in most localities because of its tendency to **self-syphon;** that is, because of the vertical drop at the outlet, the slug of draining water has a tendency to pull the water, which should remain to maintain the seal, along with it. However, the plumber will still find many existing installations of S traps

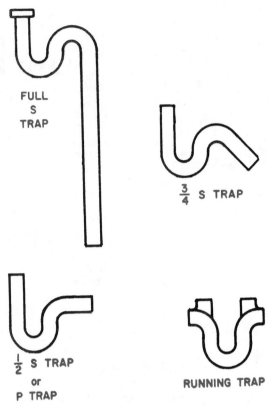

Figure 23–2 Drainage traps

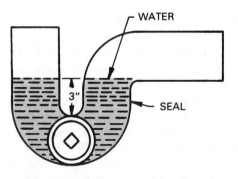

Figure 23–3 1/2 S or P trap

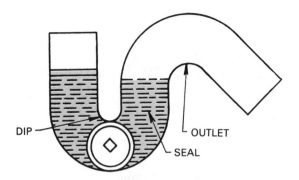

Figure 23-4 3/4 S trap

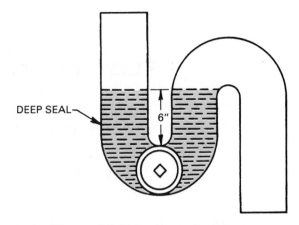

Figure 23-5 Full S deep seal trap

because codes provide a "grandfather" clause in which plumbing in place before the newer code was made law, is permitted to remain as it is. And so the plumber will not replace old worn through S traps with new ones on occasion.

Soil pipe traps usually have a 3-inch seal, but **deep seal traps** with a 6-inch seal may be obtained for rain conductors, Figure 23–5. This feature prolongs the effectiveness of the seal during droughts when ordinary seals would evaporate.

The running trap, Figure 23–2 and Figure 23–7, may be obtained with one or two vent openings to which the fresh air inlet may be connected when it is used as a trap next to a curb. A brass plug is installed in the other side.

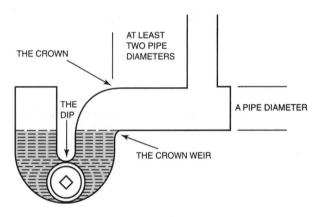

Figure 23-6

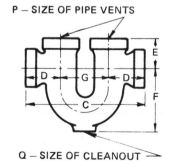

Figure 23-7 Running trap with double vent

Size	1 1/4	1 1/2	2	3	4
C	7 5/8	8 3/8	10 1/8	13 3/4	17 1/4
D	2 1/4	2 1/2	3 1/16	4 3/16	5 1/8
E	2	2 1/4	2 9/16	3 1/2	4 3/16
F	4 1/2	4 15/16	5 11/16	7 3/4	9 3/8
G	3 1/8	3 3/8	4	5 3/8	7
P	1 1/4	1 1/2	2	3	4
Q	3/4	1	1	1 1/4	2

◆ ◆

TEST YOUR KNOWLEDGE

1. What keeps sewer gases from escaping into the living spaces of a home?

2. What kind of trap has both the inlet and the outlet in a horizontal plane?

3. Do the vent(s) on a running trap violate the rule about crown venting?

4. Why has the full S type trap been outlawed in most jurisdictions?

5. How many moving parts does a bathroom P trap have?

6. Why would you need a deep seal trap?

UNIT 24

◆ ◆ ◆ ◆ ◆ ◆ ◆ ◆ ◆ ◆ ◆ ◆ ◆ ◆ ◆ ◆ ◆

LOSS OF TRAP SEALS

KEY TERMS

back-pressure
siphon
trap arms

OBJECTIVES

After studying this unit, the student should be able to:

❏ Describe siphonage and its effect on various types of traps.
❏ Describe back pressure and how to prevent it.
❏ Discuss capillary attraction and evaporation.

SIPHONAGE

A siphoned trap is a health problem to the occupants of any building. It is important to understand the siphonage of trap seal so that is may be prevented.

A **siphon** is a bent tube with arms of unequal length. If the short end is inserted in water and air is exhausted from the tube, atmospheric pressure will force the water up the short arm and over the crown. The water will continue to run until the receptacle is empty, Figure 24–1.

A *plumbing trap* is similar to a siphon. The trap may be siphoned when connected to a waste pipe. Water will stay in the trap when the atmospheric pressure, about 15 psi, is the same on both the inlet and the outlet, Figure 24–2.

When water descends in a waste stack, it picks up considerable velocity. This causes a reduction in air pressure at the top of the stack.

189

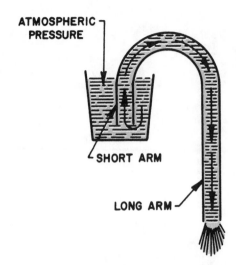

Figure 24–1

ATMOSPHERIC PRESSURE

SHORT ARM

LONG ARM

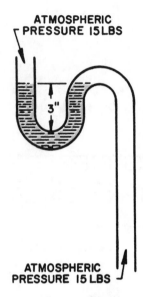

ATMOSPHERIC PRESSURE 15 LBS

3"

ATMOSPHERIC PRESSURE 15 LBS

Figure 24–2

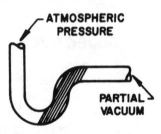

ATMOSPHERIC PRESSURE

PARTIAL VACUUM

Figure 24–3

If this air cannot be replaced, the trap will be siphoned, Figure 24–3. The causes may be improper design or size of the stack; an obstruction by rust, scale, or hoar frost; or the omission of a vent pipe.

Water and sewage descending a soil stack tend to form slugs. These slugs cause an air pressure greater than the atmospheric pressure, called a *plenum,* at the bottom, and a lower air pressure or *partial vacuum* at the top of the stack. This condition expels the water from the lower traps and siphons the trap seal on the upper floors. Both conditions admit drainage air to rooms, thereby presenting a health hazard.

A properly installed vent pipe furnishes air at the top and relieves air at the bottom. This equalizes the atmospheric pressure within the stack. Notice the slug of water in Figure 24–4 as it passes the upper trap outlet. The slug lowers the air pressure in the branch, and atmospheric pressure drives out the trap seal. At the lower trap, it creates an excess pressure that drives the water out of the trap.

When water passes the trap outlet, Figure 24–5, air pressure from the vent pipe relieves the partial vacuum. The seal remains in the trap.

The average depth of a seal in an S trap is about 3 inches. One foot of water exerts a pressure of .434 psi; 3 inches will exert 1/4 of .434 or .108 psi of pressure. This shows what a small amount of pressure a trap seal can withstand. On waste pipes that drop vertically for some

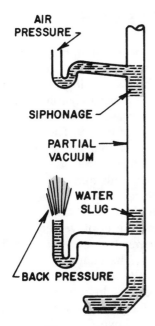

Figure 24–4

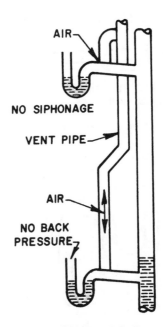

Figure 24–5

distance below a trap, the discharge of a fixture may self-siphon its own trap.

Siphonage of traps can be prevented by checking the sizes and lengths of the branch waste pipe and installing the correct vent pipes. The best trap and waste pipe connection is the 1/2 S trap connected to the continuous vent, as shown in Figure 24–6. The vent pipe supplies all the air needed, and siphonage is prevented. However, the branch should be kept short so that the top of the branch outlet is never below the dip of the trap, Figure 24–7.

The vent pipe should never be attached to the crown of the trap because grease and other matter will soon clog the vent pipe, Figure 24–8.

Oval-bottomed fixtures discharge water more quickly than flat ones. Therefore, the former are more likely to be siphoned. In the latter, the seal is retained by the trickle of water to the

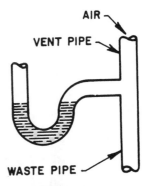

Figure 24–6

trap after the main discharge has passed. The *drum trap*, in the form of the Philadelphia regulation bath trap, cannot be siphoned, but it does not possess the self-cleaning qualities of the S trap.

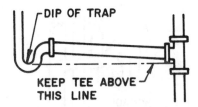

DIP OF TRAP

KEEP TEE ABOVE
THIS LINE

Figure 24–7

Figure 24–8

BACK PRESSURE

The opposite action of siphonage is back pressure. Back pressure usually occurs at the base of the stacks when water slugs, which form in soil or waste stacks, push air before them. This has a tendency to increase the air pressure at the base of the stack, Figure 24–9. The air pressure will increase if the horizontal drain is partially filled from other fixtures, or if it receives rainwater during a storm. This may not be as harmful as siphonage since, in most cases, the water will run back into the trap.

Back pressure is prevented by installing a vent pipe or relief pipe in the stack below the lowest fixture. The main vent or relief vent is connected to the soil stack with a wye connection. This prevents the collection of rust scale at that point. This pipe is usually reconnected to the soil stack above the highest fixture. Back

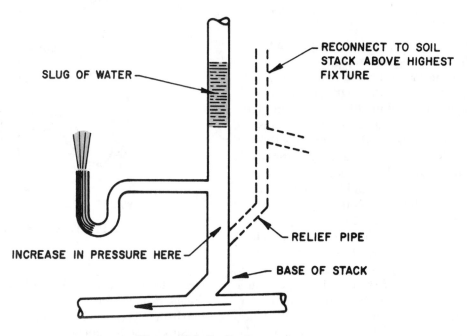

SLUG OF WATER

RECONNECT TO SOIL
STACK ABOVE HIGHEST
FIXTURE

RELIEF PIPE

INCREASE IN PRESSURE HERE

BASE OF STACK

Figure 24–9 Back pressure

pressure is then eliminated because the air can escape through the relief vent.

The fitting used at the base of the stack affects the pressure at that point. A tee is the poorest fitting. A combination wye and eighth bend, which is most often used, is better. A long-sweep bend is best.

CAPILLARY ATTRACTION

Capillary attraction is the power that small tubes, string, and other porous materials have to lift water above their own level. For example, sap rises in trees through the small tubes in the wood.

If a piece of string or lint catches in a trap, and one end hangs over into an outlet, it may act as a wick and carry the water over by drops until the seal is broken, Figure 24–10. This would allow sewer gas to enter the room. The inside of traps should be smooth to prevent material from catching.

EVAPORATION

Evaporation is the change of water to vapor. Evaporation is more rapid at high temperatures.

Air currents increase evaporation. The trap seal of an unused plumbing fixture may evaporate and admit sewer air to the house. There is no danger when a fixture is used frequently. A vented trap is more likely to evaporate than an unvented one.

Trap seals may be lost by evaporation if unused for long periods. The length of time depends upon the amount of humidity contained in the air, the temperature, whether the air is still or moving, and the surface area of the water.

The traps affected most are rain conductors in summer and floor drains in winter. Long periods of drought may cause rain conductor traps to lose their seals. For this reason, deep seal traps are required on all rain conductors. The seal of a 4-inch trap should be 6 inches deep and maintain a seal for about 3 months.

When floor drains are installed inside buildings, provision must be made to supply them with water by placing a faucet nearby, Figure 24–11. Yard and area drains are also subject to evaporation, the result being objectionable odors. However, since they are outside, they are not dangerous. The seal may be replaced by adding water.

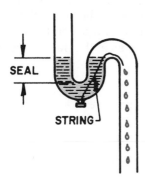

Figure 24–10 Capillary attraction

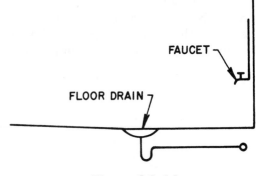

Figure 24–11

SUMMARY

Traps are devices that hold a small amount of water in a U-shaped bend. This plug of water acts as a barrier and separates the air of the building from the fumes in the sewer. The balance of pressures that keeps the water in the trap is a delicate one. Considerable engineering time has been expended by experts in the field to design plumbing systems that will protect, that is, equalize, that balance.

Water descending a stack does not always form a slug, which is a quantity of water that completely fills the pipe, with water. More often the water does not completely fill the pipe, and the air in the stack is only marginally disturbed. Under other circumstances, the water *does* fill the pipe and then the air below it is pressed downward, and air above the slug must flow into the vent pipe, if it can, to equalize the pressures above. The air being pushed down the stack ahead of the slug usually has places where it can go, and this will minimize the pressure buildup. If there is a vent on the main drain at some other point in the system, the air can go there. If there is a yard vent, the air can rush up there. If the stack and drain below the slug are clear, nothing detrimental will happen. However, if there is a blockage, either temporary (another large fixture elsewhere in the building is filling the drain), or permanent (the drain may be partially closed but is still operating after a fashion), then the pressure buildup below the falling slug of water will pressurize the traps. A bubbling sound will be heard by the occupants of those rooms with fixtures served by the stack. Sewer gas will bubble up through the traps because the seals of the traps have been broken.

The space above the falling slug of water is at a very slight negative pressure (below 15 psi). If the stack is blocked by a bird's or squirrel's nest or perhaps a year or two of leaves falling on spider webs and air cannot freely enter the top of the stack, then the air pressure may go more negative, and air needed will be drawn in through the traps on the floors above. This will make a gurgling sound, and some of the water in the traps will be drawn along with the air into the branch. This condition is more serious because the water pulled out of the traps above will not be replaced until the fixture is used. In the case of the lower fixtures that had air pushed up through them, the trap seals will be intact, because the water pushed up from below has nowhere to go but to fall back down into the trap from which it came.

In general, so long as crown venting is avoided the closer the rap is to its vent the better. The length of trap arms, which is the lateral distance between the trap and its vent, is governed by the individual plumbing codes. A usable rule-of-thumb would be to never allow the bottom of the pipe's inside diameter (I.D.) at the trap end, be higher than the top of the pipe's I.D. at the vent stack end. The bigger the trap arm's pipe size the farther it can run unvented.

◆ ◆

TEST YOUR KNOWLEDGE

1. What is the purpose of a trap?

2. What maintains the trap's seal?

3. In a properly designed system, what could still go wrong with the balance of pressures?

4. How deep is the normal trap seal?

5. What does the shape of the fixture's retaining bowl have to do with trap siphonage?

6. Which is worse healthwise, siphonage or trap blowout?

◆ ◆ ◆ ◆ ◆ ◆ ◆ ◆ ◆ ◆ ◆ ◆ ◆ ◆ ◆ ◆ ◆

SPECIAL TRAPS

KEY TERMS

garage sand trap
grease trap

OBJECTIVES

After studying this unit, the student should be able to:

❏ Explain how garage sand traps work.
❏ Describe the grease trap.

GARAGE SAND TRAPS

The drainage from garages is handled differently than the drainage from homes and other buildings. This is because the drainage from garages contains sand, gasoline, and oil. The sand that falls from or is washed from cars, for instance, may clog a regular drain or sewer.

The purpose of the garage sand trap is to prevent sand, gasoline, and oil from entering regular sewer drains. It consists of a catch basin 18 inches deep with a cover. The outlet is taken from

an elbow turned down below the water surface, Figure 25–1. The top of the trap is level with the cement floor. Since gas and oil are lighter than water, they float to the surface and do not enter the drain. The sand settles to the bottom. These traps should be cleaned periodically.

In large public garages where several floor drains are installed, a brick or concrete sand trap is used instead of the small garage sand trap, Figure 25–2. This type of trap is constructed with two compartments. The overflow from each is taken off with a sweep bend that is

197

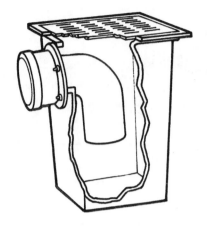

Figure 25-1 Garage sand trap

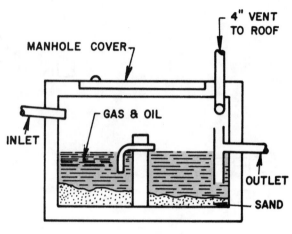

Figure 25-2

turned down. This prevents the gas and oil, which float on the top of the water, from being carried into the sewers. The dirt and sand are retained in the first chamber. Should the gas and oil overflow into the second chamber, they are checked there by the inverted outlet.

The sand trap which is vented separately to the roof, has a loose steel cover for cleaning. The outlet is connected to the main drain at any convenient location. The end of this separate drain must be vented to the roof, and the last floor drain must run to this stack to wash out the rust

scale, Figure 25-3. This sand trap can only be effective if it is periodically cleaned.

GREASE TRAPS

A **grease trap** is a receptacle placed in the waste pipes of sinks to separate and retain grease from the water. These traps are used mostly in hotels and restaurants because of the quantity of grease in the waste water. While the water is hot, the grease is in liquid form and floats on top of the

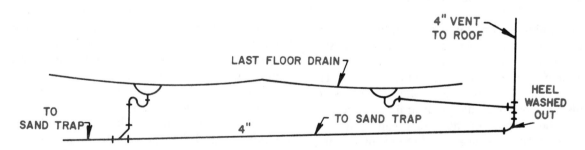

Figure 25-3

water. The grease thickens when cooled and sticks to the sides of water pipes, clogging them.

In rural sewage systems, the grease clogs the joints of loose cesspool and subsoil irrigation tile. Grease becomes as hard as soap in time and will eventually fill the pipe.

Figure 25–4 shows an example of a grease trap with baffle plates. The trap should be large enough to hold twice the capacity of the fixture. This allows the contents to cool before the next flush. The baffle plates are placed so they separate the grease. The grease, since it is lighter than water, rises to the top. The outlet is taken from the bottom to prevent grease from passing into the waste pipe. It has a large cleanout cover held in place by thumbscrews, which may be removed for cleaning.

Grease traps that use a potable (drinkable) water supply for cooling are illegal.

SUMMARY

There are some things that septic systems cannot handle. Fats, petroleum products, and indigestible solids like sand and gravel are a few things that need to be kept out of any plumbing system. Plumbers use the properties of these products to keep them from either blocking the system, making it flammable, or destroying the bacteria without which no sewage disposal system can operate.

Fats from the cooking process coagulate as they cool. From the time they leave the kitchen sink, the cooling process starts. In many cases, the fats stick to the cool pipe walls and eventually close it off. A branch waste full of thickened sticky fat can be difficult to clear with conventional drain clearing tools. The fat opens up to allow the tool to pass through it and closes up again as the sewer tape is withdrawn. This is an

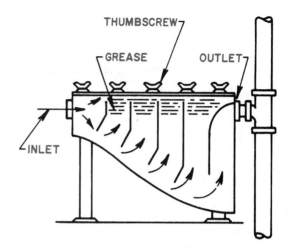

Figure 25–4 Grease trap

especially big problem in restaurants whose kitchens may be in operation for 18 hours a day. The solution is to rapidly cool the fats and heavy oils in a grease trap, forcing them to harden and accumulate in a place easy to reach, and then to clean out that trap on a regular basis.

Repair garages use specially manufactured products to soak up unavoidable spills of engine and lubrication oils. Some of this mixture is bound to be swept into the drainage system. Road sand clinging to the vehicles being serviced will be washed or swept into the drain. On occasion, oils and even gasoline find their way to the nearest floor drain. Sand is heavy, whereas petroleum products are light. If the flow of the waste is slowed by directing it into a larger reservoir, the sand will settle to the bottom where it may be scooped out. The petroleum products will float to the top where they may be skimmed away. Here again the safe operation of the system depends on timely human intervention to prevent the trap from filling and then overflowing.

◆ ◆

TEST YOUR KNOWLEDGE

1. Where might you expect to find grease traps?

2. Why would careful venting be important with garage traps?

3. Contrary to most plumbing system parts, garage and grease traps do not operate automatically. What is an important consideration of their proper operation?

4. Why is it important with drains running to garage sand traps that one of the floor drains be hooked to the lower end of the vent stack?

UNIT 26

◆ ◆ ◆ ◆ ◆ ◆ ◆ ◆ ◆ ◆ ◆ ◆ ◆ ◆ ◆ ◆ ◆ ◆ ◆

RAIN LEADERS

KEY TERM

rain leader

OBJECTIVES

After studying this unit, the student should be able to:

❑ Discuss the purpose of rain leaders.
❑ Explain where and how rain leaders are installed.

RAIN LEADERS

A **rain leader** is a pipe that carries rainwater from the roof or gutter of a structure to the correct disposal point.

Running rainwater into sewage systems is not allowed in today's ecologically-minded society. Although there are still many combination sewers in use, these are being replaced by separate storm water systems. The argument that a heavy storm may flush out the sewage system is no longer valid. This is because most systems today run well above their normal capacity. The discharge of surface water into the system does little other than overload the system. However, material that is difficult to remove may enter the system in this manner. Most codes require a separate system of piping for rain leaders and surface drains.

Outdoor, above-ground rain leaders are usually installed by the roofing contractor. If the underground portion discharges at the curb, there is no need for a trap.

Indoor rain leaders are subject to the same regulations as drainpipes. They must have a cast-iron, deep-seal trap with a cleanout that is accessible for cleaning. Indoor rain leaders have the advantage of not freezing since they are located in heated buildings. In cold buildings, such as warehouses, they are protected from freezing by steam pipes or thermostatically controlled, electric heat tapes.

Figure 26–1 shows a method in which cast-iron pipe may be used. The horizontal run of pipe tied into the vertical leader provides for some expansion in any direction. Cast-iron strainers are used to keep leaves and other debris from closing the drain. The strainer should extend at least 4 inches above the surface of the roof.

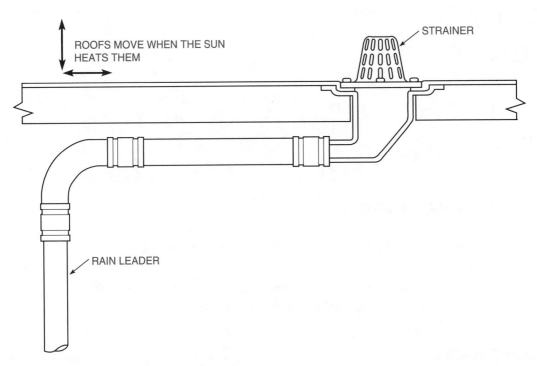

Figure 26–1 Providing for expansion

TEST YOUR KNOWLEDGE

1. Does a rain leader ever carry soil or waste?

2. Does the building storm drain terminate at the curb or a storm sewer?

3. Why would you need a strainer on a roof?

4. Why would there be a need for a trap in a storm drain leading to a storm sewer?

5. Give an advantage to running rain leaders inside of the building they serve.

◆ ◆ ◆ ◆ ◆ ◆ ◆ ◆ ◆ ◆ ◆ ◆ ◆ ◆ ◆ ◆

TESTING DRAINAGE SYSTEMS

KEY TERMS

air-pressure drop test   proving plug
hydrostatic   roughing-in tests
manometer   smoke test

OBJECTIVES

After studying this unit, the student should be able to:

❑ State how drains are tested.
❑ Compare the air test, the water test, and the smoke test.

Drainage systems are tested for leaks by using either air, water or smoke. These tests are known as **roughing-in tests**. A final visual test is made after fixtures are set and in working order.

THE AIR METHOD

To test a drainage system by the air method, all openings are plugged. Proving plugs are inserted in soil pipe openings. With copper tubing, a

testing cap is temporarily soldered on. With plastic pipe, the material may be left a little long and a test cap glued on. The proving or test plug is illustrated in Figure 27–1. In practice, air is pumped into the system by a hand pump or with an air pressure tank. Leaks or defects are visually detected by applying a soap solution to the joints. Leaks will cause bubbles to appear. Defective material must be removed and replaced. With cast-iron pipe and fittings, a careful examination should be made for defects before they are installed. It is not unusual for a soil-pipe fitting to have a sand casting hole in it. After the testing, air should be released by slowly removing the pipe cap or by using a nail to poke a hole in the copper cap before the plugs and caps are removed. Most code rules require that an air pressure of 5 pounds per square inch be maintained for 10 minutes in the presence of the inspector (an **air-pressure drop test**). A **manometer** is a measuring device that shows small fluctuations in pressure, Figure 27–2. This gauge can be attached to the system at any convenient outlet. The air pressure method of testing is time-consuming and can be expensive. But it can be used for tall buildings without fear of an excessive pressure buildup, and it also can be used in cold-weather conditions. In existing buildings where much of the piping may be concealed by building materials, the air-pressure drop test is handy for determining the presence of leaks but not necessarily the location.

THE WATER (HYDROSTATIC) METHOD

In the water (hydrostatic) test, the outlets, with the exception of the tops of the stacks, are closed, in the manner described previously, and the system is filled with water. The lower section is tested first. The testing proceeds upward as each section is checked. In this test, the lower section is under greater pressure than the top,

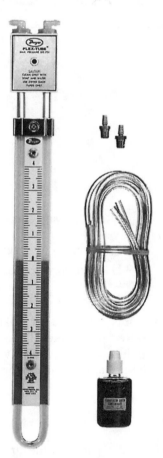

Figure 27–2 A manometer (*Courtesy of Dwyer Instruments Inc.*)

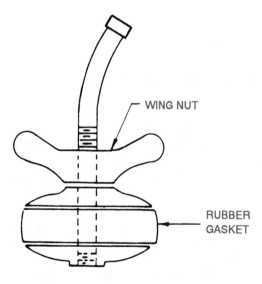

WING NUT

RUBBER GASKET

Figure 27–1

due to the head (height/pressure) of the water. Water will add nearly a half pound of pressure to every foot of elevation because of its own weight. To avoid excessive pressures, taller buildings may be tested two or three floors at a time. In cold weather, damage may be done by freezing water left in the system for too long.

THE SMOKE TEST

The smoke test is conducted according to rules outlined by local or state code, regarding where the test can be used, what procedure is to be followed, and how the test is to be conducted. The smoke test is mostly used in older buildings where it would be difficult to expose all of the piping for a visual examination. In the smoke test, a smoke machine is used to fill the system. At least three persons are involved in the test: one to run the smoke machine, one stationed in the building, and one who is stationed on the roof. Because the smell of smoke lingers, and a great deal of it may confuse the worker trying to find the exact location of a leak, the worker inside must shout a warning upon detecting smoke as the system is filled, so that the smoke machine can be shut off and the leak investigated. The worker on the roof warns when the system is filled and plugs the stack. The workers doing the checking should avoid getting smoke on their clothing as this will complicate leak detection.

SUMMARY

A soil pipe test or **proving plug** is shown in Figures 27–1 and 27–3. In use the large wing nut is loosened to relax the rubber gasket, then the plug is inserted in the pipe and the wing nut is

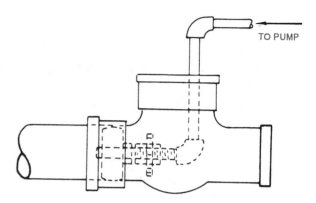

Figure 27–3 Plug-in test tee

retightened. The test plug in Figure 27–3 has a nut that looks like a rough gear instead of a wing nut. This type requires a special spanner wrench to tighten, although a large slip-joint pliers will also do.

There are other methods of testing drainage systems, such as the flow test and the peppermint oil test. These methods, however, are seldom used.

In practice, the plumbers test the system and repair any leaks. Then, with the system filled with water or pressurized, the plumbing inspector is called. The plumbing inspector checks that indeed there is water or air pressure in the system before going from joint to joint if it is a water test, or the air gauge, with a watch, if it an air-pressure drop test.

Repairing leaks on a drainage, waste, and vent system can be time-consuming and troublesome. Most codes require that leaking fittings be replaced. Some plumbing contractors require that whoever installed the leaking fittings should be the one who replaces them. This practice gives strong encouragement to careful joint making in original installations.

◆ ◆

TEST YOUR KNOWLEDGE

1. What does the roughing-in test accomplish?

2. How long should the pressure gauge remain on 5 PSI in the air-pressure drop test?

3. If the temperature is below zero, which testing method should be avoided?

4. How are leaks located with an air-pressure drop test?

5. How should leaks be repaired?

6. What is a manometer?

◆ ◆ ◆ ◆ ◆ ◆ ◆ ◆ ◆ ◆ ◆ ◆ ◆ ◆ ◆ ◆ ◆ ◆

CAST-IRON SOIL PIPE AND FITTINGS

KEY TERMS

bell and spigot	sanitary tee
bends	sweeps
combination soil fitting	test tee
double wye	upright wye
inverted wye	vent branch fitting

OBJECTIVES

After studying this unit, the student should be able to:

❏ Describe the strengths and weaknesses of case-iron material.
❏ Identify the various fitting patterns.

Cast-iron pipe is among the oldest pipe materials, along with hemlock log pipe and clay pipe. It has the advantage of being a very durable material. After some initial corrosion, the corrosion itself acts as a protective coating. There is cast-iron water supply pipe that has been in constant use for over 300 years in Europe. And so, it has an established reputation for durability. Installing cast-iron pipe by the older methods requires

some special tools called ladles, yarning irons, packing irons, and caulking irons (pronounced "corking irons" by older plumbers), Figure 28–1.

Some of the fitting nomenclature of cast-iron pipe may sound different and strange to the apprentice plumber. The elbows are called **bends** or **sweeps**. The joint style of the most common type of connection is called **bell and spigot**. More modern terminology for the bell is

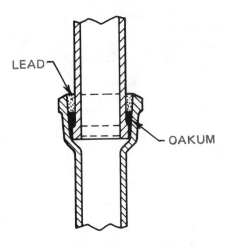

Figure 28–1 Cast-iron soil pipe

hub. Cast-iron soil pipe is made in single- and double-hub and no-hub with lengths that are 5 feet and 10 feet long. Notice that the pipe is measured from the back of the bell to the spigot end, Figure 28–2.

Cast-iron soil pipe is manufactured in two weights: service and extra heavy. Since it is lightweight, the service pipe is more likely to have sand holes or to split in handling or cutting. Soil pipe is made in 2", 3", 4", 5", 6", 7", 10", 12", and 15" inside diameter sizes.

Soil pipe, fittings, and joints may be coated with asphalt paint only after the inspector has had an opportunity to inspect the installation. All cast-iron drainage systems must be tested at an air pressure of 5 psi for a period of 10 or 15

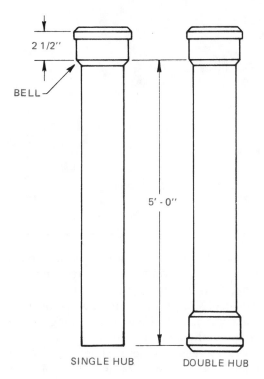

SINGLE HUB   DOUBLE HUB

Figure 28–2

WEIGHT OF SOIL PIPE

Service			
Size	Lbs. per Ft.	Size	Lbs. per Ft.
2"	4	6"	15
3"	6	7"	20
4"	9	8"	25
5"	12		

Extra Heavy			
Size	Lbs. per Ft.	Size	Lbs. per Ft.
2"	5	8"	30
3"	9	10"	43
4"	12	12"	54
5"	15	15"	75
6"	19		

minutes or it may be *hydrostatically* tested. Hydrostatic tests are tests in which the pipe is filled with water and allowed to sit overnight. If the system has been filled all the way up to the top of the stack, large leaks can be detected by observing if the water level has dropped. The plumbing inspector may want to examine each joint for seepage, however. Such testing is done to discover poor joints, sand casting holes, or split pipes and fittings which may allow sewer gas to escape into the building.

It should be noted here that the term "soil pipe" is applied to pipe materials other than cast-iron pipe. "Soil" refers to what the piping system is carrying.

When using neoprene gasket (described later) or lead caulked-type fittings, care must be taken to ensure that the flow within the system is *into the hub end and out of the spigot end* of each fitting. In other words, a length of *double-hubbed* pipe without one of the bells or hubs cut off would be illegal. This is because one of the hubs would have to have the flow coming out of it instead of going into it. The theory is that flow going in the "wrong" direction would have an easier time penetrating the joint assembly material and could leak.

The easiest and preferred method to cut cast-iron soil pipe is with a tool called a chain snap cutter. See Unit 29, *Soil Pipe Tools.* While cast-iron pipe is very rugged and durable and is often required in buildings for industrial use, repeated blows with a hard object will break it. Putting this property of cast-iron to practical use, soil pipe may be cut with a hammer and a cold chisel by placing a block of wood or lead directly under the cut line and striking the line with a chisel while rotating the pipe until it breaks. With a little practice and care, a neat cut can be accomplished when the chain snap cutter is not available.

Hint:
Safety glasses should always be worn on the job site, *especially when cutting cast-iron soil pipe.*

Cast-iron soil pipe may also be joined by several other methods, including the use of neoprene rings, Figure 28–3. The rings are placed in the bell of the pipe, which does not have a beaded end, and lubricated. The pipe is then forced into the ring, with the use of several different tools. This joint is approved in many areas for underground use and is flexible and watertight. Another advantage is the ease and speed with which this type of joint can be made.

One other type of joint that has become popular is the one used on soil pipe that does not have hubs or beads. The pipe is fastened with a neoprene sleeve and a stainless steel clamp, Figure 28–4. It is easy to install, and since there are no hubs, the pipe will fit in a smaller space.

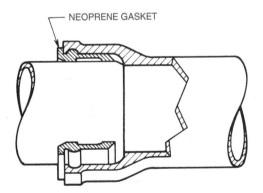

NEOPRENE GASKET

Figure 28–3

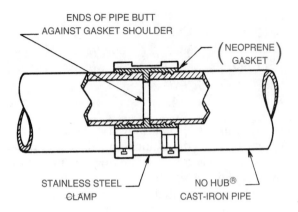

ENDS OF PIPE BUTT
AGAINST GASKET SHOULDER

$\left(\begin{array}{c}\text{NEOPRENE}\\\text{GASKET}\end{array}\right)$

STAINLESS STEEL
CLAMP

NO HUB®
CAST-IRON PIPE

Figure 28–4

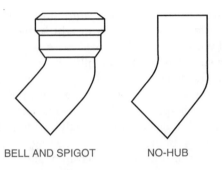

BELL AND SPIGOT   NO-HUB

Figure 28–5 1/8 bends

SOIL PIPE BENDS

To change the direction of soil pipes, several degrees of bends are made. The bend usually takes its name from the angle it turns. A 1/6 or a 1/8 bend means that there are 6 or 8 bends in a circle.

To find the degree of a bend, simply divide 360 degrees by the number of bends in a circle, such as $360° \div 8 = 45°$. Therefore, a 1/8 bend will change the direction of the pipe 45 degrees, Figure 28–5.

In Figures 28–5 to 28–8, note that:

❏ The length of a bend is the distance from the intersection of the hub and spigot center-lines to the end of the spigot, dimension D, Figure 28–6.

❏ Dimension D and X are both laying lengths.

❏ All dimensions are given in inches.

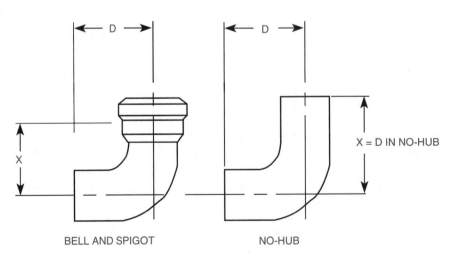

BELL AND SPIGOT   NO-HUB

Figure 28–6 1/4 bends

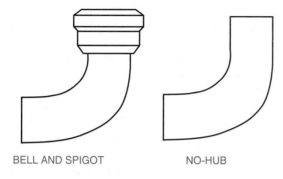

BELL AND SPIGOT   NO-HUB

Figure 28–7 Short sweeps

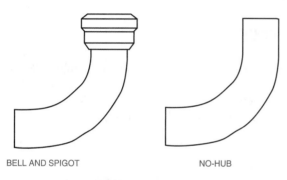

BELL AND SPIGOT   NO-HUB

Figure 28–8 Long sweeps

The sizes of different bends produced are 1/16, 1/8, 1/6, 1/5, 1/4 bend. In the 1/4 bends, there are three patterns having different radii. A 4-inch 1/4 bend has a radius of 4 inches; a short sweep bend, 6 inches; and a long sweep bend, 9 inches.

Soil pipe bends are made in the same size as soil pipe and in medium and extra-heavy weights. Extra-heavy pipe and fittings are required below buildings. All soil fittings are made with bell and bead ends and will fit any length of fitting of equal size and weight. To specify a soil bend, mention the size first and angle next.

Bends may have cleanouts, as in Figure 28–9, or the cleanouts may be placed in either side, as in Figure 28–10. The outlet in a soil bend is always smaller than the diameter of the bend.

The greater the radius of the bend, the less friction is caused by the flow of water in the drain. One-quarter (1/4) bends may be used only on vent lines because the turn is so short that in many cases the flow of sewage would be obstructed.

A return bend is used on a fresh air inlet to keep out dirt and other foreign matter.

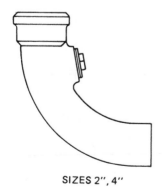

SIZES 2", 4"

Figure 28–9

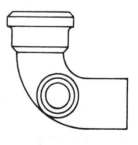

SIZES 2", 3", 4", 5", 6", 8"

Figure 28–10 1/4 bend right inlet

Y-BRANCHES

In a properly designed drainage system, the sewage should flow uniformly through the pipes without breaking up. On all horizontal pipes, it is necessary to connect branches at an angle no greater than 45 degrees.

Several forms of wyes are made for this pur-

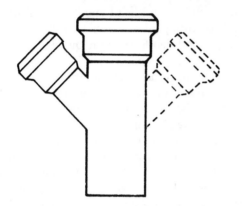

Figure 28-11 Single and double Y-branches

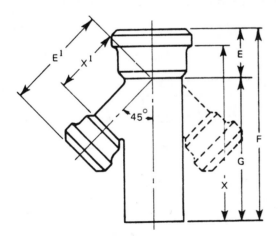

Figure 28-12 Single and double inverted Y-branches

pose. The wye is a fitting with a branch at 45 degrees, Figure 28-11. It is made in the same sizes as soil pipe with a variety of branch sizes.

A **double wye** is used where branches are to be made on both sides of the pipe at the same point.

Tees and wyes are made with only the branches reducing. **Inverted wyes** are used only at the top of vent stacks and are made in sizes from 2 inches to 6 inches, Figure 28-12.

Upright wyes are used at the bottom of vent lines, Figure 28-13.

The bell end of a wye is always the inlet. The bead end is never placed against the flow.

TEES

A *tee* is a 90-degree branch fitting, Figure 28-14. It is made in straight sizes from 2" to 15" and in reducing sizes from 12" × 3" to 15" × 12".

In the tables which follow, notice that:

❑ All dimensions are given in inches; weights are given in pounds.
❑ Tee branches (E^1) are approved for venting and cleanout purposes only, and tee openings are not recommended for use as waste inlets.
❑ Dimensions X and X^1 are laying lengths.

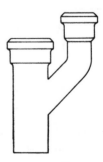

Figure 28-13

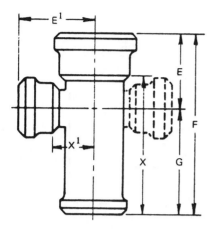

Figure 28–14

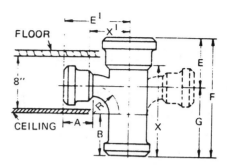

Figure 28–15 Single and double sanitary T-branches

The **sanitary tee,** shown in Figure 28–15, may be used only on vertical stacks. Notice the slight curve of the branch. This allows for a natural downward curve of the water as it enters the stack.

The sanitary tee is comparatively short so that it may be installed between floors and ceilings. These tees are also made with additional side branches. See Figure 28–15.

How to Read a Tee

A bell end of the run is never reduced. Therefore, state that size of the run first and give the size of the branch last. If the tee has an additional branch, as in Figure 28–16, hold the branch inlet toward you and read "left or right." The tee in Figure 28–16 is a 4 inch × 4 inch sanitary tee with a 2-inch right-hand inlet.

A **test tee,** shown in Figure 28–17, is placed on the main drain just inside the wall of a building. It has a large branch for the purpose of inserting a proving plug to test the drainage system. After the test is completed, a brass trap

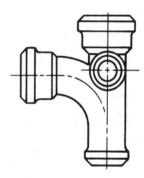

Figure 28–16 Sanitary T-branch with right-hand side inlet

screw ferrule and plug is caulked into the branch. This plug is later used as a cleanout.

COMBINATION SOIL FITTINGS

A **combination soil fitting** is a fitting that does the work of two separate fittings. For instance, the combination wye and 1/8 bend serves the same purpose as both the wye and the 1/8 bend.

Combination soil fittings have the following advantages:

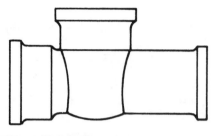

Figure 28–17   Test tee

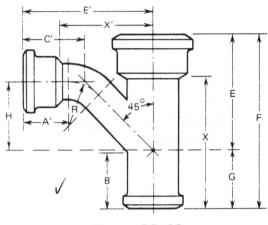

Figure 28–18

❑ They require fewer joints.
❑ They make the pipe more rigid.
❑ They take less space.
❑ They save labor.

To connect a branch or a stack to a main drain at a 45-degree angle, a combination wye and 1/8 bend is used, Figure 28–18.

The **vent branch fitting**, Figure 28–19, is used at the top of the vent stacks where they reenter the soil stack. This fitting eliminates the need for one 1/4 bend and one lead joint.

SUMMARY

Cast-iron soil pipe has an unparalleled history of durability. The newest pipe materials have a life expectancy of 50 years, whereas soil pipe installations of 300 years are still operating. Because of its cost, students can expect to find cast-iron soil pipe only in commercial buildings and in more expensive private homes. The ways of joining this material have changed greatly from earlier methods. Originally, the only style of joining was called bell and spigot. The unadorned spigot end would be slipped into the bell end and a lead and oakum joint would be made up. A piece of cast-iron pipe was nearly useless without a bell, and so, to avoid waste the

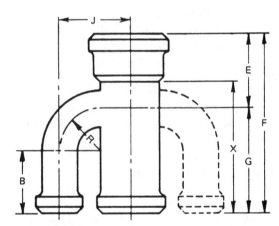

Figure 28–19   Single and double vent branches

plumber would buy double hub. This meant that when the plumber made a cut, both ends left after the cut would be useful.

Hubbed cast-iron soil pipe is still sold, but in most cases it is now joined by using a compressible neoprene gasket inside the bell and over the spigot end. When hubbed pipe is used, special tools are necessary. In the case of the gasket, a large inserter is used to pull the joint together. Often the plumber will still make up the last few immovable or hard-to-get-at joints with the old lead and oakum method. This will be explained in greater depth in Unit 29.

There is also another kind of cast-iron soil joint that is hubless. With this system, all that is needed is a tool to cut the pipe and a small ratchet-type wrench.

The old system (still in use) of naming elbow-type fittings described the change in direction that the fitting would provide in fractions of a complete circle. For instance, a half-bend would look like half of a circle and change the direction of the pipeline 180 degrees, that is, the line of pipe would make a U turn (also called

a U bend). A sixth-bend would change the direction 1/6th of a circle or 60 degrees.

Previously, in Unit 3, Fitting Specification and Identification, you were told to read a tee by giving the size of the large end of the run first, then the small end, and last, giving the size of the branch. Occasionally, branched fittings have side inlets, see Figure 28–16; the size of the side inlet is given last. But there is another factor to take into account: Will the side inlet be a left-handed one or a right-handed one? To determine this, set the branched fitting on its spigot or outlet end with the run vertical; now, turn the fitting until you are looking into the major branch. What side of the fitting is the side-inlet on? If it is the right-hand side, it is a fitting with a right-hand side inlet.

◆ ◆

TEST YOUR KNOWLEDGE

1. How long will cast-iron soil pipe last?

2. Name three tools used to join cast-iron soil pipe with lead and oakum.

3. What kind of threaded fitting can a bend be compared to?

4. What lengths does cast-iron soil pipe come in?

5. What is the smallest soil pipe size?

6. Can lengths of cast-iron soil pipe be installed backward?

7. Which kind of end preparation for cast-iron soil pipe requires the lightest tool kit?

8. How many sixth bends would it take to make a complete circle?

9. A return or U bend turns the line of pipe how many degrees?

10. Can an inverted wye be used to accept the waste branch of a kitchen sink?

CAST-IRON SOIL PIPE TOOLS

OBJECTIVES

After studying this unit, the student should be able to:

❏ Identify the soil pipe tools and describe their proper uses.
❏ Make a proper vertical bell and spigot joint.
❏ Make a proper horizontal bell and spigot joint.
❏ Cut cast-iron pipe.

LEAD AND OAKUM JOINTS

A common method of making a bell and spigot joint (cast-iron soil pipe joint) follows:

1. Light the furnace and fill the lead pot.
2. Set up the joint to be made.
3. Yarn the joint and drive it.
4. Yarn the joint again and drive it a second time.
5. Place the lead runner, if the joint is horizontal.
6. Warm the ladle.
7. Pour the joint.
8. Caulk the joint.

FURNACE AND LEAD POT

The lead heating furnace most commonly used is the *plumber's propane furnace,* Figure 29–1. This tool consists of a 20-pound propane tank to which the furnace head is attached.

Much care should be taken when attaching the furnace head to the tank. The gasket inside the furnace head, Figure 29–2, must be checked to see if it is broken or missing each time the

Figure 29–1 Plumber's propane furnace

Figure 29–2 Furnace head and gasket leaning on heat shield

head is assembled. Failure to do this may result in flames shooting out in all directions.

The best way to light the furnace is with a rolled-up piece of paper or oakum. Some plumbers use a flint striker, but this causes the hand to be very close to the flame when the furnace ignites. Treat the lighting process carefully. Once the furnace is burning, the shield may be put in place, Figure 29–2.

The purpose of the shield is to concentrate the heat around the lead pot. The pot may then be put on the furnace using a tool with a long handle through the bail of the pot. Always warm the ladle thoroughly before placing it in molten lead. The smallest amount of moisture on the ladle can cause an explosion. Most plumbers test the lead for the proper pouring temperature by pouring lead from the ladle into the lead pot. By observing how thin the last remaining sheet of lead in the ladle is, the plumber can tell if it is ready. The best way to learn this is by experience.

Yarning and Driving the Joint

Oakum expands when exposed to water. This makes it ideal for sealing pipe joints that must contain water. Oakum may be purchased in a number of different conditions. It may be untarred, tarred, or it may be what is called *white oakum*.

The oakum is placed in the joint by twisting it with one hand and forcing it down into the joint with the yarning iron in the other hand, Figure 29–3. After two circles of oakum are placed in the joint, it is driven down using a *driving iron* and a *driving hammer*, Figure 29–4. The driving hammer should be fairly heavy and firm strokes should be used. Very hard strokes with the driving hammer, however, may pump the pipe up out of the joint. The plumber places more oakum into the joint and then drives it down again. This continues until the oakum is about 1 inch from the top. If the joint has been kept straight during this process, it is ready to be poured.

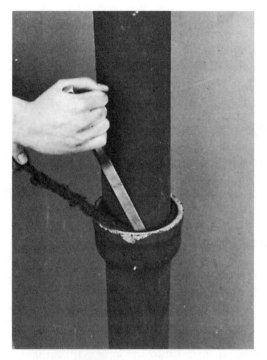

Figure 29-3 Placing the oakum

Figure 29-4 Driving iron and hammer

Horizontal Joints

Horizontal joints require the use of a *lead runner,* Figure 29–5. The lead runner is sometimes called a pouring rope. The lead runner is placed tightly around the opening in the joint and a small piece of oakum is placed into the gap where the two ends meet. The opening at the top of the pipe where the ends of the runner meet is called the *gate.* Molten lead is poured through this opening. When the lead has had an opportunity to cool the runner is removed. The lead is then caulked down firmly in three or four spots and the gate, which is now a piece of lead, is removed using a gate chisel, Figure 29–5.

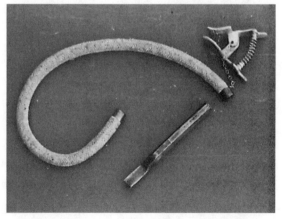

Figure 29-5 Lead runner and gate chisel

Caulking the Joint

After the head joint has cooled to the touch, it is ready to be caulked. The tools used are the caulking hammer and the inside and outside caulking irons. The caulking hammer should be light and is used with light, rapid strokes. The inside and outside caulking irons are shaped so that when the iron is held against the pipe, the working end strokes the lead at the correct point.

SOIL PIPE CUTTING TOOLS

Cast-iron soil pipe may be cut with a chain pipe cutter or a cold chisel. The chain cutter, sometimes called a snap cutter, Figure 29–6, is simply wrapped around the pipe at the point where the cut is to be made, fastened, and the handle is moved back and forth in ratchet fashion until the pipe snaps in two. The cold chisel is used with the driving hammer. The pipe is firmly supported directly under the point where the cut is desired and a continuous line of chisel marks is made around the pipe until the pipe breaks. With practice, the chisel method takes only a few minutes. The chisel used for cutting soil pipe should not be too sharp.

Special Lead Joint Tools

Most tools in this category are used to work in confined places. *Ceiling irons* work joints close to the ceiling. *Stack irons* work joints close to walls, Figure 29–7.

Insertable Gaskets

These joints are made by using neoprene gaskets in place of the lead and oakum.

These are easy to install and fairly watertight. The gasket is installed on the spigot end of the pipe. The joint is coated with a lubricant glue and forced into the joint with an installation tool, Figure 29–8. There are a number of different kinds of installation tools. A common one

Figure 29-6 Chain pipe cutter

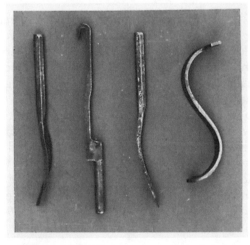

Figure 29-7 Ceiling and stack irons

works by clamping one chain around the fitting and another chain around the pipe and ratcheting them together by moving the handle back and forth, Figure 29–9. These tools may also be used for taking joints apart.

Hubless Joints

A special kind of cast-iron pipe is used for this kind of joint. The pipe and fittings have no bells or hubs. A neoprene sleeve is slipped over both halves of the joint and is then covered with a stainless steel clamp, Figure 29–10. Special tools are not required for this type of joint, but a special ratchet-type wrench is available to make the job even easier, Figure 29–11.

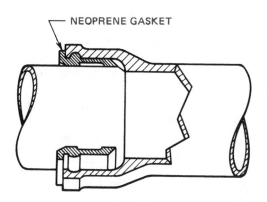

Figure 29–8

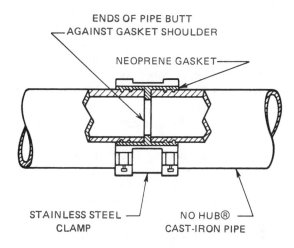

Figure 29–10

Figure 29–9

Figure 29–11

SUMMARY

This unit has briefly described the three different joining methods for cast-iron soil pipe in use today. These are:

❏ Lead and oakum. This is the traditional method and requires the most experience. Even with the more advanced joining techniques in use today, the plumber may occasionally set a closet flange and use the lead and oakum system, or "cut" a new fitting into an older line of pipe and use the older system of caulked and leaded joints. When there are only one or two joints to do in this fashion, the plumber may just melt scrap pieces of lead into a ladle and then "pour" the joint. This saves carrying the heavy and cumbersome furnace around. The plumber should always avoid breathing in the fumes given off by melting and overheated lead.

❏ Gasket insert assembly. This method requires that an insertion machine be carried from work site to work site. It is still not as heavy as the furnace and lead pot. Plumbers will make a heavy slugging tool by standing a piece of threaded pipe upright in the top of a dry, tinned can and then pouring lead into the space between the pipe and can. The result is an odd-looking but heavy mallet, which will not break cast-iron fittings. This mallet is used to bump joints together in circumstances where the assembly machine will not fit.

❏ Hubless cast-iron assembly. This is the easiest and most convenient method of joining cast-iron soil pipe. Special hubless fittings must be used. Proper and frequent pipe support is a critical consideration to prevent sagging and even joint pullout.

In all systems, the chain snap cutter is used to make pipe cuts. This is another ungainly, but convenient tool, which makes clean pipe cuts if used properly. It is indispensable for making a cut on an installed stack because the stack is usually close to a wall and the hammer and cold chisel cannot be used. Another consideration is that using the hammer and chisel method may result in a piece of cast-iron pipe dropping inside the stack and creating a difficult-to-remove obstruction. The chisel and hammer method, however, is portable and simple. The difficulty of cutting cast iron with a chisel is overrated. With practice, a clean cut on a 4-inch, service-weight cast-iron pipe can be accomplished in 2 minutes and with less than 60 hammer strokes.

◆ ◆

TEST YOUR KNOWLEDGE

1. What is the best way to light a lead furnace?

2. What does the furnace head shield accomplish?

3. What special lead and oakum tool is used on horizontal joints?

4. What can happen if the head gasket is not seated properly and the propane fuel is lit?

5. Which kind of pipe joining method requires a stainless steel sleeve clamp?

6. What kind of installation requires the application of a glue/lubricant?

7. What is a gate?

8. Why should you warm the ladle thoroughly before dipping it into the hot lead?

♦ ♦ ♦ ♦ ♦ ♦ ♦ ♦ ♦ ♦ ♦ ♦ ♦ ♦ ♦ ♦ ♦ ♦

PLASTIC PIPE AND FITTINGS

KEY TERMS

acrylonitrile butadiene styrene (ABS)
chlorinated polyvinyl chloride (CPVC)
polyethylene (PE)
polyvinyl chloride (PVC)

OBJECTIVES

After studying this unit, the student should be able to:

❑ Describe the differences between plastic pipe materials.
❑ Assemble the different kinds of plastic pipe.
❑ Explain the uses of plastic pipe materials.

PLASTIC PIPE AND FITTINGS

Plastic refers to a family of chemical compositions. While these chemical compositions may look the same, their uses in a plumbing system may be very different.

New plastic pipe materials are being introduced regularly. Many plumbing codes now permit the use of plastic pipe. Some companies are guaranteeing their plastic pipe and fittings for a fifty-year period.

Plastic piping has certain advantages and disadvantages when compared to metal piping.

Advantages

❑ It is lightweight.
❑ It is less expensive.
❑ It resists corrosion.
❑ It is fast and easy to install.
❑ It can be made with special characteristics.

Disadvantages

❑ It is less resistant to heat.
❑ It expands and contracts more when heated and cooled.
❑ It needs more support because of its flexibility.
❑ It has less crush resistance.
❑ It withstands less internal pressure.

Three common types of plastic pipe and fittings used in plumbing are *ABS, PE, PVC* and *CPVC*. All have different applications in the plumbing system.

ABS PIPING

Acrylonitrile butadiene styrene (ABS) piping is used extensively for drainage, waste, and vent piping. While ABS withstands heat well, a 10-foot section of steel pipe is recommended on commercial dishwasher drains. This dispels the heat before the waste water enters the plastic.

ABS pipe does not thread satisfactorily. An adapter is necessary to connect it to a threaded outlet. Horizontal piping must be supported at 4-foot intervals or less to avoid sagging. A lead-poured joint may be used to join plastic to a cast-iron bell. The lead will cool and solidify before the plastic begins to melt.

When connecting threaded joints, petroleum jelly or Teflon® tape is used for pipe dope. A strap wrench is used for assembling threaded joints.

Pipe wrenches scar the pipe fitting and may crack it or cause a weak spot. The tools necessary for roughing-in with plastic pipe include a rule, a tubing cutter with a special wheel, a half-round file for deburring, and a narrow paint brush for applying cement. A piece of sandpaper for cleaning dirty ABS and PVC pipe is also helpful.

ABS End Preparation (Solvent Weld)

1. Cut the pipe with a fine-tooth saw in a miter box, or with a tubing cutter with a special wheel, Figure 30–1.
2. Deburr the pipe inside and out with a half-round file, Figure 30–2.
3. Clean the pipe end of dirt, paint, or grease.
4. Apply solvent cement to both pipe and fitting, Figure 30–3.
5. Quickly insert the pipe all the way into the fitting. Twist the pipe 1/4 turn to evenly coat the joint, Figure 30–4. There should be a full bead around the socket with no gaps. Wipe off any excess cement, leaving a small fillet between the pipe and fitting.
6. Wait 3 minutes before handling.

Figure 30–1 Cutting pipe with a tube cutter

Figure 30–2 Deburring with a half-round file

Figure 30–3 Apply solvent cement to pipe and fitting.

Figure 30–4 Insert pipe into fitting and twist.

The pipe will be ready to water test within 2 hours. Because the cement dries so quickly, angles between fittings must be set quickly or the joint will be impossible to turn. Plastic fittings cannot be reused without expensive machining.

PE PIPING

PE, or **polyethylene**, pipe is a flexible plastic pipe used for underground installations, such as wells, sprinkler systems, and water supply systems. It is available in long coils of 500 feet. Maximum operating pressures vary with the particular quality and compound materials of a given product, but PE pipe may be obtained with guaranteed pressures of 80 to 160 pounds per square inch (psi).

PE piping is normally used outside of the foundation walls of a house. It is not used for hot water. The piping lies in a trench that is at least 12 inches below the frost line. The trench must be free of rocks and other sharp projections where the pipe is to be laid.

All plastics have a high expansion rate compared to steel pipe. Therefore, plastic pipe should be allowed to snake or wind down the trench. Temperatures underground are relatively constant, however, and expansion problems are rare.

Backfilling the trench must be done carefully to a depth of 6 inches above the pipe. Before backfilling further, the pipe should be pressure-tested to 150 percent of its working pressure.

PE End Preparation (Insert Filling Joint)

1. Cut the pipe squarely with a saw or tubing cutter. Ream the end with a half-round file.
2. Slip two hose clamps onto the pipe end.
3. Press the pipe onto the fitting without any pipe dope. If the fit is too tight, dip the pipe

end in a weak soap solution. Dipping the pipe end into very hot water will also help.

4. Tighten the hose clamps. Note that the screws are 180 degrees apart to distribute tension evenly, Figure 30–5.

Polyethylene pipe may also be cold flared with a special flaring tool. It is then assembled the same as any flared fitting. Insert fittings for polyethylene pipe are made of nylon or PVC.

PVC AND CPVC

Polyvinyl chloride (PVC) and **chlorinated polyvinyl chloride (CPVC)** plastic pipe materials are used for high-pressure applications (up to 400 psi). The dimensions of schedule #40 PVC (standard weight) are the same as those for standard steel pipe. CPVC is also designed to handle hot liquids up to 180°. Both come in rigid form.

PVC is used above and below ground level in such installations as underground pipelines, factories, and multiunit housing. It is threaded in schedule #80 or heavier form. Standard pipe dies may be used. They must be sharp and only used for plastic pipe. CPVC is used in the home for water supply piping and for smaller water distribution piping anywhere.

Both materials are tested with the pipe exposed. Allow sufficient drying time before pressure testing. Installing one day and testing the next day is usually sufficient. Because of long setting times, PVC and CPVC require more skill and planning to install than other kinds of plastics. The pipe fitter must consider the properties of the material and the temperature and weather conditions at the time of installation.

PVC sags at a sustained internal or external temperature of 110°. To install pipe horizontally at this or higher temperatures, it must be supported along its entire length. All plastic pipe

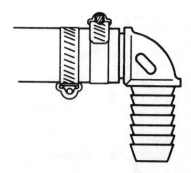

Figure 30–5

materials have a much higher expansion rate than steel pipe. CPVC expands 1/2 inch in a 10-foot length for a temperature rise between 73° and 180°. Contraction occurs at the same rate for a drop in temperature. If a pipeline is installed in midsummer and fastened securely at both ends, the joints might pull apart when the temperature drops below 32° the following winter.

PVC, like most plastics, must be well supported when it is stacked for storage, or it may take a permanent set. Dragging the pipe across a concrete floor will wear flat spots on the pipe and weaken it. Pipe hangers should not hold the pipe tightly. Strap hangers 3/4 inch or wider are best. Do not use wire hangers. The pipe should be free to slide in the hangers. Place hangers at every other joint or every 1 1/2 to 3 1/2 feet.

When joining PVC and CPVC to metal pipe, a special adapter, called a transition fitting, is used rather than a standard adapter. The transition fitting allows for the expansion difference between the two materials.

PVC and CPVC End Preparation

The assembly of PVC, CPVC, and ABS is similar with two exceptions. PVC and CPVC use a primer before the solvent cement is applied.

They are also assembled with great care.

ABS solvent cement is not interchangeable with CPVC or PVC solvent cement. The instructions on the can must be followed exactly. After assembly, avoid disturbing the joint. With PVC, a bevel is often filed on the outside end of the pipe so that cement is not pushed inside the fitting.

WELDING PLASTIC PIPE

When leaks occur on plastic pipe joints, the fitting is usually cut out with the plastic tubing cutter and a new fitting is installed. The result is an unsightly and very obvious repair. Where there was only a fitting, there now are two very out-of-place couplings and a fitting. Leaks in plastic pipe and fittings can be repaired by plastic welding.

The "torch," Figure 30–6, is actually a device like a soldering iron through which low-pressure air is pumped. The air is heated to a high temperature. The hot air brings the plastic base material to the melting point. A filler rod is then added and the fitting is welded to the pipe.

The leaking fitting should be clean and dry.

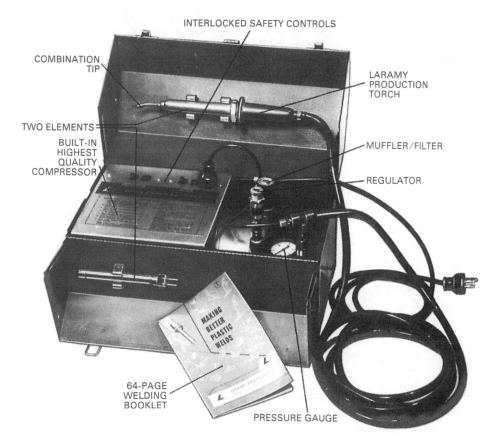

Figure 30–6 Torch for welding plastic pipe

A good plastic solvent will remove the glaze and give a roughened, clean finish to accept the filler rod. The filler rod must be of the same material (ABS, PVC, etc.) as the pipe and fitting. As the heat from the warm air torch is directed into the crack between the pipe and fitting, the filler rod is gently pressed into the crack.

By moving the torch slowly ahead of the filler rod and by pressing the filler rod at a slight angle from the vertical, the joint is welded in a smooth and continuous motion.

SUMMARY

Plastic pipe has grown to be the predominant material used for most applications in home construction. Many clever new pipe joining schemes are reducing the number of tools that a plumber must carry along. The materials are light and quite durable, if installed within certain limits. It is hard to think of a pipe and fitting material that does not have some limitations.

Plastic pipe will burn or melt easily. When it does, the resulting smoke is injurious to health. It is easily abraded and must be buried carefully so that no rocks or other hard materials lie against it. For the same reason, in places like garages where heavy equipment and vehicles might bump into it, plastic pipe must either be shielded or exchanged for galvanized steel at the likely points of contact. Remember that plastic pipe has a high coefficient of expansion. If the temperatures within the pipe change more than a few degrees, "crawling" will take place. The plumber should allow suspended pipe to move, yet at the same time, protect it from rubbing points.

TEST YOUR KNOWLEDGE

1. Which of the plastic pipe materials mentioned in the unit may be threaded?

2. Because ABS will sag if very hot water is allowed to run through it, what is recommended for kitchen sink drains?

3. If ABS cannot be threaded, then how is it joined to threaded connections?

4. What is another name for solvent welding?

5. Which of the pipe materials is best for indoor water supply?

6. How much pressure will a good grade of polyethylene pipe withstand?

7. Are there any plastic pipe materials that can be flared and joined to flare fittings?

8. Why should plastic pipe hangers be installed in such a way that the pipe is not held securely?

◆ ◆ ◆ ◆ ◆ ◆ ◆ ◆ ◆ ◆ ◆ ◆ ◆ ◆ ◆ ◆ ◆ ◆ ◆

PLASTIC SOIL
PIPE MATERIALS

KEY TERMS

DWV
soil pipe
transition fittings

OBJECTIVES

After studying this unit, the student should be able to:

❑ Define the differences in plastic pipe materials.
❑ Select the correct material for a specific job.
❑ Recognize the various fitting patterns.

PLASTIC PIPE USED IN DWV WORK

The acronym DWV means drainage, waste, and vent. Because of the many advantages of plastic pipe materials described in the previous chapter, plastic pipe has found its way into most new homes, Figure 31–1. Plastic DWV piping is not permitted in all localities and the plumber must be aware of code provisions for the particular municipality in which he or she is working. When we speak of soil pipe, we are referring to any pipe that carries human waste. This pipe may be made of many different materials. PVC

and ABS plastic pipe are often used in soil pipe applications, Figure 31–2.

PVC PLASTIC PIPE

PVC (polyvinyl chloride) pipe is a high-quality pipe material and can be used for many plumbing, pipe fitting, and air-conditioning applications. The schedule #40 weight (standard) and heavier may even be threaded. The pipe that lies within the foundation walls of the building and to a joint 5 feet outside of these foundation walls at the exit point (in other words, the main house

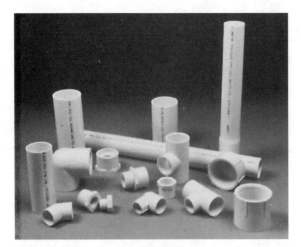

Figure 31–1 *(Courtesy Genova Products Inc.)*

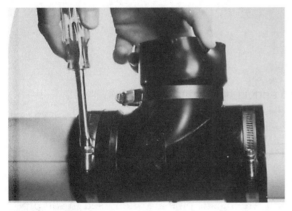

Figure 31–2 *(Courtesy Genova Products Inc.)*

drain), should be of schedule #40 weight. The joint preparation usually consists of chemical welding (gluing). Beyond this point, the main house sewer, a light grade of PVC pipe is often used and the fittings are simply pushed into the pipe without any jointing compound whatsoever. In the case of the private sewage system, the drainage field utilizes the same lightweight pipe but with perforations located at the bottom to allow the effluent to seep back into the soil.

ABS DRAINAGE WASTE AND VENT PIPING

ABS (acrylonitrile butadiene styrene) plastic pipe is rapidly becoming the material of choice for drainage and venting rough-in. It is glued and installed very easily. Indeed the plumber must take care to position the fittings accurately and quickly or the glue will set up and the fitting will cease to turn. Once the glue joint hardens, the only remedy is to cut the fitting out and install a new one with couplings. Expansion and contraction of plastic pipe is quite pronounced. The plumber should make sure that hangers are not tight and openings through which the pipe passes are large enough so that a hole will not be worn into the pipe.

FITTINGS FOR PLASTIC DWV PIPING

Most of the fittings available for copper pipe work and steel pipe work are also available for plastic pipe. Plastic pipe may be joined to copper, steel, cast iron, or other kinds of pipe with a variety of adapters, Figure 31–3. A **transition fitting** is the name given to one kind of adapter used, Figure 31–4. Another adapter fitting resembles an automotive hose clamp, except that the "hose" is heavy wall neoprene that has an internal diameter on each end specifically sized for the pipe for which it is intended. This fitting is commonly referred to as a *mission* coupling.

It is sometimes difficult to obtain T-branch fittings in the long turn pattern or TY configuration. If this is the case, the plumber should use a Y-branch fitting along with a 45-degree elbow when adding a branch drain to a horizontal run of pipe, Figure 31–5. Likewise, 90-degree elbows in plastic pipe are of the short turn pattern. While the inside of plastic pipes is smooth and aids the flow of material internally, the short turn fittings

Figure 31–3 Adapters for plastic pipe. *(Courtesy Genova Products Inc.)*

can slow the flow at the turns and cause a stacking effect. This may be good for stacks in tall buildings, as it slows the fall of the discharge and prevents a hammering effect at the base of the stack. In horizontal drains, however, the water may pile up at the turns, causing the air space over the water to close up. This airspace in partially filled pipes (drain and waste) provides a kind of internal venting, which assists drainage.

Figure 31–4 Transition fitting

SUMMARY

Soil pipe means pipe that carries human waste. The internal pressures are low, and the pipe should operate in a partially filled condition. In other words, the drainage of a bathing tub or a water closet should not be enough to completely fill the pipe. This allows the passage of venting air to pass over the drainage. In practice, bends and offsets in the line of pipe may cause a stacking effect and form slugs of water, which completely fill the inside of the pipe for varying amounts of time. Plastic soil-pipe materials have smooth interior surfaces that aid the rapid flow of waste and minimize slug formation.

Adapters, sometimes called transition fittings, allow the plastic pipe materials to be joined to any other kind of pipe. The transition fitting must allow for the different dimensional expansions between the two fitting materials.

With some of the fitting systems in use, the plumber hardly needs more than a rule, a marking tool, a cutting tool, and perhaps a nutdriver or a screwdriver. With some slip-hubbed, PVC, drainage, waste, and vent pipe (very lightweight), the plumber can dispense with the screwdriver and just push the lengths of pipe together.

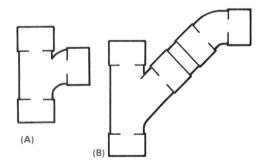

Figure 31–5 (A) Short turn TY or sanitary tee, (B) Y-branch with 45° elbow for smoother flow

◆ ◆

TEST YOUR KNOWLEDGE

1. What does DWV mean?

2. What does the term soil pipe mean?

3. How may plastic pipe, meant for drainage field construction, be identified?

4. May the glued fittings that leak be reused?

5. How can a long-turn, combination fitting, be made from short-pattern plastic fittings?

DRAINAGE FITTINGS

OBJECTIVES

After studying this unit, the student should be able to:

❑ List the special features of drainage fittings.
❑ Properly use drainage elbows.
❑ State the sizes of drainage elbows.

Drainage fittings are made of galvanized cast iron to prevent rusting. Two features of drainage fittings are not found in other types of fittings:

❑ All 90-degree drainage fittings are tapped to include the branch pipe 1/4 inch per foot,

Figure 32–1. This permits the drainage water to flow by gravity.

❑ Drainage fittings are recessed inside so that the inside of the pipe lines up with the inside of the fitting. This provides a smooth surface and allows the water to flow without obstruction, Figure 32–1.

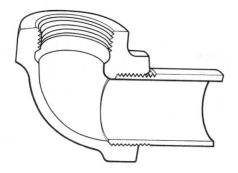 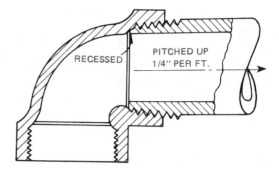

RECESSED   PITCHED UP 1/4″ PER FT.

Figure 32–1 Short turn

Drainage elbows are made in the following sizes and angles:

Sizes

1 1/4″, 1 1/2″, 2″, 2 1/2″, 3″, 4″, 5″, 6″, 8″, 10″

Angles

5 5/8°, 11 1/4°, 22 1/2°, 30°, 40°, 60°, 90°

The 90- and 45-degree elbows are made in long and short turns. Short turns are used for vent lines whereas long turns are better for drains, Figure 32–2.

Reducing 45-degree and 90-degree elbows can be obtained in sizes from 1 1/4 inches to 2 inches. The 45- and 90-degree *street* or *service* elbows are also available but should be used with care and only where the local code permits.

SUMMARY

The installation of galvanized steel drainage, waste, and venting systems with galvanized cast-iron fittings requires the use of heavy and somewhat complex machinery. While almost any of the drainage fitting pipe sizes (1 1/4″ – 10″) can be cut and threaded by hand by using ratcheting and leveraged manual threading machinery, the process is slow compared to the tools used with other pipe materials. For smaller jobs, the plumber might have most of the cut pieces machine pre-threaded at a larger shop and delivered to the job site. Pipes 2 inches and smaller can be worked fairly efficiently with portable pipe threading equipment. One might ask, why ever use this kind of material if it requires such heavy, expensive, and complex machinery? There are reasons:

❏ The material has unequaled vibration resistance.
❏ It also stands up well in places that are exposed to heavy machinery.

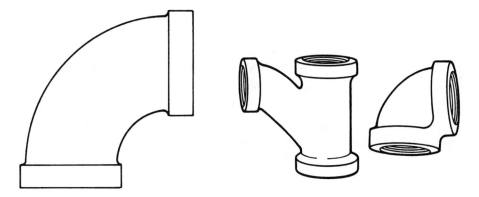

Figure 32-2

◆ ◆

TEST YOUR KNOWLEDGE

1. Other than the fact that drainage fittings are made from different materials, what distinguishes the drainage fitting pattern?

2. What precautions should the plumber consider when using short-turn fitting patterns?

3. What are the primary advantages of drainage fitting installations?

4. What are the disadvantages of this kind of pipe?

◆ ◆ ◆ ◆ ◆ ◆ ◆ ◆ ◆ ◆ ◆ ◆ ◆ ◆ ◆ ◆ ◆

FLOOR AND AREA DRAINS

KEY TERMS

area drains
deep seal trap
floor drains

OBJECTIVES

After studying this unit, the student should be able to:

❏ Discuss the types and uses of floor and area drains.
❏ Calculate the size of area drains.

FLOOR AND AREA DRAINS

Cast-iron floor and area drains are available in a wide variety of styles and patterns. **Floor drains** carry away any leakage or spilled water in such places as elevator pits, garages, boiler rooms, and laundries. **Area drains** are placed in driveways and other paved surfaces. The top is removable for cleaning purposes.

Floor and area drains installed in driveways or garages are placed flush with the pavement or floors and are extra heavy to withstand traffic, Figure 33–1. Floor drains with flanges are required in some areas of the United States, Figure 33–2.

All floor and area drains must have a trap on the waste pipe that can be cleaned in case of stoppage, Figure 33–3. The trap must be below

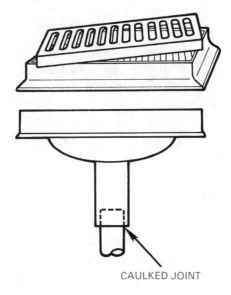

CAULKED JOINT

Figure 33-1 Catch basin frame and gate

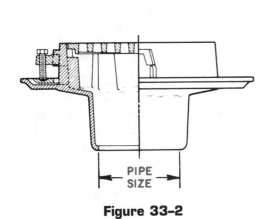

PIPE SIZE

Figure 33-2

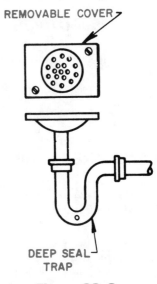

REMOVABLE COVER

DEEP SEAL TRAP

Figure 33-3

the frost line to prevent freezing. If the trap is directly under the drain, it may be cleaned by removing the grate. If the distance is not too great, the area drain trap is placed in the basement of the building. This prevents freezing or having to place it below the frost line.

Area drains have a deep seal trap to prevent the seal from evaporating in periods of drought. Floor drains must have a water supply handy in order to replace the water in the seal that evaporates from the heat in enclosed areas.

SIZING AREA DRAINS

The size of area drains for large surfaces may be determined by using Table 33–1. A 2-inch drain is the minimum size that may be used. This is sufficient for areas up to 100 square feet.

Table 33–1 is designed for rainfall rates of 4 inches per hour. *Divide* the number of square feet in the area to be drained by the local rainfall rate, then select the square feet from the column that represents the slope of the proposed drain. The size of the drain may be read at the left.

Diameter of Drain Inches	Maximum Projected Area for Drains of Various Slopes					
	1/8 in. Slope		1/4 in. Slope		1/2 in. slope	
	Square Feet	gpm	Square Feet	gpm	Square Feet	gpm
3	3288	34	4640	48	6576	68
4	7520	78	10600	110	15040	156
5	13360	139	18880	196	26720	278
6	21400	222	30200	314	42800	445
8	46000	478	65200	677	92000	956
10	82800	860	116800	1214	165600	1721
12	133200	1384	188000	1953	266400	2768
15	238000	2473	336000	3491	476000	4946

Based on rainfall rate of 1″ per hour

Table 33–1 Size of horizontal storm drains.

♦ ♦

TEST YOUR KNOWLEDGE

1. What are floor and area drains used for?

2. What things should be considered before installing a trap in an area drain inside of the building?

3. Why install a deep-seal trap?

4. If you have to drain 2000 square feet and you will have a 1/2 inch per foot slope on your drain line, and a 2 inch annual rainfall, how large of a drain line would you use?

♦ ♦ ♦ ♦ ♦ ♦ ♦ ♦ ♦ ♦ ♦ ♦ ♦ ♦ ♦ ♦ ♦

CLEARING STOPPAGES

KEY TERM

jobbing

OBJECTIVES

After studying this unit, the student should be able to:

❏ Describe how to locate a stoppage.
❏ Properly select and use stoppage clearing tools.
❏ Describe how fixture traps are cleared.
❏ Demonstrate how stoppages are removed.

Locating a stoppage in a sanitary (drainage, waste, or vent) system requires a basic understanding of the way waste systems are designed under a variety of conditions and the materials and methods used to install them. Repairing, maintaining, and enlarging existing plumbing systems is often called jobbing. The jobbing plumber must have good deductive reasoning powers along with superior plumbing systems knowledge.

Listening carefully to the customer's description of the problem is the first step in solving the stoppage location and cause. Asking good questions is the second. Some helpful questions to ask are:

1. What are the observable symptoms?

A laundry tray filled with water which recedes slowly or not at all gives us a pretty clear starting point. Water running out of a curb or yard vent is another. A cellar floor which is flooded may not be so obvious. Is there a floor drain under the water? Has it rained heavily recently? Does the water have a sewage smell? Are other homeowners on the street having the same problem? Simple questions asked in the beginning can save much fruitless exploration.

2. **Is one fixture involved or is more than one involved?**

If only one fixture is involved, the task may be limited to only one drain. Often, however, blockages are only partial and the fixture closest to the stoppage is backing up first. The homeowner may not be aware that flushing the water closet on the floor above will cause the water level to rise in the accused fixture. Gurgling sounds in other fixtures can be a tip-off that the blockage is in a main line. The plumber should run some of the other fixtures just to eliminate this possibility before turning complete attention to only one drain line.

3. **Has this particular blockage occurred before and what was the cause at that time?**

A poorly designed or installed plumbing system may have blockage problems which recur at the same place in the system. The customer will often be able to supply a history of blockages from the past and the places at which they occurred in the system. Recurring blockages at the same place call for some redesigning of the materials or routing of the system. Sink waste lines, for instance, often block up from an accumulation of grease. If the pipe

is cast iron or galvanized steel, rust and scale accumulation inside of the pipe may be providing "hooks" for grease to build up on. Replacing the old steel or cast iron pipe with smooth plastic or copper DWV tubing will often solve the problem.

4. **Does the homeowner or building manager have any knowledge of what the blockage might be?**

Children sometimes put toys in toilets and then flush the fixture. With some gentle prompting an apartment dweller will often recall that there is a fork or a washcloth missing. It is not that the plumber will have a difficult time locating the stoppage without the homeowner's help. The question may be: Is the blockage just an accumulation of hair and soap or is it a broken drinking glass? If it is the latter, it is best to remove it rather than to push it further down the pipeline to cause problems later.

5. **Has the homeowner attempted to clear the blockage and if so what was used?**

There are on the market a number of stoppage clearing chemicals that the homeowner can buy. If the fixture trap is full of "Plumbing Doctor" or some other caustic chemical the plumber could easily be severely burned when the trap is disassembled. *Always take extreme care when cutting into or disassembling drainage piping to clear a stoppage.*

SAFETY HAZARDS

As has been mentioned, caustic chemicals in a pipeline or trap may result in severe burns. It should also be kept in mind that the affected pipe is filled with human or decomposed kitchen waste. They have the potential to cause serious illness. Plumbers who do not show reasonable caution may find themselves endangered when they:

1. Cut into a pipeline which is full of backed-up sewage.
   a. Effluent may be sprayed all over, including on the plumber.
   b. An electrically powered tool may "short out" through the plumber's body.
2. Take pipe apart which is filled with sewage.
   a. Effluent or caustic chemical products may get on the skin or, worse, in the eyes.
3. Fail to clean stoppage clearing tools carefully after use.
   a. Exposure to air and sunlight kills most harmful bacteria. But built up "crud" on tools harbors the bacteria within these lumps and prevents their exposure to air and sunlight.
4. Fail to remember to shower, and change clothing before eating.
5. Fail to wear appropriate protective safety glasses, face shields, rubber gloves, boots, and aprons where required.
   a. Follow directions on containers of commercial drain cleaners meticulously.
   b. Wear protective outerwear and safety equipment when opening a pipeline below its internal water line.
   c. Pushing a rod into a plugged pipeline will often result in a violent back spray of the pipe's contents.

Often something as simple as draping an old rag over a fitting before it is opened will direct all contents down rather than outward. Of course if the line is not totally plugged, waiting for the water to drain down is the best and safest course. In larger buildings it is wise to turn off the water supply to affected fixtures so that they will not inadvertently be used while the drain line is being worked on.

PRACTICAL STOPPAGE CLEARING

Locating a stoppage in the main drain should be guided by certain conditions on the job, Figure 34–1.

If the stoppage is in the main house trap (A), the fixtures will not drain properly and water will overflow at the fresh air inlet or the laundry tray. There should be a check valve on the laundry tray waste to prevent water from backing into the trap. The stoppage may be removed by inserting a grappler through the fresh air inlet.

It may sometimes be moved by using a plunger in the fresh air inlet, Figure 34–2. This acts as a piston. The up-and-down movement causes a surging of the water within the drain, often dislodging the obstruction.

If the stoppage is between the trap and the sewer, the cleanout plug on the sewer side of the main trap may be removed and a coil wire inserted. Care must be taken when removing the cleanout plug since back pressure from the stoppage may flood the area with raw sewage.

Stoppage at point B is usually grease from the sink waste. The location of a stoppage may be found by tapping the pipe with a hammer. It will have a solid sound when the stoppage is tapped on. It may be dissolved by pouring drain solvent into the sink, or by using a cable with a cutter through the test tee.

With a stoppage at point B, the bathroom drain would not be affected. This would show that the stoppage was near the rear of the stack.

A blockage at point C may be cleared with a rod inserted in the vent stack or a hole may be cut.

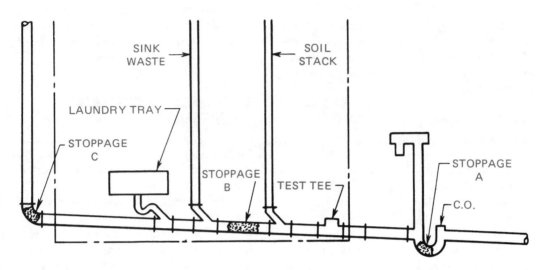

Figure 34–1

Fixture Traps and Trap Arms

Fixture traps and waste pipes are sometimes clogged by hair, lint, grease from washing dishes, or articles dropped into the toilet. Grease is the most difficult to remove because as it comes into contact with a cooler pipe, it tends to separate from the water and cling to the sides of the pipe.

Caution:
Always wash your hands and tools after working on clogged drains.

To remove grease, the *coil wire with head* is inserted in the cleanout and forced through the entire pipe, Figure 34–3. If the grease has collected for some time, it may be necessary to use a coil wire with a power-driven cutter on the end.

Figure 34–2

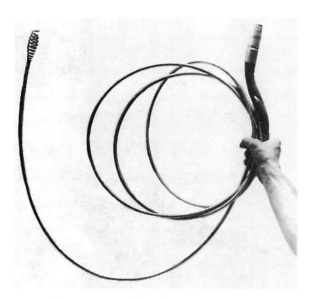

Figure 34–3 Coil wire with head

Hair and lint collect on the stainers of lavatories and laundry trays. They may be removed by using a bent wire or by removing the trap and cleaning out the pipe.

Caution:
Before taking a trap apart to clear it out, the plumber should discover if the homeowner has added any chemical drain cleaners to the trap in an effort to clear it. If this is the case, a large bucket should be placed under the trap in case of a spill. Rubber gloves and eye protection should also be worn.

Pencils, brushes, and other articles dropped into toilets may be removed by using a *closet auger,* Figure 34–4. This consists of a coil of wire within a tube. The coil has a hook at one end that is inserted in the trap of the toilet. By carefully pushing and revolving the coil, obstructions may be dislodged.

Stoppages in small fixtures may sometimes be removed by using a *force cup,* Figure 34–5. Alternatively pushing up and down creates a suction that causes a surging of the water in the pipe, dislodging the obstruction.

Fixtures with overflow outlets must have those outlets blocked off to make the force cup effective. A damp rag held over the outlet with one hand works well, Figure 34–6. Two examples of fixtures with overflow outlets are bathtubs and

Figure 34–4 Closet auger

Figure 34–5 Force cup

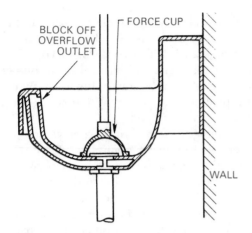

Figure 34–6 Cutaway view of lavatory

Figure 34–7 Ball-type plunger

Figure 34–8 Electric drain cleaner

lavatories (basins). A *ball-type* plunger may sometimes be more effective than a force cup, Figure 34–7. An electric hand drain cleaner is helpful for small but stubborn jobs, Figure 34–8.

SUMMARY

Clearing stoppages generally falls under the purview of the jobbing plumber. It is the responsibility of the jobbing plumber to understand the design and operation of the entire plumbing system, and furthermore must deduce this in older buildings from sketchy evidence–the locations of fixtures, vents, etc.

Most times the customer's problem can be deduced from a few polite and well informed questions asked by the plumber upon arrival at the problem site.

There are some personal health considerations about working with stopped and overflowing drains. The student should realize that these are manageable problems and knowledge and reasonable caution are all that are necessary to be safe.

TEST YOUR KNOWLEDGE

1. Where would the stoppage be (in general terms) if the customer describes a particular toilet as always backing up?

2. What kind of branch is most likely to be clogged with grease?

3. Can a closed vent cause a fixture to refuse to drain?

4. What is the first step in determining the location of the stoppage?

5. What must be done before using the force cup in a fixture with an overflow outlet?

6. What is the closet auger for?

7. If one fixture in a bathroom group is backed up and no other, what does this information tell us?

8. Why would you shut off the water supply to a bathroom group before working on the drain?

FASTENINGS

OBJECTIVES

After studying this unit, the student should be able to:

❏ State the purpose of hanger fastenings.
❏ Choose the correct hanger fastening for a specific job.

FASTENINGS

Plumbers as well as all construction craftsmen need to be able to select and install fastening devices. Fixtures must be fastened to walls and floors. Pipes must be hung from joists of wood and beams of steel. Air-conditioning and heating units must be fastened to roofs and concrete basement floors. In all cases, the plumber must make some important decisions.

1. Is the fastening going to be permanent? Will it be there when the job is complete?
2. Will the piece of equipment need to be moved from time to time?
3. How much weight must be supported by the fastening? Vibration counts as weight that is intermittently applied.
4. What are the characteristics of the material being fastened?
5. What are the characteristics of the material being fastened to?

Nails

Nails are used in the construction of wooden buildings to fasten wooden members of the structure together. The plumber will use them often.

The most frequently used nails in building construction are called appropriately enough, **common nails.** *Finishing nails* and *roofing nails* are regularly used, also.

They are purchased by the pound for quantity and by the *penny,* abbreviated with a small "d," for size. A 16d common nail, for instance, would be 3 1/2 inches long, Figure 35–1.

Common nails have a medium-large head and are used in places where the head would not normally be seen. A plumber might build a partition at the head of a bathtub, put up tub supports, and install backing boards of various types using common nails.

Nails, while being very versatile, also have some serious failings. Because they are made of steel, they rust easily and, if subjected to moisture for long enough, they will break off. The cure for this is to buy galvanized or other rust-proofed nails for use where this might occur.

Because they are normally smooth, nails pull out easily with vibration or shrinkage of the wooden members. Where this is likely to occur, the plumber might use cement-coated or threaded nails. These nails are very difficult to pull out even with a claw hammer, once they are driven in.

Finishing nails have small heads, Figure 35–2. As is indicated by their name, they are placed in exposed places like the trim around an access door. In use, they are eventually concealed by *setting* them below the surface of the surrounding wood. Setting is done by driving them slightly with a small punch known as a *nail set.* The indentation that is left is filled with wood putty and then painted over when the rest of the trim is painted.

Plumbers will often use galvanized **roofing nails** to install small strap-type hangers and to secure roof flashings in place. Roofing nails, Figure 35–2, have large heads and good holding power. Holding up a water supply line, a copper-plated pipe strap, and a small roofing nail with one hand while striking the nail with a hammer takes some practice.

Screws

Screws are generally defined as fasteners that make their own threads and do not require a nut to hold the parts of the assembly together, Figure 35–3.

This is a useful definition although the student should be aware that there is a classification of machine screw that does not quite fit. They are purchased by the box, which contains 100 or 144 screws. They are classified for size by a number between 2, the smallest, and 24, the largest. The length of a screw is given in inches.

Screws can be divided into *wood screws* and *sheet metal screws.* Each of these classifications can be further roughly divided into *roundhead, oval head,* and *countersunk head,* Figure 35–4.

The **roundhead screw** is used where appearance is unimportant and the head will not interfere with moving parts. The oval head screw is an attractive screw head and is used where the material is not thick enough to countersink and the head will be exposed. The countersunk head screw is the workhorse of the screw world and is as the name implies, finished flush or below the surface. This can then be covered or plugged and will not show after the job is complete.

Screws are also defined by the type of tool used to install them. There is the slot, clutch, Phillips, and allen head screw. For most applications, the slot head screw is used.

Common Nails

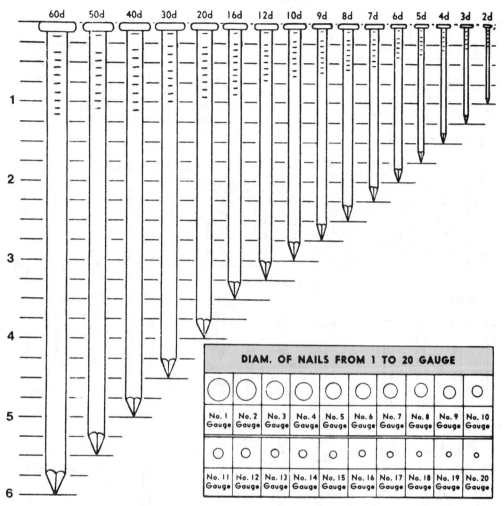

	DIAM. OF NAILS FROM 1 TO 20 GAUGE									
	No. 1 Gauge	No. 2 Gauge	No. 3 Gauge	No. 4 Gauge	No. 5 Gauge	No. 6 Gauge	No. 7 Gauge	No. 8 Gauge	No. 9 Gauge	No. 10 Gauge
	No. 11 Gauge	No. 12 Gauge	No. 13 Gauge	No. 14 Gauge	No. 15 Gauge	No. 16 Gauge	No. 17 Gauge	No. 18 Gauge	No. 19 Gauge	No. 20 Gauge

COMMON NAILS REFERENCE TABLE							
SIZE	LENGTH AND GAUGE	DIAMETER HEAD	APPROX. NO. TO POUND	SIZE	LENGTH AND GAUGE	DIAMETER HEAD	APPROX. NO. TO POUND
2d	1 inch..........No. 15	$^{11}/_{64}$	845	10d	3 inch..........No. 9	$^{5}/_{16}$	65
3d	1¼ inch..........No. 14	$^{13}/_{64}$	540	12d	3¼ inch..........No. 9	$^{5}/_{16}$	60
4d	1½ inch..........No. 12½	¼	290	16d	3½ inch..........No. 8	$^{11}/_{32}$	45
5d	1¾ inch..........No. 12½	¼	250	20d	4 inch..........No. 6	$^{13}/_{32}$	30
6d	2 inch..........No. 11½	$^{17}/_{64}$	165	30d	4½ inch..........No. 5	$^{7}/_{16}$	20
7d	2¼ inch..........No. 11½	$^{17}/_{64}$	150	40d	5 inch..........No. 4	$^{15}/_{32}$	17
8d	2½ inch..........No. 10¼	$^{9}/_{32}$	100	50d	5½ inch..........No. 3	½	13
9d	2¾ inch..........No. 10¼	$^{9}/_{32}$	90	60d	6 inch..........No. 2	$^{17}/_{32}$	10

Figure 35–1 Nail reference table. *(Courtesy Do-It-Yourself Retailing)*

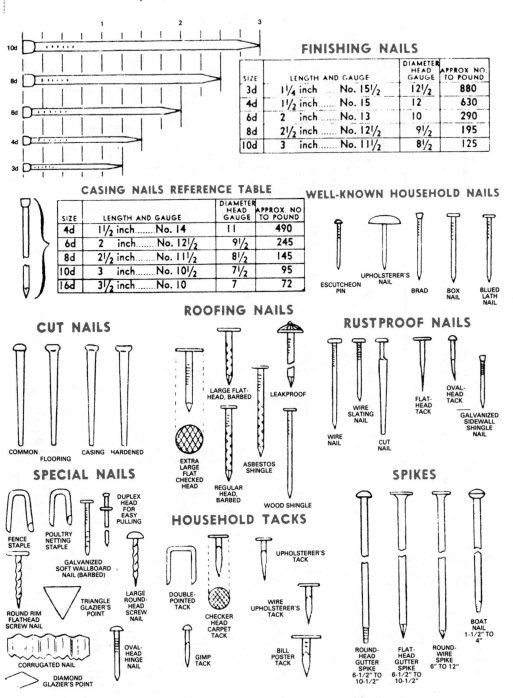

FINISHING NAILS

SIZE	LENGTH AND GAUGE		DIAMETER HEAD GAUGE	APPROX NO. TO POUND
3d	1¼ inch	No. 15½	12½	880
4d	1½ inch	No. 15	12	630
6d	2 inch	No. 13	10	290
8d	2½ inch	No. 12½	9½	195
10d	3 inch	No. 11½	8½	125

CASING NAILS REFERENCE TABLE

SIZE	LENGTH AND GAUGE		DIAMETER HEAD GAUGE	APPROX. NO. TO POUND
4d	1½ inch	No. 14	11	490
6d	2 inch	No. 12½	9½	245
8d	2½ inch	No. 11½	8½	145
10d	3 inch	No. 10½	7½	95
16d	3½ inch	No. 10	7	72

Figure 35–2 Nail head identification. *(Courtesy Do-It-Yourself Retailing)*

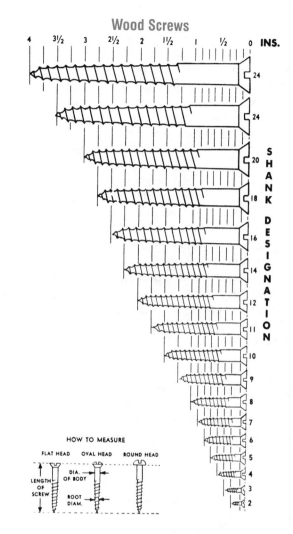

Wood Screws

Figure 35–3 Screw reference table. *(Courtesy Do-It-Yourself Retailing)*

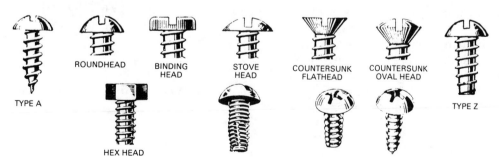

Figure 35–4 Screw head identification

Screws, especially wood screws, require considerable effort to drive them into place by hand. It is essential that the trusty screwdriver be kept in good condition. This means that the end of the blade should be square with the shank. It should also be as thick and as wide as the slot head screw in which it is going to be used.

Wood Screws. Plumbers use wood screws to fasten large hangers in place and to install wall carriers for various fixtures. It is best to use a drill bit in a drill motor to make a *pilot* hole a fraction of an inch smaller in diameter than the screw in the wood. This makes driving the screw in much easier and prevents cracking the base wood. A screw-driving bit can be obtained to go into a carpenter's brace. This allows tremendous leverage to be applied to the screw head on the larger screw sizes.

Sheet Metal Screws. Sheet metal screws have parallel sides with the exception of the starting point. This makes driving the screw all the way in much easier than the wood screw that is tapered throughout the threaded area. The sheet metal screw can also be used for lighter jobs in wood. An example of a lighter job in wood would be strap-type pipe hangers.

There is a magnetic holder for electric drills for sheet metal screws. The electric drill has enough power to drive a sheet metal screw.

In practice, a sheet metal screw is placed in the chuck. The projecting screw is pressed through the hanger against the wood and the trigger on the drill is pressed. Using the same setup with a self-tapping screw, duct work and smoke pipe can be joined in the same way without even drilling a preliminary hole.

Other Fasteners

Lead wall shields along with a lag screw can be used to fasten to concrete and block walls and floors, Figures 35–5 to 35–7. The **toggle bolt** is a simple way to fasten cabinet bases to walls, Figure 35–6. Some judgment must be exercised, however, because the wall covering may not have sufficient strength to provide support. **Molly fasteners**, Figure 35–7, are another way of fastening to plasterboard and thin wall materials.

Pipe hangers are fastened to wood, stone, brick, concrete, and metal. Study the manufacturer's catalog for other types of hangers.

Hangers fastened to wood will be more stable if screws are used instead of nails. Nails may hold small pipes, but wood has a tendency to dry out and the nails may be dislodged if there is much weight on the hanger.

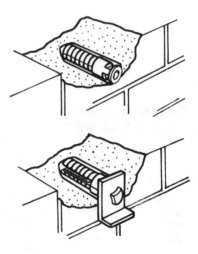

Figure 35–5 Lag screw shields

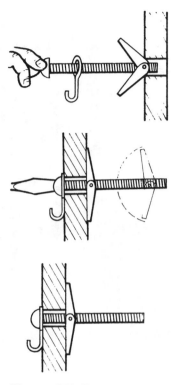

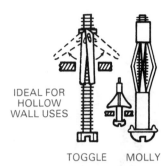

IDEAL FOR
HOLLOW
WALL USES

TOGGLE   MOLLY

BOTH BOLTS ARE SELF-ANCHORING. DRAW OF BOLT
CRIMPS SLEEVE ON MOLLY. WINGS ON TOGGLE SPRING
OPEN AFTER INSERTION IN HOLE.

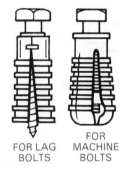

FOR LAG
BOLTS

FOR
MACHINE
BOLTS

FOR USE IN FASTENING TO CONCRETE, BRICK, OR STONE.

Figure 35–6 Toggle bolts

Figure 35–7 Anchors and shields

To fasten a clevis hanger to a wood joist, a **hanger bolt** is used, Figure 35–8.

A hole about half the diameter of the screw is bored into the center of the joist and the lag screw is inserted into the full length of the thread.

To fasten hangers to brick, stone, or concrete, a hole must be cut with a star drill or masonry drill. When using the star drill, be sure to revolve the drill during the entire cutting process. An expansion shell or bolt is then inserted in the hole. Many improvements have been made for inserting shields in concrete or masonry walls. Carbide-tipped masonry drills

that fit a regular drill motor bore holes much faster and more cleanly than star drills, Figure 35–9, but a source of power is required with this type of drill. Expansion shields or shells that are self-drilling and drive impact fasteners that may be inserted with a hammer are also available. Select the type suited for the job.

Before concrete ceilings are poured, concrete inserts are fastened to the concrete form in the proper location, Figure 35–10. Later, when

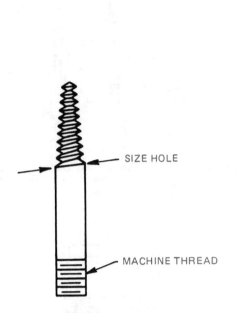

Figure 35-8 Hanger bolt

Figure 35-9 Masonry drills. *Courtesy Phillips Drill Co., Inc.)*

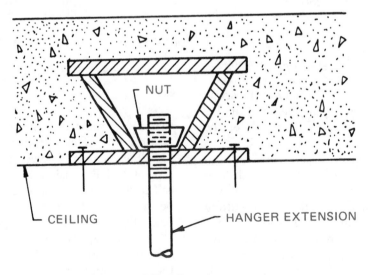

Figure 35-10 Concrete insert

the form is removed, a special nut is placed in the concrete insert into which the hanger extension is screwed.

To fasten a hanger to a steel beam, a **beam clamp** is used, Figure 35–11. Manufacturers' catalogs list many different types that are available.

As a rule, horizontal pipes are supported 12 inches to 24 inches below ceilings. Hanger extensions are used for this purpose, Figure 35–11. A vertical pipe must be supported at each floor or two, depending upon the height of the ceilings, Figure 35–12.

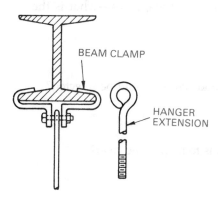

Figure 35–11

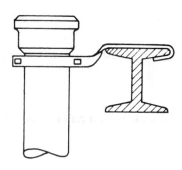

Figure 35–12 Vertical beam clamp

◆ ◆

TEST YOUR KNOWLEDGE

1. For nailing rough building construction materials—studs, joists—what is the most common fastening device?

2. When a flush finish is desired, what kind of wood screw should be selected?

3. What kind of a nail is used when the nail head is to be painted over?

4. What tools should be used to cut holes through masonry walls?

5. What is the toggle bolt used for?

6. When a fixture or pipe may be removed from time to time, would you use a screw or a nail?

FIXTURE SUPPORTS

OBJECTIVES

After studying this unit, the student should be able to:

❏ Accurately install a set of bathroom fixture supports.
❏ Box in a joist that is in the way of a water closet.

BACKING BOARD

Backing boards are anything regarded as reliable support. This includes devices for supporting and securing pipe, fixtures, and equipment to walls, ceilings, floors, or any other structural members. Backing boards are also called *supports* and *hangers*.

Backing boards may be found anywhere in a plumbing or heating system. The majority of them will be found, however, in the bathroom.

The plumber will find it necessary to install backing boards to support pipe, heating units, and fixtures as the building construction proceeds. A little care and time spent in installing backing boards correctly will make the job of fixture installation much easier later in the job.

Materials

The tools used to install backing boards are the claw hammer, the 6-foot folding rule, and the utility hand saw.

Wood used for backing boards should be a soft wood such as pine shelving material. Avoid using a wood that is hard or filled with knots. This kind of wood will split when the plumber tries to nail it. It will also be difficult to drive screws into hard wood when it is time to set the fixtures.

Installation

The backing board is usually located between two studs and is nailed to them, Figure 36–1. The proper height of the board must be determined. This can be found by looking at the rough-in drawing or taking measurements from the fixture to be installed. It is important to realize that the thickness of the finished floor must be accounted for. The work will be done with only the subfloor in place. This means that the thickness of the finished floor must be added to the height of the backing board.

Measuring from the subfloor, make a mark on the two studs located on both sides of the backing board. Include the thickness of the finished floor in this measurement. Now measure between the studs at this height. This is how long the board must be. Select a piece of soft wood at least 3/4-inch thick and 5 1/2 inches wide. Cut it to the proper length. This will be the backing board. Cut two pieces of 1 × 2 approximately an inch longer than the width of the backing board. These are the cleats. **Cleats** are short pieces of wood that support the ends of the backing board, Figure 36–2. The cleats are then nailed in place. Nail the cleat in place with two or three eight-penny nails. The center of the cleat should be placed at the same height as the mark that was made on the studs. The backing board is then centered on the marks. This means that there will be as much backing board above the marks as there is below them. The backing board is then nailed in place with sixteen-penny nails. The plumber should be careful to angle the nails so that they also go into the stud.

The distance the cleats are set back from the front surface of the stud will determine the distance that the backing board will be set in from the front surface of the stud. Sometimes, as on a lavatory hanger support, it is correct to have the backing board flush with the front of the stud. At other times it should be set back, as in a pipe support, Figure 36–3.

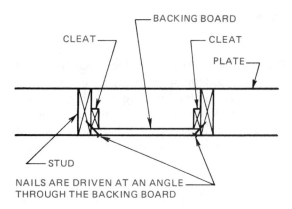

Figure 36–1

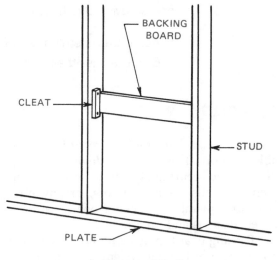

Figure 36–2

Lavatory

The lavatory is a common fixture that will require a backing board. All wall-hung lavatories have a mounting bracket. Therefore, a backing board must be provided behind the wall before the wall covering is installed. The mounting bracket that comes with the lavatory is attached to the backing board through the wall covering. The lavatory is then hung upon the bracket. Some pedestal-type lavatories do not require a backing board. The vanity should also have a backing board if it is to be installed in new construction. On old construction, the vanity is often attached to the wall with a toggle-type anchor because the vanity comes on a cabinet base. It is supported by that cabinet, but still requires additional attachment to the wall. It is poor practice to expect the pipe to hold the vanity to the bathroom wall. A backing board must also be provided behind the wall to support wall supplies and waste pipes if the job is going to have these items.

Bathtubs

Recessed bathtubs require **fixture supports** to hold the rear lip o the tub securely. This may be a horizontal 2 × 3 that runs the entire length of the tub. However, if a horizontal wooden support is used, two legs should be provided under this support so that the nails are not relied upon to carry the weight of the tub. Another way to support the tub is to cut two 2 × 3 or 2 × 4 wooden legs. These may be placed vertically upon the floor and nailed to the studs. Care must be taken to cut these supports accurately or the tub may be unsteady after it is installed, Figure 36–4. The height of the tub supports must be determined with the bathtub standing in its normal position, Figure 36–5. All cast-iron tubs should be measured individually because of differences between the castings.

Water Closet Installation

When installing a water closet in a room not originally designed for it, the plumber may find

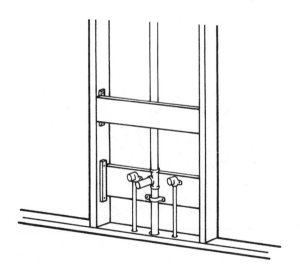

Figure 36-3 A typical backing board setup for a lavatory

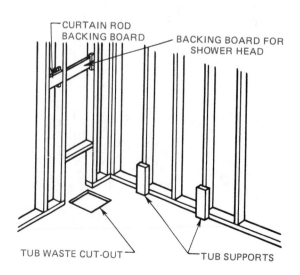

Figure 36-4 Tub fixture supports a typical layout.

that a joint interferes with the outlet to the fixture. If a section of joint must be removed to accommodate the toilet waste line, it will have to be **boxed in,** Figure 36–6. If this is not done, the flooring may be dangerously weakened. It is the plumber's responsibility to leave the structure of the house as strong, or stronger, than it was before the plumbing work was done.

Wall-hung water closets come with their own metal mounting frame enclosed in the box. It is necessary to provide a strong framework to support this frame.

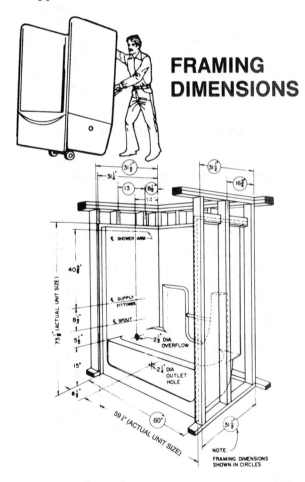

Figure 36–5 Framing dimensions

Curtain Rod Supports

Curtain rod is another name for shower curtain rod. If it is a combination shower and bathtub and the bathroom is not designed to have a tub enclosure (sliding fiberglass, plastic, or glass doors), a curtain rod must be installed. This means that a curtain rod backing board must be placed within the studded wall to support the curtain rod flanges. The curtain rod backing boards are put up in the same manner as lavatory backing boards. The distance above the subfloor should be about 6 feet, 6 inches. Because the shower curtain should hang inside the tub, the shower curtain rod must be placed accordingly.

Shower Head Supports

The shower head arm is threaded into a drop-eared elbow behind the wall. The drop-eared elbow has two screw hole lugs attached to it. This enables the elbow to be attached to a backing board. The backing board that must be installed for the shower head should be about 6 feet, 6 inches to 6 feet, 8 inches above the subfloor.

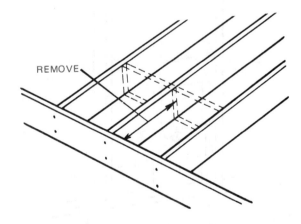

Figure 36–6 Boxing in a joist. Cut out piece marked remove. Install piece on hidden line.

Miscellaneous Supports and Structure

Occasionally, the plumber will have to provide support for overhead piping. If the piping is running across the joists (at right angles to them), it can be either hung from the joists or strapped to them. If, on the other hand, the piping is running in the same direction as the joist, the plumber may have to cut wooden pieces to fit between the joist (2 × 3s or 2 × 4s would do). The plumber would then either hang the pipe from the wooden pieces or strap the pipe to them.

A **trouble or access door** is placed at the head of the bathtub so that the plumber can get into the parts of the tub that may cause trouble. This door is usually found in a closet. To make a trouble door, the plumber places a horizontal crosspiece high enough to be slightly above the tub diverters. A long continuous cleat is nailed around the rectangle formed about 1/4 inch back from the face of the studs. A panel of 1/4-inch plywood is then cut to fit into the rectangle, Figure 36–7.

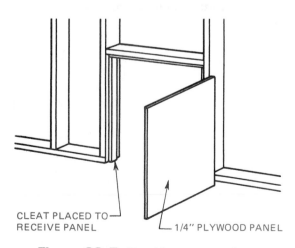

CLEAT PLACED TO
RECEIVE PANEL
└─ 1/4" PLYWOOD PANEL

Figure 36–7 Trouble or access door

SUMMARY

Construction plumbers, more so than jobbing plumbers, frequently find themselves acting as carpenters. This happens most often in the placement of wood in the existing structure to support the weight of pipe, equipment, and fixtures. Backing boards, as they are called, may need to be placed between joists and studs to provide a firm foundation for pipe, or to provide support for suspended fixtures. The plumber should be proficient with measuring, marking, and cutting new wood for this purpose. Occasionally, existing structure must be removed. When this is the case, the plumber carefully removes no more than is necessary and reinforces the surrounding area structurally so that the strength of the building is not compromised.

In new construction, the backing boards are put in place before the wall coverings. Then, after the wall coverings and finishes have been added, the plumber returns to hang and install fixtures. If the plumber has not carefully placed those backing boards, which have now been covered, it is too late; the consequences may be serious. In rough finished areas, like basements and garages, the backing boards may be moved and no harm is done.

At the time when supports are being installed, the finished floors may not have been laid or the wall coverings (wallboard, etc.) put in place. The plumber needs to know the thickness of these finishing materials and allow for them. In most cases, the backing boards will be of such a width that a half-inch here or there will not matter.

◆ ◆

TEST YOUR KNOWLEDGE

1. What is the common name for wood placed within walls to provide a firm foundation for fixture supports?

2. What do joists support?

3. What do studs do?

4. Why would a vanity, with its cabinet base, require a backing board?

5. What is the purpose of "boxing in" a joist after a section has been removed?

6. What is the purpose of a trouble door?

7. What is a piece of light wood prenailed to a joist or a stud and meant to serve as a nailing support for the backing board called?

UNIT 37

PIPE HANGERS

KEY TERMS

clevis hanger
coil hanger
F & M hanger
perforated iron strap

pipe strap
Reznor hooks
split ring hanger
spring cushion hanger

OBJECTIVES

After studying this unit, the student should be able to:

❏ List the various methods of supporting pipes.
❏ Sketch the various devices used to support pipes.

Pipes must be securely supported both horizontally and vertically under many different conditions. Pipe can be attached to various building materials.

The first consideration is the weight of the pipe and its contents. Expansion and contraction due to changing temperature within the pipes must also be considered. Pipes in some buildings are subject to vibration.

The simplest small pipe support is the Reznor hook. It is a bent piece of wire with two pointed ends that are driven into a wooden joist, Figure 37–1.

The pipe strap is used for small pipes on walls or ceilings, Figure 37–2. Roundhead screws should be used on pipe straps.

Perforated iron strap is used to support small pipes, such as water and heating pipes, Figure 37–3. It may not be used on main drainpipes.

A clevis hanger is good for large pipes. The advantage is that the top half may be installed first and the bottom half may be removed. When used with an extension eyebolt, it may be adjusted to line up the pipe, Figure 37–4.

The F&M hanger consists of two malleable iron halves hinged at the bottom and bolted at

Figure 37–1 Wire pipe hook (Reznor)

Figure 37–2 Pipe strap

Figure 37–3 Perforated iron strap

the top. This is hung on a rod or pipe support, Figure 37–5.

Ceiling coils or a number of parallel pipes may be supported by a **coil hanger,** as in Figure 37–6.

To support pipes in buildings that have considerable vibration, a **spring cushion** hanger is sometimes used.

Many special hangers may be made on the

job from bar or band iron. They may be either bolted or welded, Figure 37–7.

The **split ring hanger** is shown in Figure 37–8. A split ring hanger is a commercial hanger that may be disassembled while remaining attached to joists.

Since they are rigid, steel, brass or copper pipes require that supports be installed about every 10 feet. Cast-iron pipe, because of the soft

Figure 37–4 Clevis hanger

Figure 37–5 F&M hanger

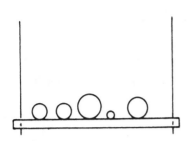

Figure 37-6 Coil hanger

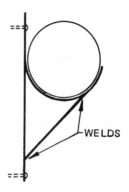

Figure 37-7

lead joints, must be supported at closer intervals. Many codes require lead joints to be supported at every joint.

Some codes specify a 5- or 10-foot distance between hangers. A 5-foot distance between hangers is usually better since the pipe is more easily kept in alignment. One hanger under each stack is required because of the extra weight at those points. Glass pipe and plastic pipe require support at varied intervals. Check the manufacturer's catalog for the recommended distance between hangers.

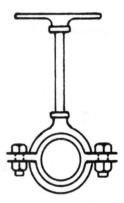

Figure 37-8 Split ring

◆ ◆

TEST YOUR KNOWLEDGE

1. Is the weight of the pipe being supported the only consideration when selecting the pipe hanger?

2. What would be the fastest form of pipe hanger to install?

3. In high vibration conditions, what might be a good selection for steam pipes?

4. Which hanger allows for an indefinite number of pipes to run in the same hanger?

WATER SYSTEMS

(Courtesy Wheeler Manufacturing)

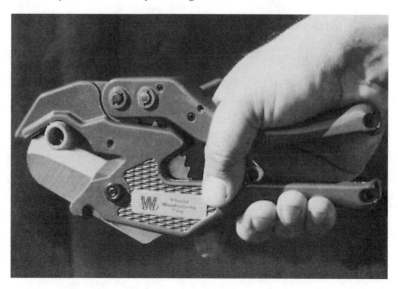

INTRODUCTION TO WATER SUPPLY

KEY TERMS

cased
cistern
drilled wells
driven wells

dug wells
potable
soft water
water table

OBJECTIVES

After studying this unit, the student should be able to:

❏ Understand the sources of potable water.
❏ Describe the treatment of water.
❏ Understand the water cycle of potable water.

WATER SUPPLY SOURCES

There are three sources of potable water. These sources are: rainwater, underground water, and surface water. Potable means drinkable.

RAINWATER

Rainwater is used in some areas where water levels in the ground are very deep or nonexistent. Water runoff from roofs is collected in cisterns and used for household purposes. A **cistern** is a holding tank for the rainwater. Although rainwater is relatively pure, it may be contaminated by the roof, the gutter and downspouts, or the cistern. It should receive some chemical treatment to kill bacteria before use.

In farm communities where water is in short supply, cisterns are used for washing water and for irrigation. This relieves the burden of supplying all of the water from the well. Because rainwater contains few minerals, it is called **soft water**.

UNDERGROUND WATER

Under the surface of the earth, at various depths for different locations, there is a layer of saturated ground. The level at which the saturated ground is found is called the **water table**, Figure 38–1.

This level does not always remain at the same depth. Depending upon the season or amount of rainfall, it may move up or down. Underground water from a well that has been

properly located and constructed is the safest. Well water is generally used untreated.

Wells

Where municipal water is not available, each home must have its own source of water supply. Wells are by far the most common source of water. The object of a well is to make the water lying beneath the water table available for use. Where the water table is close to the surface, wells are sometimes dug by hand. **Dug wells** are rarely deeper than 20 feet. Hand-dug wells may go dry during long rainless periods. When water is not added to the underground supply over a long period, the water table goes deeper and deeper. It can very easily go beyond the depth of shallow wells. Also, hand-dug wells are more subject to pollution than deeper wells are.

Another kind of well is called a **driven well**. The driven well is used in loose, sandy, or gravelly soils. A device called a well point is screwed into a length of pipe and then it is driven down into the earth with hammer blows. It is one of

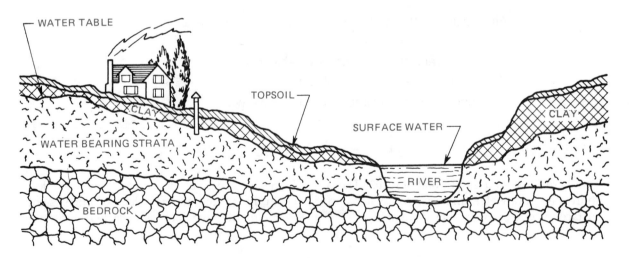

Figure 38–1

the easiest methods of establishing a well. However, it is limited to depths of less than 60 feet and to soils that are not rocky.

Drilled wells are the deepest, most dependable, and the most pollution-free. It is possible, but not practical, to drill a well with small equipment. Drilled wells are made with massive machinery under the supervision of a well drilling specialist. The wells are drilled down into the water-bearing strata. The well is then **cased.** This means that a large-diameter steel pipe is lowered into the well hole. Then, the space around the casing is filled with grout (a cement) from the surface to a depth below the polluting effects of the surface water, Figure 38–2. The well pipe itself is then placed within this casing.

SURFACE WATER

Water that runs in streams or is found in depressions, such as in lakes, reservoirs, ponds, or oceans, is called surface water.

Urban areas ordinarily use surface water for their potable water source. This source is the most plentiful. However, it is also the most easily contaminated. The water authority must treat this water before it is piped to the consumer. Depending upon the amount of impurities, the water may have to go through a number of stages. Screening removes large particles. Filtering removes smaller particles and impurities. Chemical purification removes bacteria and other organic pollutants.

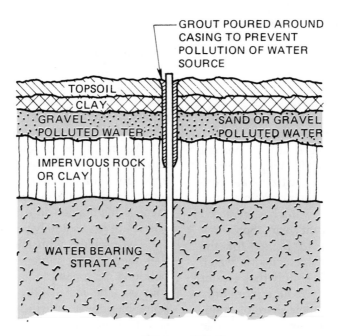

Figure 38–2

WATER CYCLE

Water that is consumed today is actually very old. Water is continually repurified and recycled by nature. The water cycle works in the following way:

1. The sun causes evaporation from bodies of surface water (lakes, rivers, streams, and oceans). The water is then held in suspension in the air.
2. The water held in suspension precipitates in the form of rain or snow.
3. a. Some of the rain or snow runs off into streams and rivers and becomes surface water again.
   b. Some of the rain or snow soaks into the ground where it eventually enters the water table. There is an underground flow of this water down toward bodies of water. This brings the water back to its beginning state.

SUMMARY

Water is not a problem in the modern world; after all the Earth's surface is still 75 percent water. The supply of safe drinking water is, however, a problem. Because citizens are concerned about the quality of their drinking water supply, the sale of "spring" water has increased by many multiples in recent years. Some say that industrialization is to blame for the reduced availability of potable drinking water. Some say that the reason is simply that greater numbers of people not only require more pure water, but also contribute their share of pollution back into the environment—a double threat. Double the number of people and dou-

ble the potable water requirements and also double pollution. If there were fewer people, so it goes, factories would produce fewer cars, refrigerators, farm tractors, etc. Whatever the cause may be, the plumber will be called upon to install more water purification and softening units than ever before. The plumber will also be called upon to render more and more advice. A heartening observation about the state of our dwindling water supply is that water is only H_2O, a molecular combination of two atoms of hydrogen and one of oxygen, and, as such, that which has become polluted can be made clean again. The cost will be high.

Water can be distilled by the addition of heat, which changes its state to water vapor or steam and then it may be condensed back into its liquid state. This process kills bacteria and leaves contaminating solids behind. The water becomes distilled water and safe to drink, in most cases.

Water can be filtered and treated, the present procedure. The more efficient form of filtration is to force the water through a membrane that is so fine-textured that only the water molecules can come through. This is even used to remove salt from salt water.

Water can be decomposed into its constituent elements, oxygen and hydrogen, by the application of an electrical current and then recombined to make water again. If the equipment is maintained in a sterile condition, this will result in absolutely pure water.

The problem with all of these methods is that they are far more expensive than water obtained in a simple fashion from nature's sources. The installation and maintenance of home-based water purification systems, even on so-called safe municipal water supply systems, will become a new frontier for plumbers.

TEST YOUR KNOWLEDGE

1. Name three sources of potable water.

2. What kind of water do cisterns hold?

3. What is generally the safest source of potable drinking water?

4. Can rainwater be contaminated?

5. What is water that contains few minerals and a neutral acid level often referred to?

◆ ◆ ◆ ◆ ◆ ◆ ◆ ◆ ◆ ◆ ◆ ◆ ◆ ◆ ◆ ◆ ◆

PRIVATE WATER SYSTEMS

KEY TERMS

a pump's prime
air control
deep-well pump
jet pump

piston-type pump
pressure switch
shallow-well pump
submersible pump

OBJECTIVES

After studying this unit, the student should be able to:

❏ Describe the parts of a typical private water supply system.
❏ Explain how the parts of the system work together.

PRIVATE WATER SYSTEMS

Private water supply systems may use springs, creeks, cisterns, and other sources for their water supply. The typical system uses a cased well for its water source. If properly constructed, this is the most reliable, pollution-free water supply. The parts of the private system are:

❏ Pump
❏ Pressure tank

❏ Pressure switch
❏ Foot valve
❏ Air volume control

How the System Works

The pump of a well either lifts or pushes water from the well into a tank until the desired pressure is reached (40 to 45 psi), Figure 39–1. A pressure-actuated electrical switch turns the pump on and off. The switch might be adjusted

283

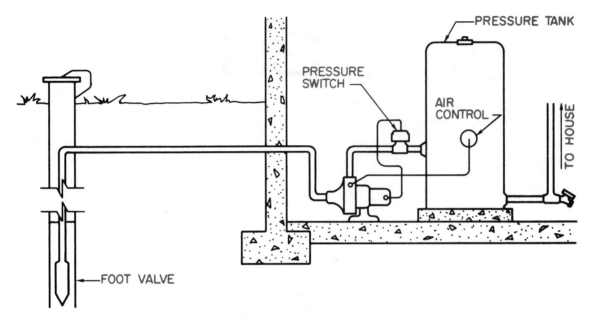

Figure 39–1 Well

to turn on the pump when the pressure in the tank drops to 20 psi, and then turn it off when the pressure climbs to 40 psi.

Since water is incompressible, the pressure tank operates with a cushion of air trapped in the top. If the tank is completely filled with water, the pressure in the tank will drop immediately when a tap is opened. The pump will then turn on to restore pressure. Because there is no air cushion, the pressure will build up immediately, and the pump will shut off again. This will repeat again and again before a small container fills with water.

Air is compressible. This property enables it to expand without a rapid pressure loss. It is important to maintain an air cushion in the top of the tank. To maintain this air cushion, a device, called an air control, is used. A foot valve is placed at the end of the pipe submerged in the well. It prevents the water being drawn up the well pipe from falling back down into the well

when the pump turns off. If this check valve is not there, the pump will lose its prime when it stops running. When a pump is primed, it is full of water and has no air in it. Water pumps will not operate when they are air-bound.

PUMPS

There are several types of water pumps and pumping systems. A pump may be a jet type, submersible, or reciprocating. These three pumps may be further classified for either a shallow well or deep well.

A shallow-well pump is limited to a theoretical depth of 35 feet. In practice, because of friction loss and pumping efficiency, depths of 25 feet are usually the limit for this type. The shallow-well pump uses atmospheric pressure (14.7 psi) to lift the water. Sucking water up a straw is a good comparison.

The more common **deep-well pump** pushes the water up the well pipe to the tank. The **submersible** deep-well pump (Figure 39–2) accomplishes this by being lowered into the water in the well. It is sealed so that water cannot enter the electric motor, Figure 39–3. The deep-well **jet pump** uses a two-pipe system within the well to push and draw the water to the surface, Figure 39–2. The jet pump diverts some of its water back down the well. This water is forced through a *venturi,* a narrow opening in the jet body, to create a low pressure. This low pressure assists the pump in lifting the water up the pipe.

The reciprocating or **piston-type pump** must use a long pump rod extending to a piston that is submerged in the well water. Because of the complexity of this system, the reciprocating pump is usually limited to shallow-well applications for the home.

PRESSURE SWITCH

The **pressure switch** is mounted on the tank or in the discharge pipe from the pump. It uses the existing pressure to activate an electrical switch. In a common application, it *makes* or turns the pump on at 20 psi, and *breaks* or turns the pump off at 40 psi.

AIR CONTROL VALVES

The job of the air control valve is to add a small amount of air to the pressure tank each time the pump cycles. Without air control the water would slowly absorb the air cushion in the tank.

There are several methods of admitting air to the pressure tank. Most use the negative pressure on the suction side of the pump to draw air in, Figure 39–4. The air control valve, or *air charger,* must add air only when the tank requires it.

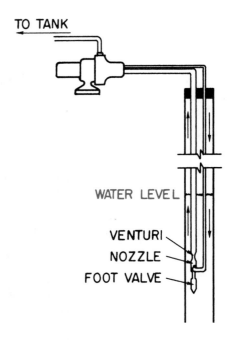

Figure 39–2 Deep-well jet pump

PRESSURE TANK

The pressure tank may simply be a cylindrical tank designed to handle the maximum working pressures of the system. Some tanks have a plastic foam float that rides on the water surface in the tank. The float separates the air and the water. This prevents the water from absorbing the air. Some tanks separate the air and water with a rubber diaphragm or air bladder.

SUMMARY

The use and application of private water pumping systems is quite varied. The information in this unit is generally limited to common applications for rural homes. In practice, the span of useful and working systems is quite large.

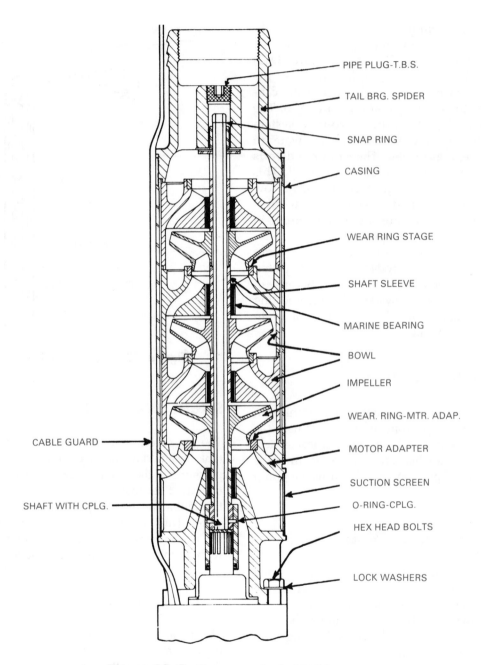

PIPE PLUG-T.B.S.

TAIL BRG. SPIDER

SNAP RING

CASING

WEAR RING STAGE

SHAFT SLEEVE

MARINE BEARING

BOWL

IMPELLER

WEAR. RING-MTR. ADAP.

MOTOR ADAPTER

SUCTION SCREEN

O-RING-CPLG.

HEX HEAD BOLTS

LOCK WASHERS

CABLE GUARD

SHAFT WITH CPLG.

Figure 39-3 Cutaway of submersible pump

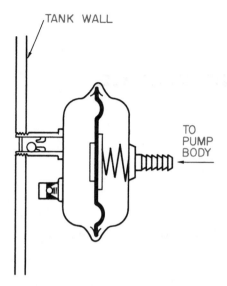

TANK WALL

TO
PUMP
BODY

Figure 39–4 Air control

Physics plays a great part in the design of private water supply systems.

Some of these systems make use of a storage tank to hold water which is collected at opportune times for use when the occupants of the building need it. Many farms in the Midwest used (and still use) windmills to lift water gallon by precious gallon into an elevated cylindrical holding tank. Once the water is in the holding tank, the elevation of the tank above the living quarters will provide the pressure of the water at the faucets. The effect of this system is to allow the pump to work day and night to fill the tank; the tank then supplies the water when needed at constant pressure by virtue of its height. A two- or even one-gallon per minute system might provide all the water that the farm family needs. The deeper the well, the less water a pump of a given size can produce. In arid regions, the water may be deep and the supply slow because of the depth.

The same concerns and solutions can be applied to skyscrapers and high-rise buildings, only in this case the water might be available from city mains at 65 psi But the height of the top floors will make water at that pressure undeliverable. It requires 0.434 psi to raise a column of water one foot. If the stories in the building are 10 feet apart, it would require 4.34 psi to raise the water one story. This means that the pressure on the second floor would be 65 psi – 4.34 psi or 60.66 psi A little calculation would show that in a relatively small building of only 15 stories, there would be no pressure available at all on the top floor. What to do?

In the case of tall buildings, they may go back to the elevated tank concept of the farm windmill with a twist. They may lift the water in stages with booster pumps every few floors somewhat like a bucket brigade until the water is delivered into a storage tank on the top floor. The storage tank will then supply water to all quarters below it at a constant pressure but again with a twist. The cumulative booster pump pressures which originally lifted the water all the way to the 100th floor, are now regained in the supply pipe below the roof tank. Now the plumber will install pressure reduction valves every few floors going downward so that occupants receive the water at usable pressures and equipment is not stressed beyond design limits. Without this, the water pressure on the tenth floor could theoretically go to 390 psi.

A less exciting consideration is that the private water supply pump be no more powerful than the well can handle. Wells are rated in gallons per minute. This means that a given well might be able to supply 10 gallons per minute on a continual basis. If the plumber installed a well pump that delivers 15 gallons per minute, the pump might literally suck the well dry and lose its prime before the controls shut off. When a pump loses its prime, the impeller housing fills with air and the flow of water stops. The homeowner or plumber may need to re-prime the pump manually by filling the well pump housing and all of the pipe coming from the well.

◆ ◆

TEST YOUR KNOWLEDGE

1. Which is more compressible, air or water?

2. Name two types of well pumps.

3. Name two well configurations that influence the kind of pump and pumping system used.

4. What is the practical maximum depth for shallow-well pump setup?

5. Which of the pumps requires that two pipes run from the house to the well?

6. What makes the pump go on and off?

7. What does a foot valve do in the system?

8. What do the terms *make* and *break* refer to?

TOOLS FOR
WATER SUPPLY SYSTEMS

KEY TERMS

annealing   flaring tools
bending spring   swaging tool
flame soldering   tubing cutters

OBJECTIVES

After studying this unit, the student should be able to:

❑ Recognize the tools used in water supply systems.
❑ Use water supply tools.

No materials in the plumbing trade have changed more than those used in water supply piping. In a period of 50 years, water piping has gone from lead to galvanized steel, to copper, and now to plastic. The gradual change to plastic pipe materials has just begun. It seems that eventually almost all piping for home use will be plastic. At present, copper piping is the most frequently used.

Plastic piping requires, for the most part, only simple hand tools. These include a wood saw and perhaps a miter box and measuring tools. Some companies make a special tubing cutter for plastic pipe.

Copper water tube, however, requires a number of special tools. These include: flaring tools, swaging tools, copper tubing and fitting cleaners, soldering tools, and copper bending tools.

TUBING CUTTERS

Tubing cutters, Figure 40–1, are lightly constructed pipe cutters. They should never be used on steel pipe, however, because they will either break or bend out of line quickly. To use the tubing cutter, it is tightened against the

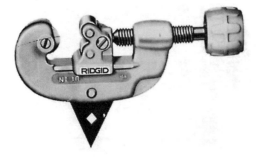

Figure 40–1 Tubing cutter. *(Courtesy The Ridge Tool Company)*

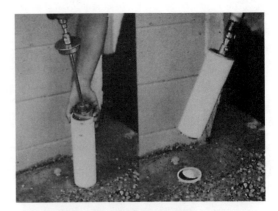

walls of the tubing and turned one complete circle around the tube and then tightened a little more. This is continued until the tubing is cut through. Turning the adjustment in slowly will result in a small burr on the inside of the tube. This can be removed with the blade provided on the back of the tubing cutter. Some tubing cutters have reamer blades with square holes in them. This hole is a key for operating the valve on acetylene tanks. Some plastic tubing cutters are shown in Figure 40–2.

FLARING TOOLS

Flaring tools are used to provide a mechanical joint in copper tubing, Figure 40–3. In order to flare, the end of the copper tube is clamped in the flaring block and a steel cone is forced into the end of it with a screw device. This cone flares out the end of the tubing to the proper dimension, Figure 40–4. The outside dimension of the flare is controlled by how much of the tubing is permitted to project through the flaring block.

SWAGING TOOLS

A **swaging tool** is used to join two lengths of tubing together. The tool is driven into the end of the

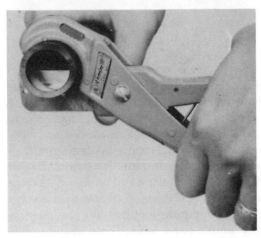

Figure 40–2 Plastic tubing cutters. *(Courtesy Wheeler Manufacturing)*

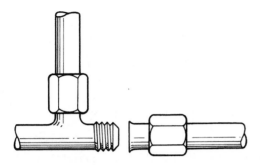

Figure 40–3 Flared tubing joists. *(Courtesy The Ridge Tool Company)*

Figure 40–4 Flaring tool

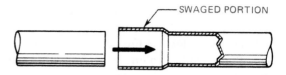

Figure 40–5 Swaged joint

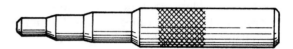

Figure 40–6 Swaging tool

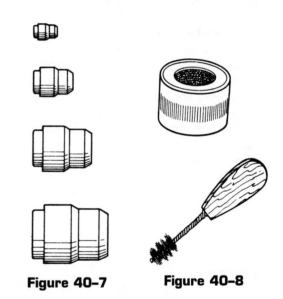

Figure 40–7   **Figure 40–8**

pipe. This forms a socket that will accept the end of the other piece of tubing perfectly, Figure 40–5. Some swaging tools have a number of different sizes on the same tool, Figure 40–6. Some swaging tools come in sets with a different tool for each size pipe, Figure 40–7. Hard copper tubing should be annealed before it is swaged.

Annealing is a process of heating and cooling that is used to avoid brittle or cracked joints. In annealing, the tubing is heated with the torch until a deep purple color is obtained. The copper tubing can then be allowed to cool and it will remain soft. It helps to put a little soldering flux on the tool before inserting it in the pipe. The tubing should be held in the hand while the swaging tool is being driven in. Placing the end of the tubing on a plank will usually result in a damaged end.

CLEANING TOOLS

In order to solder, the surfaces to be soldered together must be perfectly clean. Plain steel wool may be used for this purpose, or a specially designed tool may be used, Figure 40–8. Both the outside of the tubing and the inside of the fitting must be cleaned. Self-cleaning fluxes may be used. However, if the pipe or fittings are old or dirty, they will still have to be cleaned. Joints made with self-cleaning fluxes must be soldered soon after assembly. Failure to solder within the same day means that the joint must be recleaned.

SOLDERING TOOLS

The type of soldering commonly used to join copper pipe and fittings is called **flame soldering**. A torch flame is applied to the fitting. When the fitting and tubing reach the melting point of solder (375 to 425°F), the solder will flow into the fitting and bond the fitting to the pipe.

The torch kit consists of a tank of gas, a regulator, a flexible hose, a handle, a torch tip, and a key for the gas tank, Figure 40–9.

Tanks of gas may contain acetylene, propane, or a number of other inflammable preparations. The containers may vary in size and shape, but the use of the gases is much the same. The amount of gas left in acetylene bottles is determined by the remaining pressure. The amount of gas left in propane and similar gas containers is determined by the remaining weight. The outlets of all flammable gas bottles are connected by means of left-handed threads. This is a safety precaution.

> **Caution:**
> Never place gas bottles in a small closed space, such as a car trunk. A leakage could cause a violent explosion. Always secure bottles so that they cannot fall or roll about.

It is a wise precaution to have a carbon dioxide fire extinguisher along when using gas bottles. A broken hose or loose connection can

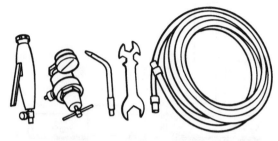

Figure 40–9 Torch kit (left to right): handle, regulator, torch tip, key, and hose. *(Courtesy The Ridge Tool Company)*

result in a fire. It is sometimes impossible to turn off the gas supply valve when the fire is near the bottle. A carbon dioxide fire extinguisher will put the fire out and enable the operator to turn off the gas.

Regulators are fastened directly to the tank. They cut down the pressure in the tank to that required by the torch tip. Some regulators are adjustable. To increase tip pressure turn the control valve clockwise.

Hoses are subject to wear and breakage. They should be inspected daily.

Different sized torch tips are required for different jobs. When larger masses of metal must be brought up to soldering heat, the tip must be longer.

Tips on Flame Soldering

❑ Move the flame around the joint in order to increase the temperature evenly.
❑ Solder will flow toward the hotter area. By concentrating the heat on the fitting the solder can be made to flow up into a socket.
❑ Because the fitting is heavier than the pipe, keep the flame mostly on the fitting.
❑ Do not feed in more solder than is necessary to fill the joint.

BENDING SPRINGS

A bending spring is a long closely wound spring, Figure 40–10. The spring is slid into position on the tubing and the tubing is bent with the spring on it. The spring prevents the tubing from flattening. Hard tubing must be annealed before bending.

Figure 40–10 Bending spring

Some Tips on Bending

❏ Place a mark about 10 inches from where the tube is to be bent. This can be used as a reference point when the spring covers the original mark.
❏ Make the bend sharper than is actually needed. Then, straighten it to the angle originally intended. This will loosen the spring and make it easy to remove.
❏ Never try to pull a spring off the pipe. This will stretch it and probably ruin it. Push it off while turning it.

SUMMARY

Efficient, well-maintained tools and new pipe joining methods make any job easier. The light, easy, simple trend in plumbing materials, while making life better, cannot be used as an excuse for sloppy workmanship. With all of the flexible pipe materials, both plastic and metallic, the plumber should keep in mind that, unlike electrical wiring, water supply systems may need winterizing, and occasionally a pipe must be located within a wall with considerable accuracy. Flexible tubing run for long distances between joists in the basement not only looks unsightly but may have water traps. A *water tap* is a plug of water that remains in a dip in the tubing after the system is drained. It cannot be predicted that the house will never be without heat, and these water-filled dips may freeze and burst the pipe. When retrofitting water supplies in finished building walls, it is convenient and understandable to feed soft tubing from one floor to the next, rather than to install pipe or hard tubing. When the building is under construction, however, straight runs of pipe ensure a certain amount of predictability for where a craftsperson may expect to find the pipe in the future. Tubing that languishes from stud to stud is more likely to receive a stray drill hole than is rigid tubing. When rigid tubing is used, leaks are found and new connections are made more easily without cutting large holes through expensive wall coverings.

It cannot be emphasized enough that the propane and acetylene fuels used to solder and braze copper and stainless steel tubing are dangerous and need to be treated with care and understanding. Many houses are burned down because a craftsperson did not use precautions when melting a lead joint or soldering fittings in close proximity to the building's flammable construction materials. When doing this kind of soldering work, an approved fire extinguisher should be handy at all times. Having a water hose with a spray head attached nearby is also a good precaution, and may be mandated. To test the connections after a soldering outfit is assembled, make a soap and water solution, turn the tank valve on, and then paint the solution on every joint with an acid brush. Large bubbles will reveal the location of leaks.

♦ ♦

TEST YOUR KNOWLEDGE

1. Is a swaged joint made to join a fitting to tubing?

2. What is the square hole in the reamer of the copper tubing cutter for?

3. Do you really need to ream tubing after you cut it?

4. Do you have to clean copper tubing and fittings when you use self-cleaning flux?

5. What does annealing do?

6. How should you test the connections on your torch kit?

◆ ◆ ◆ ◆ ◆ ◆ ◆ ◆ ◆ ◆ ◆ ◆ ◆ ◆ ◆ ◆ ◆ ◆

SOLDERING AND BRAZING

KEY TERMS

face feeding	soft soldering
flux	soldering
hard solder	soldering iron

OBJECTIVES

After studying this unit, the student should be able to:

- ❑ Select soldering materials.
- ❑ Clean metals in preparation for soldering.
- ❑ Prepare the joint.
- ❑ Solder a proper tubing connection

Soldering is the process of joining two pieces of metal using heat and the application of another metal of a different composition. Solder always melts at a lower temperature than the metal it is joining together. The technique of soldering requires a careful attention to the application of heat, the careful use of cleaning materials, and the proper soldering materials and tools. To become proficient at soldering, a good deal of practice is required. In plumbing, the technique is usually flame soldering, that is, applying heat with a torch directly to the material to be joined. There are plumbers, however, who make lead pans and roof flashings with a soldering iron. That is the indirect application of heat, by first heating an iron, and then applying the iron to the metal to be soldered. The "soldering iron," by the way, is usually made of copper, Figure 41–1.

Figure 41-1 Soldering iron

SOLDER

There are many solders. There are solders made of brass, copper, zinc, silver, aluminum, and even gold. There are hard and soft varieties of the above-mentioned solders. We will confine ourselves in this unit to those regularly used in plumbing.

There are two types of solders used in the plumbing trade: hard and soft. The term **hard solder** is generally reserved for those solders containing silver. Hard solder requires more heat than soft solder. Hard soldering can be done with standard acetylene soldering equipment if the materials being soldered are light. For joints of 1 1/4 inches and larger, an oxyacetylene torch is required.

Soft soldering is the more common operation used by plumbers to join fittings and tubing together. Solder may be purchased in many compositions. Those concerning the plumber usually contain the element lead. Some common soft-solder types follow.

50-50   This is 50% lead and 50% tin.
60-40   This is 60% lead and 40% tin.
95-5   This is 95% tin and 5% antimony.

In most states, 50-50 and 60-40 solders are only permitted on heat circulation pipes and drainage systems. Some states forbid the sale of all but lead-free solder.

In the early years of the last century, water supply pipes were often made entirely of lead. It has since been discovered from hair samples taken from this period of time that a higher concentration of lead existed in the bodies of the people of that age than exists in the bodies of the people of this age. The reason for this higher concentration could be attributed to the lead water supply piping of that period. Some authorities have recently discovered a connection between the very small ring of lead solder left inside the copper tubing in lead-soldered joints in copper pipe and fittings and low-level lead-poisoning of humans. Therefore, many plumbing codes forbid the use of soft solders with any lead content whatsoever.

A solder type that may be used from time to time for special applications is silver solder. Silver solders that contain a high percentage of *copper* are termed hard silver solders. Those that have a higher percentage of *silver* are called soft silver solders. All silver solders require great heat and great care to accomplish a reliable joint.

FLUXES

The use of a **flux** is indispensable to soft soldering. If no flux is used, the solder will not flow on the material being soldered. The air around us contains oxygen. The oxygen in the air slowly oxidizes most metals. At room temperature this occurs slowly. However, if heat is applied, this process is speeded up greatly. In fact, years of normal oxidation on copper will take place in minutes if a torch flame is applied to the copper tubing. Flux prevents the air from coming in contact with the pipe. It does this by producing a gas when heated, which takes the place of the air around the joint. Commercial fluxes often have a chemical that etches the copper when heated. This cleans the copper in addition to the cleaning that the plumber applied.

There are all kinds of fluxes. Most have been discovered by trial and error; for example, sheep fat, or tallow; rosin, the resin from a tree; and borax. Or the plumber simply uses a good grade of commercial flux. There are fluxes that require the plumber to first clean the joint mechanically. There are also self-cleaning fluxes. If the pipe and fittings are new; if the soldering takes place within a couple of hours of the application of the flux; and if the joint is carefully cleaned after soldering, the self-cleaning flux will save work. Many plumbers prefer the standard flux that requires cleaning the tubing and fittings by mechanical methods first.

CLEANING

In order to provide a base for the solder to join with the metal to be soldered, the metal must be bright and clean. This may be accomplished by various mechanical methods.

> #00 or #000 steel wool
> Emery cloth
> Special wire brushes
> Strip of screen impregnated with grit
> Electrical cleaning machine

Steel wool will provide a bright and clean surface for soldering. On small tube sizes like 3/8 and 1/2 inch, it is sometimes difficult to clean inside the fitting If the end of a dowel is split and a small amount of steel wool is inserted, this can be overcome. In all cleaning methods the plumber must not touch the metal to be soldered with the fingers. This will deposit an oil on the metal and may prevent the solder from sticking.

A better way to clean fittings in the small pipe sizes is to use a fitting brush, Figure 41–2. This is a commercial product that makes the job fast and easy. Brushes for the outside of the pipe may also be purchased. Care should be taken so

Figure 41–2 Fitting cleaning brush (*Courtesy Nibco Inc.*)

Figure 41–3 Sand cloth (*Courtesy Nibco Inc.*)

that the brushes will not be contaminated with grease and oil when not in use.

Sand cloth may be purchased in a "plumber's roll," Figure 41–3. This is a long strip of sand cloth. In use, the plumbers simply tear off what they need. When the cloth becomes filled with oxides, copper, and brass, they throw it away and tear off another piece.

On large jobs, a machine may be used that will clean and ream the tubing and fittings very quickly, Figure 41–4.

FLAME SOLDERING

All tubing should be reamed before it is soldered. The burr that is pressed inside the tubing creates a turbulence and slows down the flow. The back of most tubing cutters has a triangular-shaped piece of metal, which can do an adequate job. There are also tubing reamers available from various manufacturers that are very

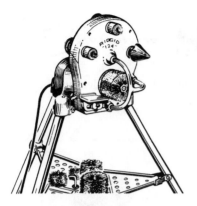

Figure 41–4 Copper cleaning machine *(Courtesy The Ridge Tool Company)*

Figure 41–5 Inner-outer reamer *(Courtesy The Ridge Tool Company)*

easy to use, Figure 41–5. Because copper is a soft metal, the steel blade of a pocket knife is preferred by many.

Clean both the inside of the fitting and the outside of the tubing. The tubing should be bright well above the depth that the fitting will slip on it. The fitting should be bright and clean all the way to the bottom of the socket. Care should be taken so that the fingers do not touch the cleaned copper and brass. Time spent in doing a careful job of cleaning tubing and fittings will be more than repaid later when the job is tested out. Leaks are often difficult and time consuming to repair.

A thin but continuous coating of flux is applied to both the tubing and the inside of the fitting. The fitting should then be turned about on the pipe to smear the flux inside of the fitting. Excessive flux should be avoided. Flux inside of the fitting will be pushed inside of the tubing and cause corrosion. Excessive flux on the outside of the joint will drip away at the first application of heat.

Heat is applied to the fitting and the pipe, not to the solder. The plumber should be con-

stantly aware of where the greatest thicknesses of metal are. This is where most of the heat must be applied. The tubing being of lighter gauge than the fitting will get up to soldering heat more rapidly than the fitting. So the flame must be kept on the fitting much longer. On 1/2 inch and smaller tubing, there is no need to heat the tubing at all. It will receive enough heat from the edges of the flame and through transference from the fitting. As the fitting gets up to the proper soldering heat level, test the opposite side of the fitting with the solder. If the solder starts to melt as soon as it is touched to the tubing, it is hot enough. Remove the flame and apply the solder to the face of the fitting. This is called **face feeding**. If the joint is no larger than 3/4 inch, and it is a horizontal joint, keep feeding solder until one drop falls. The most common mistake that beginners make is to keep feeding solder after the joint is full. This will often result in a ball of solder forming inside of the pipe. If this ball of solder does not stick to the inside of the joint, it will travel with the flow of water often ruining valves in its path. If it remains stuck to the inside of the joint, it will form an obstruction in the pipe.

On larger tubing sizes, the torch will have to be kept on the fitting while the solder is applied. The flame should never be used to melt the solder, however. Once the tubing and the fitting are up to soldering heat, the flame is kept a little ahead of the solder and both are moved around the joint. When the joint is finished, it should be inspected. There should be a thin line of solder visible all the way around. If the fitting is overheated, the flux will be exhausted before the solder is applied. The copper may turn black, or oxidize, and then the solder will not take. Often plumbers will apply additional flux and make a joint that appears to be good. Many times the solder is only around the face of the fitting and the joint will fail under pressure and time, Figure 41–6. If the joint turns black and the solder will not stick, take the joint apart and reclean it.

On 3 inches and larger tubing, it is a good idea to use two torches at the same time, one on the side feeding the solder and the other on the opposite side keeping the temperature high. Solder will flow toward the hottest area. To draw the solder down into the socket, work the torch toward the base of the socket.

To draw the solder up into an upside down joint, two forces are at work. First, as already stated, solder flows toward the hottest area. Second, liquid will flow up any narrow space, like the thin space between a fitting and the tubing in it. This natural phenomenon is called *capillary action*.

SUMMARY

Soldering and welding are methods of joining materials that leave the joint as strong as the base material. Generally, welding joins materials using the same material as the filler and solder uses a filler material that is different from the base.

Soldering is an ancient art, most of which has been developed through trial and error. Much of the nomenclature and definition of the art is filled with exceptions. One is tempted to say that these are methods for joining metals, yet plastic may be welded also. In the case of lead welding, *lead burning* is the term used, not

SOLDER PENETRATION OF ONE-THIRD THE CUP DEPTH—
BREAKING LOAD, APPROXIMATELY 2100 POUNDS.

SOLDER PENETRATION OF THE ENTIRE CUP DEPTH—
BREAKING LOAD, APPROXIMATELY 7000 POUNDS.

Figure 41–6 Depth of solder penetration (*Courtesy Nibco Inc.*)

welding, to build up nuclear reactor shielding. The terms *hard* and *soft* soldering refer only to the heat required to melt the filler material. Brazing means to solder with brass, and silver soldering means soldering with silver.

The craftsperson must develop a sense of the thickness and the heat-carrying capabilities of the materials, as well as the capabilities of the tools the craftsperson uses. In general, the plumber uses a technique called flame soldering, but in special circumstances, the plumber may use a soldering iron that has been heated in a charcoal burner. Jewelers may use a glass blowpipe held in the mouth and blown through an alcohol flame.

Brazing/engineering data

Best results will be obtained by a skilled operator employing the step-by-step brazing technique that follows.

1. The tube should be cut to desired length with a square cut, preferably in a square-end sawing vise. The cutting wheel of the type specifically designed for cutting copper tube will also do a satisfactory job. The tube should be the exact length needed, so that the tube will enter the socket of the fitting all the way to the shoulder of the cup. Remove all slivers and burrs left from cutting the tube, by reaming and filing, both inside and outside.

2. To make up a proper brazed joint, the clearance between the solder cup and the tube should be approximately .001 to .010. Maintaining a good fit on parts to be brazed insures:

Ease of application — Excessively wide tolerances tend to break capillary force; and, as a result the alloy will either fail to flow throughout the joint or may flush out of the joint.

Corrosion resistance — There is also a direct relation between the corrosion resistance of a joint and the clearance between members.

Economy — If brazing alloys are to be used economically, they, of necessity, must be applied in the joint proper and in minimum quantities, using merely enough alloy to fill the area between the members.

3. The surfaces to be joined must be clean and free from oil, grease and heavy oxides. The end of the tube need be cleaned only for a distance slightly more than it is to enter the socket. Special wire brushes designed to clean tube ends may be used, but they should be carefully

CLEAN TUBE

used so that an excessive amount of metal will not be removed from the tube. Fine sand cloth or emery cloth may also be used with the same precautions. The cleaning should not be done with steel wool, because of the likelihood of leaving small slivers of the steel or oil in the joint.

CLEAN FITTING

4. The socket of the fitting should be cleaned by methods similar to those used for the tube end, and care should be observed in removing residues of the cleaning medium. Attempting to braze a contaminated or an improperly cleaned surface will result in an unsatisfactory joint. Brazing alloys will not flow over or bond to oxides; and oily or greasy surfaces tend to repel fluxes, leaving bare spots which will oxidize, resulting in voids and inclusions.

FLUX TUBE

5. Flux should be applied to the tube and solder cup sparingly and in a fairly thin consistency. Avoid flux on areas not cleaned. Particularly avoid getting excess flux into the inside of the tube itself. Flux has three principle functions to perform:

A. It prevents the oxidation of the metal surfaces during the heating operation by excluding oxygen.

B. It absorbs and dissolves residual oxides that are on the surface and those oxides which may form during the heating operation.

C. It assists in the flow of the alloy by presenting a clean nascent surface for the melted alloy to flow over. In addition, it is an excellent temperature indicator, especially if an indicating flux is used.

6. Immediately after fluxing, the parts to be brazed should be assembled. If fluxed parts are allowed to stand, the water in the flux will evaporate, and dried flux is liable to flake off, exposing the metal surfaces to oxidation from the heat. Assemble the joint by inserting the tube into the socket hard against the stop. The assembly should be firmly supported so that it will remain in alignment during the brazing operation.

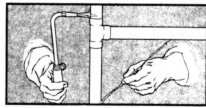

HEAT TUBE

7. Brazing is started by applying heat to the parts to be joined. The preferred method is by the oxy-acetylene flame. Propane and other gases are sometimes used on smaller sizes. A slightly reducing flame should be used, with a slight feather on the inner blue cone; the outer portion of the flame, pale green. Heat the tube first, beginning at about one inch from the edge of the fitting. Sweep the flame around the tube in short strokes up and down at right angles to the run of the tube. It is very important that the flame be in continuous motion and should not be allowed to remain on any one point to avoid burning through the tube. Generally, the flux may be used as a guide as to how long to heat the tube, continuing heating after the flux starts to bubble or work, and until the flux becomes quiet and transparent, like clear water. The flux will pass through four stages:

(Courtesy Nibco Inc.)

Brazing/engineering data

A. At 212 F the water boils off.

B. At 600 F the flux becomes white and slightly puffy and starts to work.

C. At 800 F it lays against the surface and has a milky appearance.

D. At 1100 F it is completely clear and active and has the appearance of water.

8. Now switch the flame to the fitting at the base of the cup. Heat uniformly, sweeping the flame from the fitting to the tube until the flux on the fitting becomes quiet. Avoid excessive heating of cast fittings.

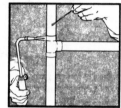

HEAT FITTING

9. When the flux appears liquid and transparent on both the tube and the fitting, start sweeping the flame back and forth along the axis of the joint to maintain heat on the parts to be joined, especially toward the base of the cup of the fitting. The flame must be kept moving to avoid burning the tube or the fitting.

APPLY BRAZING WIRE

10. Apply the brazing wire or rod at a point where the tube enters the socket of the fitting. The temperature of the joint should be hot enough to melt the brazing alloy. Keep the flame away from the rod or wire as it is fed into the joint. Keep both the fitting and the tube heated by moving the flame back and forth from one to the other as the alloy is drawn into the joint. When the proper temperature is reached, the alloy will flow readily into the space between the tube outer wall and the fitting socket, drawn in by the natural force of capillary attraction. When the joint is filled, a continuous fillet of brazing alloy will be visible completely around the joint. Stop feeding as soon as the joint is filled, using table on page 76 as a guide for the alloy consumption.

NOTE: For tubing one inch and larger, it is difficult to bring the whole joint up to heat at one time. It frequently will be found desirable to use a double-tip torch to maintain the proper temperature over the larger area. A mild pre-heating of the whole fitting is recommended. The heating then can proceed as in steps 7, 8, 9 and 10. If difficulty is encountered in getting the whole joint up to heat

HEATING TUBE

HEATING FITTING

at one time, then when the joint is nearly up to the desired temperature the alloy is concentrated in a limited area. At the brazing temperature the alloy is fed into the joint and the torch is then moved to an adjacent area and the operation carried on progressively all around the joint.

HORIZONTAL JOINTS—When making horizontal joints, it is preferable to start applying the brazing alloy at the 5 o'clock position, then move around to the 7 o'clock position and then move up the sides to the top of the joint, making sure that the operations overlap.

HORIZONTAL JOINT

VERTICAL JOINTS—On vertical joints, it is immaterial where the start is made. If the opening of the socket is pointed down, care should be taken to avoid overheating the tube, as this may cause the alloy to run down the tube. If this condition is encountered, take the heat away and allow the alloy to set. Then reheat the

REMOVE EXCESS FLUX

solder cup of the fitting to draw up the alloy.

After the brazing alloy has set, remove residual flux from the joint area as it is corrosive and presents an unclean appearance and condition. Hot water or steam and a soft cloth should be used. Wrot fittings may be chilled; however, it is advisable to allow cast fittings to cool naturally to some extent before applying a swab. All flux must be removed before inspection and pressure testing.

TROUBLE SPOTS

If the alloy fails to flow or has a tendency to ball up, it indicates oxidation on the metal surfaces, or insufficient heat on the parts to be joined. If work starts to oxidize during heating, it indicates too little flux, or a flux of too thin consistency. If the brazing alloy refuses to enter the joint and tends to flow over the outside of either member of the joint, it indicates this member is overheated, or the other is underheated, or both. In both cases, operations should be stopped and the joints disassembled, recleaned and fluxed.

(Courtesy Nibco Inc.)

The Fine Art Of Soldering

When adjoining surfaces of copper and copper alloys meet under proper conditions of cleanliness and temperature, solder will make a perfect adhesion. The strength of joint is equal to or even greater than the strength of tube alone. Surface tension seals the joint. Capillary attraction draws solder into - around - through - and - all - about the joint.

WITH 95-5 SOLDER AND INTERMEDIATELY CORROSIVE FLUX

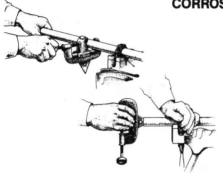

1. Cut tube end square; ream, burr and size.

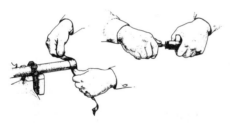

2. Use sand cloth or steel wire brush to clean tube and cup to a bright metal finish.

3. Apply solder flux to outside of tube and inside of cup of fitting carefully so that surfaces to be joined are completely covered. Use flux sparingly.

4. Apply flame to the fitting to heat tube and solder cup of fitting until solder melts when placed at joint of tube and fitting.

5. Remove flame and feed solder into the joint at one or two points until a ring of solder appears at the end of the fitting. THE CORRECT AMOUNT OF SOLDER IS APPROXIMATELY EQUAL TO THE DIAMETER OF THE FITTING . . . ⅝″ solder for ⅝″ fitting, etc.

6. Remove excess solder with a small brush or wiping cloth while plastic.

(Courtesy Nibco Inc.)

◆ ◆

TEST YOUR KNOWLEDGE

1. What is a soldering iron?

2. Why are solders containing lead forbidden for use on water supply systems?

3. What does flux do?

4. What is a fitting brush?

5. Why should the cleaned surface not be touched?

6. What is the most common mistake made by beginners?

7. What makes soft solder flow upward inside of a joint?

8. Why must two torches be used on 2-inch copper joints?

Unit 42

♦ ♦ ♦ ♦ ♦ ♦ ♦ ♦ ♦ ♦ ♦ ♦ ♦ ♦ ♦ ♦ ♦

Sizing the Water Supply System

KEY TERMS

friction loss
hardness (water)
pressure loss method
pressure reducing valve
velocity method

OBJECTIVES

After studying this unit, the student should be able to:

❑ Explain friction loss in pipes.
❑ Discuss water velocity.
❑ Size pipe by the velocity method and the pressure-loss method.

FRICTION LOSS IN PIPES

To understand the loss of pressure caused by water flowing through pipe, one must consider the length and size of the pipe, the roughness of the interior, and the number and type of fittings.

In Figure 42–1, tank A is filled with water. When valve E is closed, gauges B and C show the same pressure. This is the pressure exerted by the head H, or H × .434. However, when valve E is opened and water flows, gauge C shows less pressure than gauge B. This pressure loss is due to the rubbing action of the moving water against the sides of the pipe and is known as friction loss.

If the pipe is twice as long, the friction is twice as much. If the pipe corrodes and becomes very

305

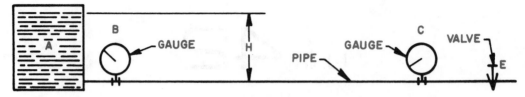

Figure 42-1

rough, the friction loss increases considerably. If the pipe size increases, the friction loss is reduced. This is because only that water that actually touches the side of the pipe causes friction.

Friction is proportional to the square of the velocity. Friction is increased by changes in the size of the pipe because of the eddies that are set up in the pipe as shown in Figure 42–2. *Eddies are air or water currents which move against a main current.*

WATER VELOCITY

Water has a number of properties that concern the plumber. **Hardness** is the total mineral content of the water and affects the scaling of the inside pipe surfaces. The *pH* of the water means the acid balance of the water. A pH of 7 is neutral. A low pH means that the water has high acid content. With the low pH, for instance, copper water tubing is liable to corrode quickly.

These factors can be controlled somewhat with water-softening equipment. The plumber can slow down the speed of the water in the pipe by increasing the size of the tubing. The slower the water flows, the less corrosion that takes place. Water that flows too fast:

- ❏ Is noisy.
- ❏ Causes water hammer.
- ❏ Causes excessive wear and corrosion, especially when water is hard, acidic, or hot (over 150°F).

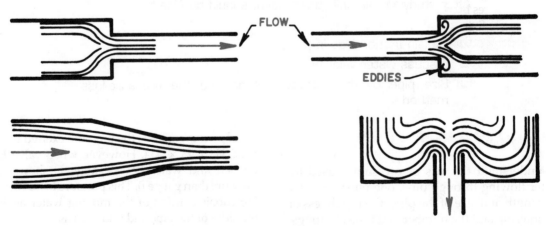

Figure 42-2

Under normal conditions velocity should be 8 feet per second (fps). If the water has a low pH (high acid), water-softening equipment, quick-closing valves, or very hot water (over 150°F), the velocity should be 4 fps.

FACTORS TO CONSIDER IN PIPE SIZING

The tasks of water supply system are to:

❏ Supply adequate water to any outlet under normal conditions.
❏ Operate in silence.

When a plumber "sizes a water supply system," essentially what he/she is doing is making a material list for the *tees* in the system. As the flow from two fixtures is joined together a tee is used. The outlets to the fixtures are large enough to supply each fixture. The inlet to the tee must be large enough to supply *both* fixtures. The same is true for tees supplying groups of fixtures. Each outlet of the tee must be large enough to supply the group supplied by that outlet and the inlet must be large enough to supply both of the groups services by the two outlets. If the plumber begins at the fixture furthest away from the water supply and works his/her way backward sizing tees along the way, he/she will eventually end up at the water service and will be able to calculate its size from the accumulated data.

On many occasions as a plumber works his/her way backward, a tee will be selected which has an inlet bigger than it needs to be because there are only so many sizes to choose from. When this happens, the inlet side of the succeeding tees may not change until the load from the branches builds up to require the next size. Also, as a practical matter, only 2 or 3 dif-ferent sizes are needed to be used in the average house, rather than buying 5 different sizes because the calculations have indicated that five different sizes would be optimum. The plumber would simply oversize some parts to avoid having a lot of leftover tubing.

There are many different methods of pipe sizing. In home construction, most experienced plumbers will not need to go through the formal mathematics of pipe sizing. They will use their past experience to size the water supply piping. In larger buildings, however, pipe sizing can make large differences in the cost of the installation and the way the system fulfills its tasks.

There are two major methods of pipe sizing, the **pressure-loss method** and the **velocity method**. There are also many other methods which are usually variations or combinations of these two. The velocity method is acceptable for buildings of three stories or less and it is easy to use and understand. This method is explained in detail later in this unit.

The velocity method makes some assumptions, however, which limit its usefulness. When the facts of the project exceed these assumptions the friction loss method should be used. The *National Standard Plumbing Code* has many charts and graphs, and the full explanation for this method.

Velocity Method Assumptions

1. **The building will be house-sized in area.**

   When the length of runs in the water supply get longer, the friction caused by the water molecules rubbing against the sides of the pipe and fittings, and the turbulence caused by rough edges and pipe fittings, will slow the flow and reduce the pressure available at the outlets.

NOTE:

Turbulence in piping systems has been much studied and is not completely understood. It does not conform to mathematical methods. It is known that turbulence, once introduced into the water flow, has a tremendous and unreasonable drag on that flow. Plumbers who wink at the practice of reaming water tubing should take note: "Suppose you have a perfectly smooth pipe with a perfectly even source of water. . . . All the rules seem to break down. When flow is smooth or laminar, small disturbances die out. **But past the onset of turbulence,** disturbances grow catastrophically. This onset–this transition–became a critical mystery in science." *Chaos, Making a New Science,* James Gleick, 1987, Viking Penguin Inc. **Un-reamed pipe creates turbulence!**

2. **The building will be three stories or less.**

For each story in height (approximately 10 feet) the pressure that was available at the water main drops 4.34 psi. If the water pressure at the main was 65 psi at 15 stories there would be **NO pressure** available. And of course, at lower levels there would be less and less available as the number of floors increased.

In taller buildings, this is compensated for by having booster pumps at some floor levels. Some very tall buildings have a water tank located at the top for fire fighting purposes. The level in this tank is maintained by booster pumps. The water to the floors of the building is then supplied by gravity. This may seem like a low pressure supply system. But now the water pressure *increases* by 4.34 psi at each lower floor level. Every tenth floor or so must have a **pressure reducing valve** installed to avoid damage to the terminal outlets and piping from excess pressure!

3. **There will be no designed continuously running outlets.**

Air-conditioning towers, manufacturing processes, and continuous irrigation processes will make the velocity system inapplicable.

4. **There will be at least 40 psi of available water pressure.**

Buildings with supply pressures under 40 psi are also to be sized using some other method than the velocity method.

SIZING PIPE BY THE VELOCITY METHOD

Sizing pipe by the velocity method applies to buildings of three stories or less. The water pressure available must be at least 40 psi.

Step 1. Obtain the available pressure at the main and the corrosive qualities of the water from the local authorities.

Step 2. Make a schematic drawing of the entire system. Show where the branches are located and which fixtures go with which branches. Identify all quick-closing valves, such as flushometers that require 4 feet per second piping.

Step 3. Refer to Table 42–1. Mark on the drawing the total water supply fixture units (wsfu) to each fixtures and the hot-and-cold water valves.

Step 4. Starting from the fixture outlets, mark minimum supply sizes down from Table 42–2.

Step 5. Working from the fixtures to the service pipe, size the system according to the velocity limitation tables in Table 42–3. If the system pressure is above 40 psi, go to the nearest pipe size. If the system pressure is very close to the minimum (40 psi), and the wsfu load falls between two pipe sizes, choose the larger.

Example:

Determine the pipe size for a small home. It has a bathroom group (tub, toilet, and lavatory), an outside hose bib (a valve that may run continuously), a small washing machine, a laundry tray, and a kitchen sink. It has city water, and the municipal authority says the

water is noncorrosive and has a minimum pressure of 55 psi.

Step 1. The pressure is above 40 psi. Because the water is noncorrosive, the system can be designed for 8 feet per second.

Step 2. The schematic drawing is shown in Figure 42–3.

Step 3. According to the load values assigned to fixtures, Table 42–1:
   a. Bathtub was 2 wsfu; 1.5 cold, 1.5 hot.
   b. Lavatory has 1 wsfu; .75 cold, .75 hot.
   c. Toilet has 4 wsfu; cold only.
   d. Outside hose bib is not on the chart. Since it may run continuously, it should be rated at 2 wsfu.
   e. Washing machine with an 8-pound capacity has 2 wsfu; 1.5 cold, 1.5 hot.
   f. Laundry tray has 3 wsfu; 2.25 cold, 2.25 hot.

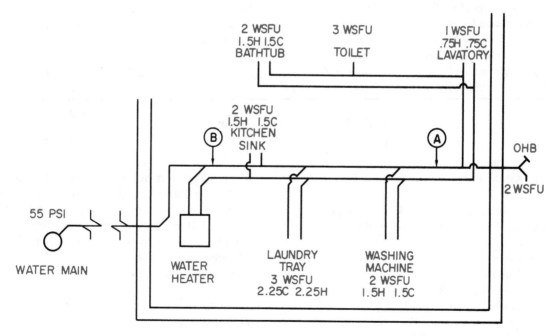

Figure 42-3

Fixture	Occupancy	Type of Supply Control	Load Values, in Water Supply Fixture Units		
			Cold	Hot	Total
Water closet	Public	Flush valve	10.		10.
Water closet	Public	Flush tank	5.		5.
Urinal	Public	1" flush valve	10.		10.
Urinal	Public	3/4" flush valve	5.		5.
Urinal	Public	Flush tank	3.		3.
Lavatory	Public	Faucet	1.5	1.5	2.
Bathtub	Public	Faucet	3.	3.	4.
Shower head	Public	Mixing valve	3.	3.	4.
Service sink	Offices, etc.	Faucet	2.25	2.25	3.
Kitchen sink	Hotel, restaurant	Faucet	3.	3.	4.
Drinking fountain	Offices, etc.	3/8" valve	0.25		0.25
Water closet	Private	Flush valve	6.		6.
Water closet	Private	Flush tank	3.		3.0
Lavatory	Private	Faucet	0.75	0.75	1.
Bathtub	Private	Faucet	1.5	1.5	2.
Shower stall	Private	Mixing valve	1.5	1.5	2.
Kitchen sink	Private	Faucet	1.5	1.5	2.
Laundry trays (1 to 3)	Private	Faucet	2.25	2.25	3.
Combination fixture	Private	Faucet	2.25	2.25	3.
Dishwashing machine	Private	Automatic		1.	1.
Laundry machine (8 lbs.)	Private	Automatic	1.5	1.5	2.
Laundry machine (8 lbs.)	Public or General	Automatic	2.25	2.25	3.
Laundry machine (16 lbs.)	Public or General	Automatic	3.	3.	4.

Table 42–1 Load values assigned to fixtures.

  g. Kitchen sink has 2 wsfu; 1.5 cold, 1.5 hot.

Step 4. Referring to Table 42–2 the pipe sizes from the tees to the fixtures will be:
  a. Bathtub—1/2"
  b. Lavatory—3/8"
  c. Toilet—3/8"
  d. Hose bib—1/2"
  e. Washing machine—1/2"
  f. Laundry tray—1/2"
  g. Kitchen sink—1/2"

Step 5. By referring to the velocity limitation tables in Table 42–3 the rest of the building can be sized. There is a separate table for each kind of pipe. Assume that L-type copper tube is used within the house and that a K-type copper tube is used for the underground service pipe.

The tables are divided into ten columns. From left to right:

Fixture or Device	Size (in.)
Bathtub	1/2
Combination sink and laundry tray	1/2
Drinking fountain	3/8
Dishwashing machine (domestic)	1/2
Kitchen sink (domestic)	1/2
Kitchen sink (commercial)	3/4
Lavatory	3/8
Laundry tray (1, 2, or 3 compartments)	1/2
Shower (single head)	1/2
Sink (service, slop)	1/2
Sink (flushing rim)	3/4
Urinal (1″ flush valve)	1
Urinal (3/4″ flush valve)	3/4
Urinal (flush tank)	1/2
Water closet (flush tank)	3/8
Water closet (flush valve)	1
Hose bib	1/2
Wall hydrant or sill cock	1/2

Table 42–2 Minimum size of fixture supply pipes.

❑ Column 1 lists the pipe size required.

❑ Column 2 lists the inside pipe diameter.

❑ Column 3 lists the flow in gallons per minute of that size.

❑ Column 4 lists the load (wsfu) for tank-type toilets.

❑ Column 5 lists the load (wsfu) for flush valve-type toilets.

❑ Column 6 lists the amount of pressure which is lost in 100 feet.

❑ Columns 7, 8, 9, and 10 are the same as columns 3, 4, 5, and 6. Columns 7, 8, 9, and 10, however, are for velocities of 8 feet per second. Columns 3, 4, 5, and 6 are for velocities of 4 feet per second.

Results: In step 4, the branches were sized from the fixtures to the tees that serve them. From the bathroom group to point A on Figure 42–3, there are 7.25 wsfu on the cold side. According to the tables under the 8 fps and flush tank toilets column, 1/2-inch tubing will handle 2.5 wsfu and 3/4-inch tubing will handle 7.3 wsfu.

Step 1. From point A to the bathroom group, there should be 3/4-inch tubing on the cold side. Therefore, there are 2.25 wsfu from the tee at point A to the tee in the bathroom group on the hot side.

Step 2. From point A to the bathroom group, there should be 1/2-inch tubing on the hot side. From point B to point A the pipe must carry the kitchen sink, the washing machine, the laundry tray, plus the bathroom group and hose bib. On the cold side there are 12.50 wsfu. On the hot side there are 7.5 wsfu.

Step 3. From point B to point A on the cold side, there should be 3/4-inch tubing. 12.5 is closer to 7.3 than it is to 22.5.

Step 4. From point B to point A, there should be 3/4-inch tubing on the hot side. From point B to the street main, the cold line must carry all of the load for the building, both hot and cold. When this occurs, the total figure per fixture must be used rather than adding the individual hot and cold figures. There are 25 total wsfu for the building. Types L and K tubing will carry 22.5 and 19.5 in the 1-inch size respectfully.

Step 5. Therefore, the water service and the main to point B will be in the 1-inch size using this method.

Copper Water Tube, Type K

Nominal Size (in.)	Actual I.D. (in.)	Flow (gpm) q	Velocity = 4 feet per second			Flow (gpm) q	Velocity = 8 feet per second		
			Load (wsfu) 1 *	Load (wsfu) 2 *	Friction (psi/100') p³ *		Load (wsfu) 1 *	Load (wsfu) 2 *	Friction (psi/100') p³ *
			Col. A	Col. B			Col. A	Col. B	
½	.527	2.7	.75	—	8.5	5.4	2.3	—	31.0
¾	.745	5.5	2.3	—	5.6	11.0	6.3	—	20.2
1	.995	9.7	5.3	—	4.1	19.4	19.5	5.8	14.4
1¼	1.245	15.2	10.8	5.0	3.1	30.4	54.0	14.0	11.1
1½	1.481	21.5	25.0	7.8	2.6	43.0	98.0	34.0	9.2
2	1.959	37.6	78.0	24.0	1.8	75.2	251.0	130.0	6.5
2½	2.435	58.2	166.0	69.0	1.4	116.4	460.0	340.0	5.2
3	2.907	82.8	289.0	161.0	1.2	165.6	725.0	663.0	4.2
4	3.857	146.0	609.0	528.0	0.8	292.0	1705.0	1705.0	3.0

Galvanized Iron and Steel Pipe, Standard Pipe Size									
½	.622	3.8	1.5	—	8.2	7.6	3.7	—	31.0
¾	.824	6.7	3.0	—	6.0	13.4	8.4	—	22.5
1	1.049	10.8	6.1	—	4.6	21.6	25.3	7.7	17.2
1¼	1.380	18.6	17.5	6.0	3.4	37.2	77.3	23.7	12.8
1½	1.610	25.4	37.0	9.3	2.9	50.8	132.3	52.0	10.8
2	2.067	41.8	93.0	29.8	2.2	83.6	293.0	171.6	8.4
2½	2.469	59.8	174.0	75.6	1.8	119.6	477.0	361.0	6.8
3	3.068	92.0	335.0	209.0	1.4	184.0	842.0	806.0	5.4
4	4.026	158.6	688.0	615.0	1.1	317.2	1980.0	1930.0	4.1

Schedule 40 Plastic Pipe, (PE, PVC & ABS)									
½	.622	3.8	1.5	—	6.8	7.6	3.7	—	24.2
¾	.824	6.7	3.0	—	5.1	13.4	8.4	—	18.0
1	1.049	10.8	6.1	—	3.7	21.6	25.3	7.7	13.2
1¼	1.380	18.6	17.5	6.0	2.8	37.2	77.3	23.7	9.6
1½	1.610	25.4	37.0	9.3	2.3	50.8	132.3	52.0	8.2
2	2.067	41.8	93.0	29.8	1.7	83.6	293.0	171.6	6.1
2½	2.469	59.8	174.0	75.6	1.4	119.6	477.0	361.0	4.8
3	3.068	92.0	335.0	209.0	1.1	184.0	842.0	806.0	3.8
4	4.026	158.6	688.0	615.0	0.8	317.2	1930.0	1930.0	2.8

*[1] Col. A applies to piping which does not supply flush valves.

*[2] Col. B applies to piping which supplies flush valves.

*[3] Friction loss, p, corresponding to flow rate, q, for piping having fairly-smooth surface condition after extending service, applying the formula:

$$q = 4.57 \ (p) \quad (d)$$

Table 42-3 Sizing tables based on velocity limitation (*Courtesy National Standard Plumbing Code*)

Copper and Brass Pipe, Standard Pipe Size

Nominal Size (in.)	Actual I.D. (in.)	Flow (gpm) q	Velocity = 4 feet per second				Velocity = 8 feet per second			
			Load (wsfu) 1 *	Load (wsfu) 2 *	Friction (psi/100') p^3 *	Flow (gpm) q	Load (wsfu) 1 *	Load (wsfu) 2 *	Friction (psi/100') p^3 *	
			Col. A	Col. B			Col. A	Col. B		
½	.625	3.8	1.5	—	6.8	7.6	3.7	—	24.2	
¾	.822	6.6	3.0	—	5.1	13.2	8.4	—	18.0	
1	1.062	11.0	6.3	—	3.7	22.0	26.4	8.0	13.3	
1¼	1.368	18.3	16.8	6.4	2.8	36.6	75.0	22.7	10.0	
1½	1.600	25.2	36.3	9.3	2.3	50.4	130.0	51.0	8.4	
2	2.062	41.6	92.0	29.5	1.7	83.2	291.0	170.0	6.2	
2½	2.500	61.2	181.0	80.0	1.4	122.4	492.0	376.0	4.9	
3	3.062	92.0	335.0	209.0	1.1	184.0	842.0	807.0	3.9	
4	4.000	158.0	685.0	611.0	0.8	316.0	1920.0	1920.0	2.9	

Threadless Copper and Red Brass Pipe (TP)										
½	.710	4.9	2.0	—	5.9	9.8	5.3	—	20.8	
¾	.920	8.3	4.2	—	4.4	16.6	13.2	5.7	15.5	
1	1.185	13.7	9.0	—	3.3	27.4	44.0	10.5	11.7	
1¼	1.530	22.9	28.9	8.3	2.4	45.8	110.0	40.0	8.5	
1½	1.770	30.6	55.0	14.5	2.1	61.2	181.0	80.0	7.2	
2	2.245	49.4	126.0	48.5	1.6	98.8	369.0	240.0	5.6	
2½	2.745	74.0	245.0	125.0	1.3	148.0	631.0	537.0	4.4	
3	3.334	109.0	421.0	305.0	1.0	218.0	1081.0	1081.0	3.5	
4	4.286	180.0	816.0	774.0	0.8	360.0	2318.0	2318.0	2.6	

Copper Water Tube, Type L										
½	.545	2.9	1.0	—	8.2	5.8	2.5	—	29.0	
¾	.785	6.0	2.5	—	5.2	12.0	7.3	—	18.7	
1	1.025	10.3	5.5	—	3.9	20.6	22.5	7.0	13.7	
1¼	1.265	15.7	11.5	5.0	3.0	31.4	58.0	15.5	10.7	
1½	1.505	22.8	28.5	8.0	2.5	45.6	109.0	38.0	8.7	
2	1.935	38.6	82.0	26.0	1.8	77.2	261.0	138.0	6.3	
2½	2.465	59.5	172.0	75.0	1.4	119.0	474.0	356.0	4.9	
3	2.945	85.0	300.0	178.0	1.1	170.0	750.0	692.0	4.0	
4	3.905	149.0	636.0	544.0	0.8	298.0	1759.0	1759.0	2.8	

Table 42–3 Sizing tables based on velocity limitation (*Courtesy National Standard Plumbing Code*)
Continued

SUMMARY

Sizing the water supply system is also a matter of experience.. In a small building, the length of the supply branches may be ignored. In a large building, the length and the layout of the branches become critical. The available water pressure may fluctuate on a rural system as the pump comes on at 20 psi and goes off at 40 psi. Another factor that must be considered is the height of the building. The pressure available in the basement water main will be reduced by 20 psi five floors above.

The usage of fresh water daily per person has steadily risen. At this time it is approximately 100 gallons of potable water per day for every man, woman, and child in the United States. To make use of the available water pressure to best advantage, the plumber must pay great attention to the location and length of the water mains and the length and size of the branches. Another consideration, as always, is the local plumbing code or, on occasion, the individual experience of the plumbing inspector. Some inspectors insist on oversized risers between floors. Some demand that water mains and all horizontal runs be pitched back toward a convenient low point for easy winterizing. The position and style of the valves used are important. Some valves, like gate valves and ball valves, place very little restriction in the flow, whereas the traditional "stop and waste" (globe type) valve causes great restriction.

TEST YOUR KNOWLEDGE

1. What factors affect friction loss?

2. Why do the gauges on the same floor in a system being tested all register the same when all of the outlet valves are turned off?

3. If friction is proportional to the square of the velocity, what would happen if the velocity of the water were doubled?

4. What pH would be equivalent to water that is neither acid nor alkaline?

5. What is wrong with pipes with high water flow speeds?

6. What is a good water flow speed to shoot for?

7. The water comes into the building at a given pressure and speed. How can a plumber control the speed of water within the individual branches?

8. Where can detailed instructions for other methods of water supply sizing be found?

WATER PRESSURE AND HYDROSTATIC PRESSURE

KEY TERMS

air chamber
head
water hammer

OBJECTIVES

After studying this unit, the student should be able to:

❏ Explain water and hydrostatic pressure.
❏ Determine water pressure and hydrostatic pressure.
❏ Discuss the importance of air chambers.

WATER PRESSURE

To understand water pressure, consider the weight of a cubic foot of water at average temperature to be 62.5 pounds. Since pressure is usually given in pounds per square inch, consider the cubic foot as 144 columns, 1 square inch in cross section and 12 inches high, Figure 43–1. The weight of each column is 62.5 pounds divided by 144, or .434 pound.

If this column is placed vertically and a gauge is attached to the bottom, it shows that the weight and pressure (.434) are equal. However, this is only true when the surface area is 1 square inch. Pressure per square inch (psi) is computed by the vertical height above a certain point and is exerted outward in all directions.

A gauge attached to the bottom of a cubic foot of water shows no more pressure than the 1-inch column described. The area above 1 inch

317

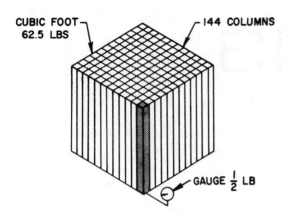

Figure 43–1 One cubic foot

does not affect the pressure, neither does the shape of the vessel.

If the 144 one-inch square columns shown in Figure 43–1 are placed vertically above each other, they will exert a pressure of .434 × 144. This equals 62.496 or 62.5 psi. Note that this is also the weight of a cubic foot.

To determine the pressure per square inch at the bottom of a tank, multiply the **head** or height of the water above the bottom by .434. For example, the pressure at the bottom of a tank that is 8 feet deep is .434 × 8, or 3.472 psi.

Study the columns in Figure 43–2. Columns 2, 3, 4, and 5 are different sizes and shapes. However, they are each 12 inches high. By opening each valve separately, the gauge will register .434 psi, the pressure exerted by a column 1 foot high.

Column 1 is 4 feet high. Therefore, the pressure on the gauge would register .434 × 4, or 1.736 psi.

HYDROSTATIC PRESSURE

Pressure on any part of an enclosed liquid is exerted uniformly in all directions. The pressure acts with equal force in all directions. Therefore,

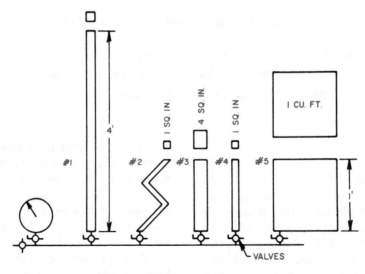

Figure 43–2

if a pressure of 1 psi is exerted over an area of 1 square inch upon an enclosed liquid, every square inch of the vessel is subjected to a pressure of 1 psi.

Example:

A cylinder having a base area of 1 square inch is connected to a cylinder having a base area of 100 square inches. Each has water-tight plungers. A 1-pound weight on the small cylinder will, therefore, support 1 pound for each inch, and 100 pounds on the large cylinder, Figure 43–3.

Study the cylinder in Figure 43–4 having pistons A, B, C, D, E, and F. The areas of the pistons are as follows:

A = 100 square inches
B = 7 square inches
C = 1 square inch
D = 6 square inches
E = 8 square inches
F = 4 square inches

Disregard the weight of the pistons and water. If a force of 5 pounds is applied to piston C whose area is 1 square inch, how much pressure must be applied to the other pistons to counterbalance this force?

Solution:

A force of 5 pounds on piston C equals a pressure of 5 psi. Therefore, multiply the pounds per square inch by the area of each piston to obtain the number of pounds supported by those pistons.

Example:

Force A is 100 square inches times 5 pounds, or 500 psi. Using the same process, find the required force on the other pistons.

If cylinder C has an area of 8.25 square inches and a force of 150 pounds is applied to it, the forces on the other pistons are found by dividing the force by the square inches and multiplying by the area of each piston.

Solution:

150 divided by 8.25 equals 18.182 psi. Piston D equals 18.182 pounds multiplied by 6 square inches, or 109.092 psi.

This is the principle upon which hydraulic

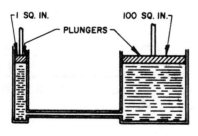

Figure 43–3

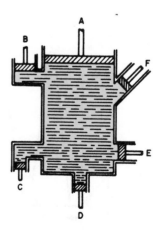

Figure 43–4

presses, hydraulic lifts and elevators, flushometer valves, and reducing valves operate.

WATER HAMMER AND AIR CHAMBERS

Water hammer is caused by a moving body of water suddenly stopping in a pipe. Water is almost incompressible. Therefore, when a valve is closed quickly, a shock is sent through the system. This shock has been known to create pressures as high as 800 psi.

The intensity of the water hammer depends upon the volume of water, the velocity at which it is flowing, and how suddenly the valve is closed. Self-closing and quick-compression faucets, valves, and pumps cause nearly all water hammer. Water hammer is also caused by suddenly opening a valve that allows water to flow under pressure into an empty pipe.

The shock and excessive pressures caused by water hammer may cause considerable damage to valves and piping. Reducing or check valves in the line make the shock less severe.

In Figure 43–5, valve A is attached to pipe B containing water at 50 psi of pressure. When valve A is opened, the pressure drops to 30 psi due to friction. When valve A is closed suddenly, the pressure jumps to 680 psi. This is a tremendous shock. Instead of having only one shock, a series of shocks occur as shown in Figure 43–5. Each succeeding shock becomes less intense and further apart until the water is at rest at the original 50 psi of pressure.

Air chambers are the shock absorbers of the plumbing system. They are used to relieve the shock resulting from water hammer. While water is almost incompressible, air is highly compressible. Therefore, air confined in a properly sized pipe will act as a cushion.

Water absorbs air and, in time, the air chamber becomes waterlogged. A *petcock* is placed near the top and a valve and drain placed at the bottom to recharge the air chamber. The air chamber is placed near a quick-closing valve or at the end of a long run of pipe. Experiments have shown that a short air chamber having a large area is better than a long air chamber with a small area. An air chamber 6 inches in diame-

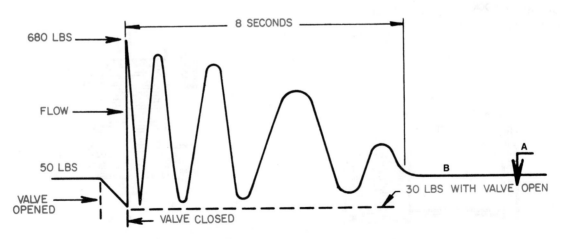

Figure 43–5 Shock waves of water hammer

ter and 18 inches long is 27 percent more efficient than one 2 inches in diameter and 64 inches long.

When the air chamber is filled with air at atmospheric pressure (15 psi) and water at 30 psi (two atmospheres) is turned on, the air in the air chamber is reduced to one-half its original volume. Likewise, with 45 psi (three atmospheres) or with 60 psi (four atmospheres), the air is reduced to one-third and one-fourth, respectively, of its original volume.

The air chamber is placed in a vertical position to receive any air from the water main, Figure 43–6. The top petcock is placed 6 inches below the estimated water level to prevent air leakage. The valve and petcock are placed at the bottom for drainage. Notice that with pressure of 60 psi (four atmospheres), the air volume is reduced one-fourth.

Since air chambers can become waterlogged, making them ineffective until recharged, other devices have been developed to absorb high-pressure surges or shocks. One such device, Figure 43–7, consists of an elastic, compressible material in the shape of a tube and insert. It is inserted at some point in the pipeline to serve the same purposes as an air chamber. At B, the tube and insert are in normal position before the faucet or valve is opened. The expansion of the tube against the insert as it absorbs a sudden shock is shown in C. As the shock recedes, the tube and insert return to their original position.

SUMMARY

The essence of this unit on hydraulics is that, with hydraulics, a pressure of 1 pound on an opening in a filled and enclosed vessel can lift 100 pounds at another opening. The amount of work that must be done to raise 100 pounds is still 100 pounds worth of work. To raise 100 pounds one inch, the 1 pound must be pressed into its particular opening 100 inches. And so the tradeoff is: light pressure over a long distance for strong pressure over a short distance.

The hydraulic jack must be pumped many times to raise the tire of the car an inch off the roadbed. The effect of the jack is to allow the stranded motorist to work easily for a longer period of time to lift the wheel, instead of the impossible task of grasping the bumper and lifting the car. A check valve on the pump piston allows the motorist to add up shorter strokes into the effect of one long theoretical hydraulic stroke. The important thing to remember is that the motorist does do all the work that is necessary to lift the car by the bumper.

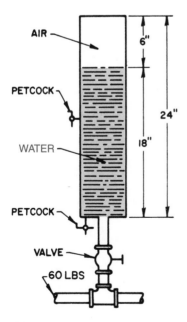

Figure 43–6 Air chamber

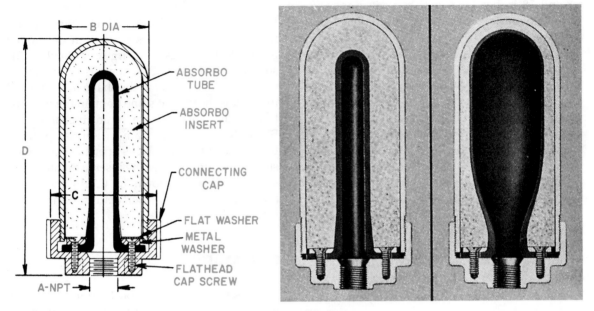

Figure 43–7

TEST YOUR KNOWLEDGE

1. What causes water hammer?

2. How does the air chamber work?

3. What does a cylinder of water having a base area of 1 square inch and a length of 12 inches weigh?

4. What does a common water pressure gauge measure?

5. If you had an oil drum 4 feet high filled with water and a 2-inch pipe 4 feet high filled with water, and you measured the pressure at the bottom of each, what would be the gauge readings?

UNIT 44

◆ ◆ ◆ ◆ ◆ ◆ ◆ ◆ ◆ ◆ ◆ ◆ ◆ ◆ ◆ ◆ ◆

THE WATER SERVICE

KEY TERMS

bursting pressure
cast copper
compression fitting
corporation ferrule
corporation stop
curb key
curb stop or cock

flared fitting
friction loss
solder or sweat fittings
water service
working pressure
wrought copper

OBJECTIVES

After studying this unit, the student should be able to:

❏ Explain the uses, sizes, and weights of copper tubing.
❏ Identify the uses of solder-type and flared fittings, and the sizes and types available.
❏ Describe the precautions that must be taken when making copper joints.
❏ Discuss house service piping and rough-in.

COPPER TUBING

Copper has certain advantages over wrought iron and steel: it resists rust; it is easy to handle in its soft form and can be bent around obstructions that would otherwise require joints; and it will withstand a severe load. It has been known to withstand freezing six times before breaking.

There are disadvantages in the use of copper tubing. These include its relatively high initial cost and its degree of expansion. Copper expands about twice as much as iron or steel; therefore, sufficient space must be allowed for this in installation.

Copper tubing is available in hard or soft tubing and in four different weights: DWV, lightweight, medium, and heavy. Hard copper tubing is usually used for exposed piping because it has better appearance. It is available in 20-foot lengths. Annealed (soft) copper tubing is used to form bends around obstructions, in inaccessible places, and underground.

Two weights of copper tubing, types K and L, are available in either hard or soft tubing. Type M and type DWV are available in hard form only. Type K has the thickest wall and, in soft tubing, is usually used for underground water service pipes, or whenever high pressure may dictate its use. Type K is manufactured in 60-foot and 100-foot coils in sizes up to and including 1 inch. The 60-foot coils are also available in sizes of 1 1/4 inches and 1 1/2 inches; 40-foot coils are available in a 2-inch size. Soft copper tubing is not usually furnished in sizes larger than 2 inches.

Type L has a thinner wall and is used for general heating and plumbing purposes. It is also used for underground water distribution piping. Types K, L, M, and DWV are manufactured in sizes from 3/8 inch to 12 inches. Type DWV is the lightest, followed by M, L, and K. It is used for interior water distribution, for heating, and for drainage.

Joints on hard and soft copper tubing, which is used inside buildings, are sweated with solder or brazed. In some areas, soft copper joints placed underground are flared since solder is attached by chemicals in the ground. Underground pipes should never be laid in cinders unless properly protected by a layer of sand, rust-resistant clay, limestone, or special wrapping.

The sizes, weights, and thicknesses of copper tubing are found in Table 44–1.

COPPER TUBE FITTINGS

Solder Fittings

The fittings used on copper tubing may be wrought or cast in copper, brass, or bronze. When copper is **wrought,** it has been hammered or formed by pressure into a desired shape; when it is **cast,** it has been shaped with the use of a mold. The wrought-brass sweat elbow is shown in Figure 44–1. Figure 44–2 illustrates a wrought-copper elbow with an iron pipe thread (IPT). In general, there are threes of fittings for use with copper tubing: the **solder or sweat fitting,** the **compression fitting,** and the **flared fitting.**

Copper tube fittings are made in a large variety of sizes and shapes, similar to malleable

Standard Water Tube Size	Actual Outside Diameter	Nominal Wall Thickness			Theoretical Weight		
		Type K	Type L	Type M	Type K	Type L	Type M
Inches	Inches	Inches	Inches	Inches	Lb./Ft.	Lb./Ft.	Lb./Ft.
3/8	.500	.049	.035	.025	.269	.198	.145
1/2	.625	.049	.040	.028	.344	.285	.204
5/8	.750	.049	.042	. . . .	.418	.362	. . . .
3/4	.875	.065	.045	.032	.641	.455	.328
1	1.125	.065	.050	.035	.839	.655	.465
1 1/4	1.375	.065	.055	.042	1.04	.884	.682
1 1/2	1.625	.072	.060	.049	1.36	.114	.940
2	2.125	.083	.070	.058	2.06	1.75	1.460
2 1/2	2.625	.095	.080	.065	2.93	2.48	2.03
3	3.125	.109	.090	.072	4.00	3.33	2.68
3 1/2	3.625	.120	.100	.083	5.12	4.29	3.58
4	4.125	.134	.110	.095	6.51	5.38	4.66
5	5.125	.160	.125	.109	9.67	7.61	6.66
6	6.125	.192	.140	.122	13.9	10.2	8.92
8	8.125	.271	.200	.170	25.9	19.3	16.5
10	10.125	.338	.250	.212	40.3	30.1	25.6
12	12.125	.405	.280	.254	57.8	40.4	36.7

Table 44-1 Sizes and weights of copper water tubing

iron fittings. Wrought-type solder fittings are made in sizes from 1/8 inch to 4 inches. Cast-type fittings are made in sizes from 1/8 inch to 12 inches.

It is often necessary to connect copper tubing to threaded pipes or equipment. Fastenings for this purpose have one opening that is threaded and are generally referred to as *adapter*

Figure 44-1 Copper sweat elbow

Figure 44-2

fittings or *fittings to copper* in specifications. To specify such a fitting, write: "one 1/2-inch 45-degree elbow fitting to copper." To specify a fitting with one end threaded, write as follows: "one 1/2-inch copper to 1/2-inch IPT elbow, or one 3/4-inch copper to 3/4 copper to 1/2-inch IPT." The branch is mentioned last.

To secure a tight solder joint, the fitting and tubing must fit to a close tolerance. The solder flows into the joint by a capillary action. This action can take place only when the tolerance is correct. The tube and the fitting must be thoroughly cleaned and fluxed to assure a perfect joint. *Flux* is an acid or rosin mixture that is applied to metal surfaces to remove oxide film before soldering.

Flared Fittings

Flared-type copper tube fittings are made of cast brass or bronze in sizes from 1/8 inch to 3 inches. They are used to join soft copper tubing in underground service pipes or where pipes must be cleaned often.

Flared fittings are similar to ground joint unions, Figure 44–3. The machined end of the fitting and the collar are then tightened onto the fitting. This ensures a tight joint between the tube and the fitting.

An adapter is used to join copper tubing to iron pipe. Elbows and tees are also manufactured so that tubing may be attached to any type of threaded pipe, including galvanized, brass, wrought iron, stainless steel, or plastic, with either inside or outside threads.

Flared fittings, Figure 44–4, are specified the same as galvanized fittings. A tee used for copper tube and galvanized pipe is specified as 1/2-inch copper to 1/2-inch copper to 1/2-inch IPS (iron pipe size).

COPPER TUBE JOINTS

Joints on copper tubing are easy to construct. However, some simple rules must be followed carefully to prevent leaks.

To make a sweat joint follow these steps:

1. Use either a tube cutter or a hacksaw with 24 teeth per inch to square the end of the tube. if the tube is not cut square, solder may run inside the pipe. Ream any burrs.
2. Thoroughly clean the outside end of the tube and the inside of the fitting. Use steel wool or a cleaning tool. Solder will not stick to dirty or oxidized metal.
3. Apply solder paste sparingly to the outside end of the tube and the inside of the fitting. Insert the tube into the bottom of the fitting and revolve it so as to distribute the flux.
4. Apply the flame of a torch to the fitting cup and heat the fitting and the tube. When the solder melts, press it to the fitting; this draws solder into the joint. When the solder ring shows all around the joint, it is finished. Do not use any more solder; it will run inside the pipe.

Caution:
On large pipes, two torches may be used or one section of the joint may be completed at a time. Care must be taken to prevent the heat of the flame from coming in contact with wood or any other inflammable material.

5. Gently wipe the excess solder from the joint while it is still in a plastic form. This makes a neat job.

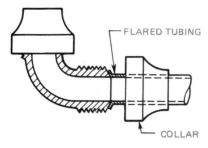

Figure 44-3

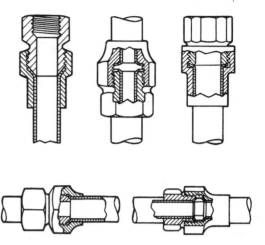

Figure 44-4

Soft solder is made in combinations of 50 percent tin and 50 percent lead to 95 percent tin and 5 percent antimony alloy. It may be used on pipes where the temperature will not exceed 250°. Solders with lead content may not be used on potable water piping.

HOUSE SERVICE PIPE

The house water service carries water from the water main, which is usually under the street, into the building. A device called a corporation ferrule or corporation stop is inserted into the water main with a threaded connection into the wall of the water main. This is usually done by the city water department if the city water supply is maintained by the city. Many municipalities allow private companies to supply the city's water needs. In this case the corporation stop would be supplied by that company.

A special drilling machine is used that can drill and tap (or, put threads in) and then install the valve without turning off the street water pressure.

The corporation stop is a valve. This valve is of the *cock* type. The cock-type valve can be turned completely on and completely off by turning the blade on the valve body only 90 degrees, see Figure 44–5. This cock is installed with a short run of pipe, usually "K" type soft copper, running into the property which is

being served by this valve. At that point, just inside the property line, another cock-type valve is installed. This is called the curb cock. The curb cock can then be turned off, the corporation stop turned on, the trench in the street can be backfilled, and the street patched with asphalt or whatever the street paving material may be. The corporation stop spends its life covered, under the street and inaccessible without great effort.

The size of the water service pipe to the building is determined by the number and kinds of fixtures in the building which it must supply. The smallest water service allowed is 3/4" I.P.S. This would be for a small home. Houses with 3 or 4 bathrooms could well use a larger size. It should be noted here that soft coiled copper water tubing can be obtained in sizes up to 2" I.P.S. Some plumbing codes do not permit soldered joints to be used underground; in these cases compression or flare-type copper joints will be required. The BOCA and the National Standard Plumbing Codes do not object to this kind of jointing. The complaint is that acids or chemicals in the soil react with the soldering materials and break down the joint integrity.

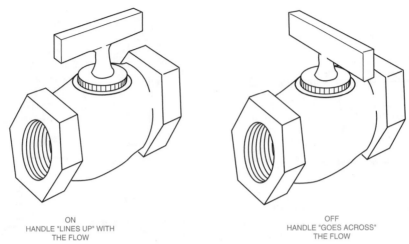

ON
HANDLE "LINES UP" WITH
THE FLOW

OFF
HANDLE "GOES ACROSS"
THE FLOW

Figure 44–5

Silver soldering or brazing could be used with reasonable safety in these kinds of soils.

The water service should be in a separate trench from the sewer lateral and displaced horizontally at least 10 feet. The local authority should be consulted about this requirement, however, as codes often give local authorities the latitude to waive this rule. In these cases, if a superior sewer lateral material is used and a shelf is cut in the side of the trench above the sewer lateral the water service may be allowed in the same trench with the sewer lateral. The service must touch no large rocks or stones and must be deeper than the frost line.

There should be three separate shutoff valves in the water service pipe as indicated by Figure 44–6 and Figure 44–7. One is buried under the street paving material, one is placed at the curb and must be reached using a special curb key, and one is just after the point where the pipe enters the building. The valve placed just inside of the building is followed immediately by a water meter. This allows the water meter to be serviced without turning off the water supply at the curb.

The cocks used at the water main and at the curb have a very low friction loss. Water flowing through a pipe is slowed and its pressure is reduced somewhat by the action of the water molecules rubbing against the inside wall of the pipe. This is called friction loss. It is expressed in equivalent feet of pipe. The cocks used at the street main and at the curb slow down the flow of water as much as 2 feet of pipe each. Globe-type valves and compression valves have a friction loss equivalent to 30 feet of pipe! Where a valve is to spend its service life almost always in an open state, install gate-type valves or ball-type valves inside of the building. These valves have a low friction loss (2 feet of pipe). If a valve is used to control the amount of flow going through a line on a more or less regular basis install a globe or a compression-type valve. Otherwise, install gate-type or ball-type valves. The savings in reduced pressure loss can be considerable.

Materials for water service

Plastic pipe materials permitted for water service piping are:

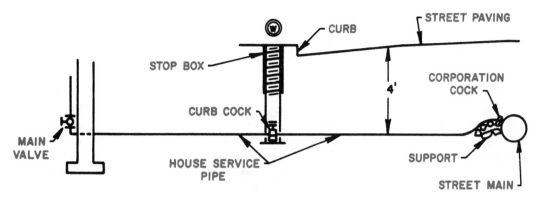

Figure 44–6

PVC,   Polyvinyl chloride
CPVC, Chlorinated polyvinyl chloride
ABS,   Acrylonitrile butadiene styrene
PE,   Polyethylene
PB,   Polybutylene

Other allowable materials include:

Cast-iron water pipe
Copper pipe and tubing
Brass pipe
Galvanized steel

Before using thermoplastic pipe materials, it should be noted that water service piping must have a *minimum* working pressure capacity of 160 psi at 73.4°. Working pressure is different from bursting pressure. Working pressure is the expected internal pressure when in use, whereas bursting pressure is the pressure at which a pipe and fitting may be expected to fail. Make sure that the materials used have the correct working pressure.

Galvanized steel water services should be used only in projects with a short expected life

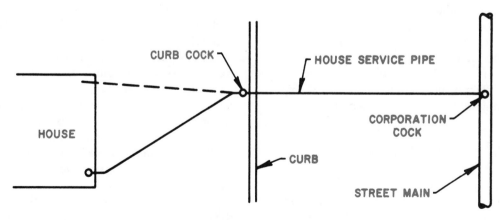

Figure 44–7

span. Even though the zinc protective coating that galvanizing provides is effective, it should not be expected to have a service life of over 15 years.

> **Caution:**
> When replacing metallic water service piping in older buildings with thermo-plastic piping, make very sure that the water piping in the building did not serve as the only electrical ground.

BASEMENT WATER MAINS

The basement water main is that part of the water supply system that extends through the basement of a building. The branches to all fixtures are connected to the basement main, Figure 44–8.

The basement main is attached to the house service pipe where it enters the building. Brass, copper, galvanized iron, or thermoplastic pipe and fittings may be used. It should run directly with as few bends as possible to avoid friction.

The basement main must be large enough to supply all the fixtures that may be used at one time There should be at least 15 psi of pressure at the highest fixture. Failure to provide this may cause back siphoning, particularly in high buildings, and may also pollute the water supply.

Basement mains should be supported every 10 feet. Reznor hooks are used for small pipes. Pipe hangers are used for large mains.

SWEATING OF PIPES

Moisture appearing and dropping from the surface of cold water, ice water, or brine pipes is caused by warm, damp air condensing on the cold pipe. The drops appear along the bottom of the pipe as shown in Figure 44–9. This sweating should be avoided since moisture dropping from pipes can damage floor coverings, furniture, or merchandise.

Two conditions must be present to cause sweating of pipes. First, the air must contain a high percentage of moisture. Second, the pipe must be cold. This usually happens in the summer during hot, humid weather. It may also take place in a warm building in which considerable water and steam are used, such as in laundries, dye houses, or dairies. In these buildings, the water pipes are kept cold by continuous use. Under these conditions, sweating occurs. Similarly, in a home a leaking faucet or tank ball cock will cause the cold water pipes to sweat.

To prevent sweating of pipes, cover the pipes to prevent the moisture-laden air from coming in contact with the pipes. This is done by covering the pipes and fittings with insulation or other special covering and painting the outside of the coverings with several coats of paint. This must be done carefully or moisture may penetrate.

SUMMARY

Copper tubing is very durable in most installations. It is affected by a high acid (low pH) content in the water and may develop leaks in time if the high acid problem is not corrected by water treatment. This may be revealed by blue stains on the fixture porcelain. The plumber must take care that the solder used on the potable water supplies contains no lead. Use hangers and straps that are either copper or copper flash-coated or made of some nonmetallic material. Remember that dissimilar metals may corrode where they are in contact. Also hang your pipe in such a way that it can expand noiselessly.

The house water service carries water from the curb to the house. Because of chemical interaction with the soil, solder joints underground should be avoided. Flare fittings are the most frequently used joining method. If the potable water supply is a well, the tubing used most often is polyethylene (PE) with insert fittings and stainless steel tubing clamps. If the pump is in the house, insert fittings with clamps are per-mitted to hook up the pump. After that point, copper, steel, PVC, or CPVC pipe should be used. In general, copper and plastic pipe mate-rials have become the materials of choice for roughing-in new home water supply.

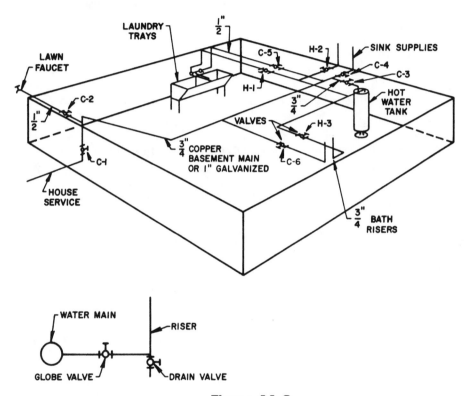

Figure 44-8

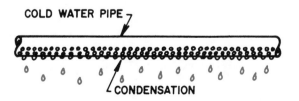

Figure 44-9 Sweating pipe

◆ ◆

TEST YOUR KNOWLEDGE

1. Which weighs more, 1 1/2-inch type L, or 1 1/2-inch type DWV copper tubing?

2. Which has the thinnest walls, type K, L, or M?

3. Which two types will not be found in the soft (annealed) form in any size?

4. What kind of solder must be used on domestic water supply lines?

5. Why is the fit between tubing and fitting important in a solder joint?

6. What causes cold water supply pipes to sweat and drip?

7. What is the minimum pressure that should be supplied to each outlet?

8. In regard to friction loss, a globe valve will provide the restriction to flow that is equivalent to how many feet of pipe?

Unit 45

◆ ◆ ◆ ◆ ◆ ◆ ◆ ◆ ◆ ◆ ◆ ◆ ◆ ◆ ◆ ◆ ◆ ◆

Roughing-In

KEY TERMS

bypass
coppering
hose bibs
main (the)
pressure-reducing valves

roughing-in
safety valve
spreader
strainers
water meters

OBJECTIVES

After studying this unit, the student should be able to:

❏ Explain basement layout.
❏ Select the proper materials for the job.
❏ Locate equipment properly.
❏ Describe the safety aspects of heating equipment.

Roughing-in is the installation of all of the pipelines and parts of the plumbing system that can be put in before the placement of the fixtures. The measurements must permit the easy installation of the fixture. The pipes must be of the proper size and material.

This unit mainly concerns the rough-in of those things that might be found in the basement or utility areas of a home. The rough-in of other individual devices and fixtures will be explained in the units that address those particular devices and fixtures.

MATERIALS

The term coppering mean the installation of copper water pipe in the basement.

The copper water pipe should be no less than type L in weight. Hard tubing looks neater and seems to have greater resistance to electrolysis.

Plastic water pipe should be at least schedule #40. Schedule #40 is standard weight. It should be supported at closer intervals than copper tubing. It may be a better choice than copper for use with well installations. Sometimes, well water is so acidic that copper tube is slowly eroded away from the inside. Plastic pipe is not harmed by acid.

Basement Layout

The best layout for basement piping is usually the one that uses the least pipe. The main is the pipe that leads from the water meter or pressure tank. Running the main between the appliances to be served results in branch lines of equal length, which is good. A 3/4-inch copper main will serve the average home having up to two bathrooms. From the 3/4-inch main, the branches may be 1/2-inch lines to the various areas of the house. Each area, such as the kitchen, should have its own shutoff valves. However, the plumber should try to avoid having too many shutoff valves on the route to any particular outlet. Globe-type valves greatly restrict the flow of water.

Water Heaters

It is best to run the pipe full size to the water heater from the main. This will provide more equal flow between the hot and cold water lines. A valve should be placed on the cold water side of the heater. Avoid soldering close to the heater because this may melt the plastic drop tube inside the heater. All water heaters must be provided with a temperature and pressure relief valve, Figure 45–1.

Outside Hose Bibs

Outside hose bibs should have a valve with a bleeder on it placed in the line. This should be placed in a way that will not allow freezing. The bleeder on this shutoff valve allows the draining of the line during the winter months. An anti-siphon device should be placed in the line also. This prevents contaminated water from entering the water system. This is a possibility when a garden hose is left attached and a pressure failure occurs.

Testing

All water lines should be capped and tested before drywall or any other wall covering is applied. When testing, full service pressure is applied to the lines. A visual inspection of every joint should then take place. The pressure should be left on for at least 24 hours before wall coverings are installed. If the water is on during the installation of drywall, leaks will be immediately discovered when a nail is driven into the pipe.

WATER METERS

Water meters are instruments for measuring the quantity of water used in buildings. They are placed on the main waterline before any branches are connected. In small dwellings, they are sometimes installed in a pit at the curb so they can be read at any time, Figure 45–2.

Meters are installed in an upright position and where they may be easily read. Since they cause considerable friction, a meter one size larger than the pipe is sometimes used. Meters

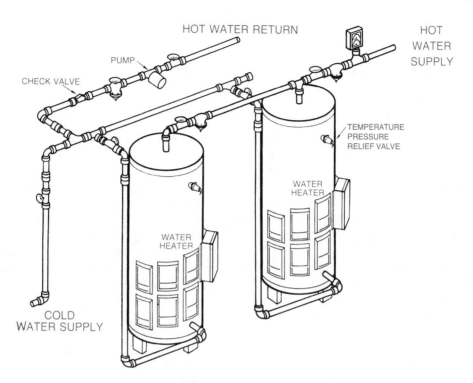

HOT WATER RETURN

PUMP

CHECK VALVE

HOT WATER SUPPLY

TEMPERATURE PRESSURE RELIEF VALVE

WATER HEATER

WATER HEATER

COLD WATER SUPPLY

Figure 45–1

must be protected from freezing. If freezing is still possible, the meter should have a breakable bottom. This protects the vital meter parts.

A valve is placed within 12 inches of the meter on the pressure side of the waterpipe. On 2-inch and large-size meters, a valve must be placed on each side along with a tee for testing. This tee is placed on the house side of the meter. A bypass is sometimes installed on large meters to permit uninterrupted service in case repairs must be made to the meter. Dirt, grit, and pipe dope should be blown out of the line before meters are installed. An arrow on the meter indicates the direction of flow. In new buildings, a temporary connection is installed in place of the meter to prevent damage to the meter.

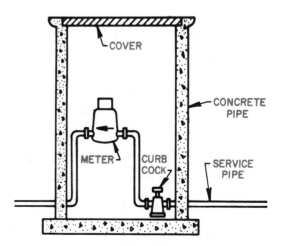

COVER

CONCRETE PIPE

METER

CURB COCK

SERVICE PIPE

Figure 45–2 Meter pit

Ordinarily, pipes are strong enough to support small meters, but large ones require masonry support.

Hot-water swells the hard-rubber discs and prevents proper registering. It is, therefore, prevented from backing through the disc meters. This is done by installing a check valve and a safety valve. A bypass may be installed to allow hot water to pass around the meter, Figure 45–3. However, the use of the bypass is regulated by some codes.

Some meters register in gallons; others register in cubic feet. Some have a straight reading dial; others have circular reading dials, Figure 45–4. To read a circular face dial, start at the highest numbered dial and write the last number that the hand has passed. Read the next lower dial and place this number to the right of the previous number. Continue until the lowest dial is read. This is the number of gallons or cubic feet used. A previous reading subtracted from this reading will give the amount of water used since that reading. The 1-foot dial shown on the right in Figure 45–4 is used for testing the meter accuracy, but not for recording

CURB BOXES

In order to turn the curb stop handle, an extension curb box is installed. The box is placed directly over the curb cock so the stop key will easily engage the tee handle, Figure 45–5.

The box is set on a brick or stone support to prevent settling. If the valve is not in the center or not in an upright position, it will be difficult to engage the handle with a stop key, Figure 45–6.

The curb stop is placed about 4 feet deep to prevent freezing. The top of the box is flush with the pavement for pedestrian safety. If the box is covered, its location is determined by locating the water main and tracing it back to the box. A flashlight or a mirror in the sun may be used to see the bottom of a curb box.

STRAINERS

In passing through water mains and pipes, water picks up grit, rust, scale, and other harmful materials. If this grit is caught under the washer of a valve or faucet, it may destroy the

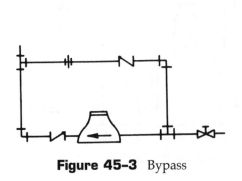

Figure 45–3 Bypass

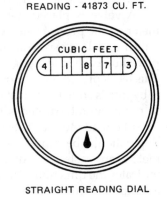

READING - 96872 CU. FT.

READING - 41873 CU. FT.

CIRCULAR READING DIAL

STRAIGHT READING DIAL

Figure 45–4 Reading of water meter

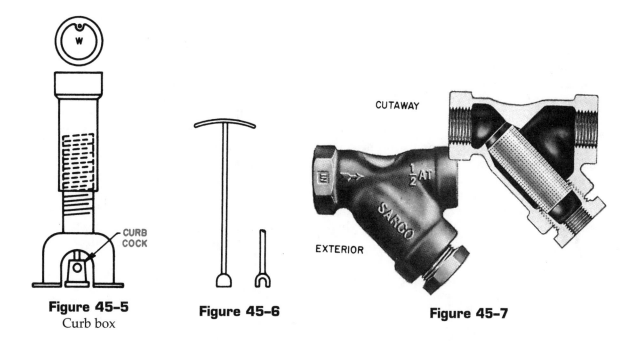

Figure 45–5
Curb box

Figure 45–6

Figure 45–7

washer and cut the seat. If it is caught in a regulating valve, it prevents the valve from controlling the water pressure. This, in turn, may cause a tank or piece of equipment to fail and do considerable damage. To prevent this, strainers are installed in the lines ahead of such valves and equipment.

Manufacturers of regulating valves, realizing their value, placed strainers on their equipment. Strainers must always be placed on steam-reducing valves. Failure to do so may cause an explosion. The proper location for a strainer is between the shutoff valve and the regulating valve or piece of equipment that is to be protected, Figure 45–7.

Good strainers have an effective area twice the pipe size area. The strainer should be the same size as the pipe.

On large valves, a bypass is installed so that water or steam may be temporarily supplied while repairs to the regulating valve are being made.

PRESSURE-REDUCING VALVES

Pressure-reducing valves are installed to protect water supply systems from excessive pressure (over 60 psi). The valve is located near the point of entrance when the whole system is to be protected. In large buildings they are used in certain zones or floors. They are also used to protect special fixtures or equipment. They must be placed on supplies to heating systems in small buildings. This is because heating boilers and radiators are guaranteed to stand only 30 psi of pressure.

When pressure-reducing valves are installed on the supply to any heating device, a safety valve must also be installed. The safety valve is placed between the pressure-reducing valve and the heater to guard against an explosion.

Pressure-reducing valves for water supplies are the diaphragm type, Figure 45–8. Diaphragm valves work on the principle that low pressure on a large area (the diaphragm) will overcome a high pressure on a small area (the valve disc.)

The valve consists of a body with the seat facing down. The stem, with a washer on the bottom, extends upward and is attached to the diaphragm. As the pressure in the house side reaches the desired number of pounds, the pressure pushes up on the diaphragm. This

pulls the washer against the seat and closes the valve. The spring is placed on top of the diaphragm to assist in opening the valve and for smoother operation.

The pressure at which the valve closes may be increased by tightening an adjusting screw on top of the valve. The washer is renewed by removing the cap at the bottom of the valve. The diaphragm may be renewed by removing the bolts around the rim.

Better results are obtained when the valve operates completely open. The line is always blown out before connecting the reducing valve. A strainer is always placed ahead of the reducing valve. Large valves are bypassed.

BYPASS

When reducing or thermostatic-control valves are installed on steam or waterlines, they usually have a bypass around the valve. A bypass permits water or steam to be used temporarily when the reducing or thermostatic valves are out of order, Figure 45–9.

If the reducing valve requires repairs, the gate valves in the 2-inch line are closed, and the 1 1/4-inch globe valve on the bypass is opened. This provides a temporary supply. Watch gauge B on the low-pressure side to prevent excessive water or steam pressure.

When there is a thermostatic-regulating valve where the steam pressure is high, the corresponding steam temperature may heat the water to a dangerous degree. A safety valve must be placed on the low-pressure side of all regulating valves. This prevents pressure due to expansion from backing through such a valve and causing an explosion.

A gauge should be placed on each side of the valve so that both pressures may be readily seen. A strainer must also be placed in front of a

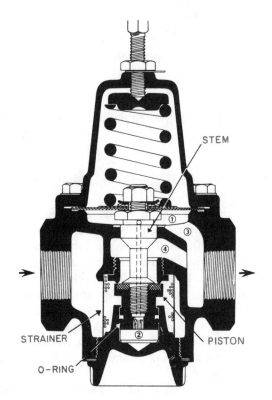

STEM

STRAINER

O-RING

PISTON

Figure 45–8 Diaphragm pressure-reducing valve

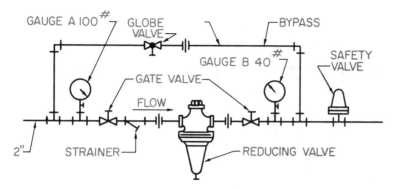

Figure 45–9 Bypass

reducing valve to prevent chips or rust scale from entering. If chips get caught beneath the seat washer, the valve will not shut off tightly and pressure will increase on the low-pressure side. The safety valve should be set about 25 percent higher than the desired low pressure. Unions are placed on each side of the reducing valve to make its removal easier.

SUMMARY

When roughing-in, the plumber will bring all of the plumbing, both water supply and waste, into position so that the final terminal equipment and fixture installations will be simplified and secure.

If the house is only one of a number of houses being built on the same basic plan, it may be that the locations and preliminary carpentry work have already been decided. If it is one-of-a-kind or the first of a series, the plumber will find that many decisions will have to be made even though the estimates have been rendered and the architect's drawings have been examined.

After the water service pipe and the main house sewers are installed, the plumber might install a spreader where the meter will be installed by the water authority. Often, arrangements are made by the contractor with the water authority to supply unmetered water at a set rate until the building is completed. The plumber will turn the water on and off at the curb cock using a curb key until the first valve on the inlet side of the spreader is installed.

The basement (or utility room) contains the plumbing and heating machinery that assure the home or building owners a comfortable and trouble-free residence in the structure. Here we will find the water meter, the house heater and circulating pipe, the water heater, the washer and dryer, the water pump and compression tank, the water softener, water strainers if necessary, the central-vacuum pump, system control valves, and all of the primary plumbing equipment.

Equipment with small spaces or openings for fluid to get through require strainers to protect their ability to function. Even minute flecks of hard material can disable them. Most pressure-reducing valves fall into this category. A common fitting in a bypass assembly with a regulating valve is a strainer.

◆ ◆

TEST YOUR KNOWLEDGE

1. What is meant by the term *water service*?

2. What does schedule #40 mean?

3. Does everything that could be described as roughing-in happen in the basement?

4. Why must every water heating device have a pressure/temperature relief valve?

5. What is the purpose of a water meter?

6. What is installed to keep dirt and grit from damaging valve seats?

7. What could go wrong if the curb cock is not installed exactly upright?

8. What is the purpose of a bypass around a pressure-reducing valve?

9. Which side of the reducing valve should the temperature/pressure relief valve be placed?

10. To increase the pressure on the low pressure side of an adjustable reducing valve, which way should the adjustment screw be turned? Clockwise or counterclockwise?

SECTION **5**

HEATING
WATER

(Courtesy A.O. Smith Water Products Company)

PRINCIPLES OF CIRCULATION

OBJECTIVES

After studying this unit, the student should be able to:

❏ Describe how water circulates.
❏ Discuss the factors that affect circulation.
❏ State how circulating pipes are installed according to the principles of circulation.

CIRCULATION

Water is composed of small particles called *molecules*. When water is heated, these molecules expand (increase in size) and currents are set in motion. This movement is called circulation or convection. It is caused by the difference in weight between hot and cold water.

A cubic foot of water at 60°F weighs 62.4 pounds. A cubic foot at 212°F weighs 59.7 pounds.

If containers of water at these temperatures were placed on a balance sale, Figure 46–1, the cubic foot of 60°F water would sink, since it is 2.7 pounds heavier, forcing the 212°F water to rise. This is the reason water circulates between heaters and hot-water tanks and in hot-water heating systems.

In Figure 46–2, notice how water circulates when heated and how the water is expanded on the hot side of the loop. It is heavier on the cold side of the loop, causing the water to sink. It is claimed that water will circulate with as little as 7° difference in temperature, but the greater the difference in temperature, the faster it circulates.

Figure 46–3 shows water circulating to a radiator. Notice how the water travels up from the heater when heated and descends to the heater after being cooled in the radiator.

The pipes that carry water from heaters to hot-water storage tanks are called circulating pipes. The top, or hottest, pipe is called the flow pipe, and the bottom, or cooler, pipe is called the return pipe, Figure 46–4.

Natural Laws

Natural laws must be observed in all circulation work.

1. **Hot water is lighter than cold water and tends to rise.**

Therefore, all flow pipes must have an upward construction from the heater to the hot-water tank or radiator. Cold water is heavy and sinks to the bottom. Therefore, all return pipes must slant down toward the heater.

2. **Air is driven out of water when the water is heated.**

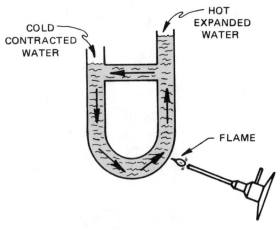

Figure 46–2

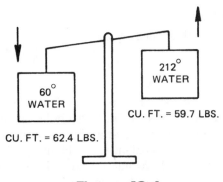

Figure 46–1

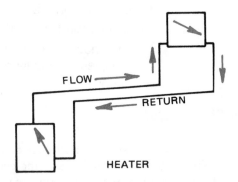

Figure 46–3 Circulation to radiator

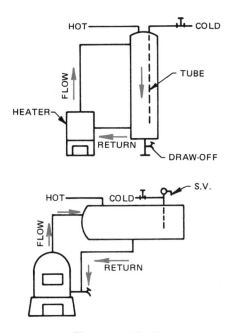

Figure 46-4

If it gathers in pockets in the pipes, the job will be noisy and circulation may stop. When pipes are properly installed, the air is relieved at high points in the system.

- ❏ Circulating pipes should be short.
- ❏ They should be as straight as possible.
- ❏ They should have few fittings.
- ❏ They should be carefully reamed.
- ❏ They should be the same size as the tappings of the heater.

Caution:
It is dangerous to place valves in the circulating pipes since the closing of these valves may cause an explosion. Every heating device should have a safety valve installed in the tank or heater to relieve expanded water or steam.

SUMMARY

Heat rises and cold falls. In a given quantity of a liquid or gaseous substance, the warmer, expanded molecules are less dense and therefore float to the top of that quantity. Suppose there were no gravity, would there be circulation?

Heated liquid or gas expands. Therefore, in a closed system, heat means pressure. More heat produces more pressure. Safety demands that some safety device be installed to dissipate excessive pressure. It is also a fact that increasing the pressure will increase the heat, and decreasing it will reduce the heat. This is an important fundamental of refrigeration and air-conditioning.

Heat is energy and moves toward cold. Cold dissipates heat. Energy has a tendency to move toward a lesser state. These last three statements are also about movement but a movement of energy. If someone holds a bar of iron by one end and heat the other end, eventually his/her hand will begin to feel very warm. This happens because the heat applied flows toward the cooler end of the iron bar as it attempts to bring all energy into equilibrium.

In a boiler we have a closed water heating system. The water in that closed system will eventually lose all of its free oxygen. Whatever metals or corrodible substance the oxygen finds, it will attack until there is no more free oxygen. Water is composed of two atoms of hydrogen and one atom of oxygen. The oxygen that it takes to make a molecule of water is bound. It is no longer free to attack corrodible metal surfaces. So when the free oxygen, which is introduced with the water during the initial fill-up, and some occasional "make-up" water is used up, corrosion within the closed heating system stops. That is why galvanized pipe is unnecessary for heating systems, and that is why old steam and water heating pipe is still usable after it is removed years later.

Any water heating system should be treated as a closed system and a proper heat and pressure relief valve should be installed.

◆ ◆

TEST YOUR KNOWLEDGE

1. Which is lighter, a bucket of warm water or a bucket of cold water?

2. Will circulation take place if you heated a single place on one side of the bucket?

3. What are pipes that carry the water in closed heating systems called?

4. What is the pipe that carries the water back to the boiler called?

5. What is the pipe coming up out of the water boiler called?

AUTOMATIC STORAGE GAS WATER HEATERS

KEY TERMS

drip leg
TPR valve
try handle

OBJECTIVES

After studying this unit, the student should be able to:

❏ Describe the construction and use of the automatic gas water heater.

WATER HEATERS

The automatic storage gas water heater is a vertical storage tank enclosed in a sheet metal case that is insulated to reduce heat loss, Figure 47–1. One type has a preheater screwed into the bottom of the tank. the gas flame is directed against this preheater. The bottom of the tank and the flue also absorb heat.

The gas is automatically controlled by a thermostat, Figure 47–2. This thermostat is placed in the side of the tank so that the incoming cold water turns on the gas. When the water is heated to 140° or 160° (the thermostat setting), it lowers the gas flame. This type of water heater has a graduated thermostat so the gas can be reduced gradually to a small flame. The flame maintains the heat of the water and acts as a pilot for the next operation.

A draw-off cock is installed at the bottom. A draft hood is placed on the flue pipe to prevent down draft that may put out the flame.

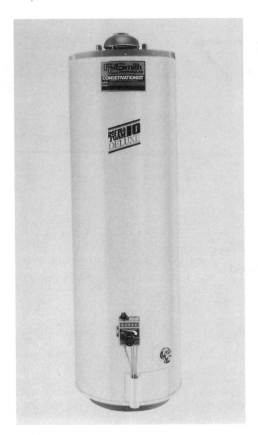

Figure 47-1 *(Courtesy A.O. Smith Water Products Company)*

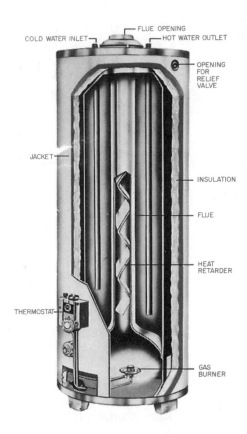

Figure 47-2

The flue is connected to the chimney so gas fumes can escape in case the pilot light goes out. Some of these heaters are equipped with Bunsen burners with fixed openings; others have adjustable air shutters.

Another type of gas water heater has no preheater but is arranged so the whole tank is a heating surface. The tank and the outer jacket are separated about 3/4 inch. This space is used as the flue.

Automatic gas water heaters are insulated with 1 1/2 inches of spun glass to conserve heat.

Some heaters have a heat trap to prevent hot water from circulating in the supply line.

Caution:
Every water-heating device must have a reliable relief valve to prevent explosions in case the thermostat fails to shut off the gas (see Figure 47-3, Figure 47-4, and Figure 47-5).

Figure 47–3

Figure 47–4

The heaters are equipped with an automatic pilot control. If the pilot goes out, the gas supply will not resume until the pilot is relighted.

After installing a gas water heater, the piping should be checked for leaks before lighting the pilot.

Tips

When a new heater is installed, it is often the case that the gas supply line has filled with air. Because fuel gas is often supplied at very low pressures, it may take a long time for the gas to reach the pilot orifice when the plumber is attempting to light the pilot for the first time. To assist the lighting process, many plumbers first make sure that there are no open flames or electrical sparks. Then, they open the union at the heater and turn on the gas until the first odor of gas is detected. The union joint is then made up tightly. After the air has completely cleared, the pilot is lighted. In this way, the plumber will only have to hold the pilot valve open for a short period of time.

When a water heater is first filled with cold water in humid weather and then the burner is lighted, condensed moisture may drip down from the flue pipe into the Bunsen burner for quite a long while. This may give the false

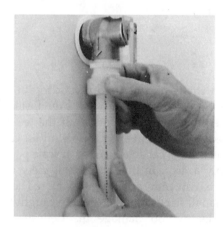

Figure 47–5

impression that the tank is leaking.

When the heated water may back up into a cold water supply branch or even back to the water meter in some cases, it is a good practice to place a check valve on the upstream side of the water heater. However, remember that every water-heating device must have a temperature/pressure relief valve installed, especially one that has a check, or regulating, valve.

These devices trap a quantity of water on the downstream side and have the potential for an explosion.

COMMERCIAL WATER HEATERS

Commercial heaters often are joined to a storage tank. This ensures that a large quantity of hot water will be available. The heavy-duty burner unit will be under a more continuous load. By avoiding long periods of idleness with attendant cycles of cold and hot burner parts, greater efficiency is attained, Figure 47–6. This set-up also provides very hot water for sterilizers and dishwashers and also water at 140°F for ordinary use.

SUMMARY

When changing from a vertical drop to a horizontal run in gas piping, use a tee instead of a 90-degree elbow. The bottom of the tee is then closed with a short pipe nipple and a cap and forms what is called a drip leg, a scale pocket, or simply a drip. In time this small piece of pipe and its cap may accumulate water. The water and small flecks of scale that are caught by this drip leg would damage the delicate control valves and regulating valves typically found on gas-heating systems. When gas water heaters are hooked up, one of these drip legs should be installed at the water heater's control valve.

> **Note:**
> The water that may collect in one of these is very flammable.

Figure 47–6 *(Courtesy A.O. Smith Water Products Company)*

Figures 47–3, 47–4 and 47–5 show the installation of a blow-off pipe on a TPR (temperature pressure relief) valve. A TPR valve opens on excessive temperature or pressure. The plumber installs a transition adapter so that plastic pipe can be used with the brass threaded fitting. When something goes wrong and the TPR valve "blows," the exiting water is very hot. A piece of tubing should be led down from it to within 1 to 2 inches from the floor. The TPR valve shown has a try handle on it. This should be pulled up briefly every few months to test the valve's functioning and to clear out any scale or dirt that may have collected.

TEST YOUR KNOWLEDGE

1. Where is the flue pipe from a water heater hooked up?

2. What is the purpose of the heat trap in the outlet pipe within the jacket of the water heater?

3. What device prevents downdrafts from putting out the water heater's flame?

4. (Extra credit question.) What would make you think that calling a water heater a "hot water heater" might be incorrect?

5. What happens when the pilot light goes out? Is there any danger?

6. What is a TPR valve?

◆ ◆ ◆ ◆ ◆ ◆ ◆ ◆ ◆ ◆ ◆ ◆ ◆ ◆ ◆ ◆ ◆ ◆ ◆

ELECTRIC WATER HEATERS

KEY TERMS

dip tube
heating elements
sacrificial rod

OBJECTIVES

After studying this unit, the student should be able to:

❏ Describe the electric water heater and how it operates.

ELECTRIC WATER HEATER

Electric water heaters, Figure 48–1, are usually the storage type. They may be equipped with a time switch to take advantage of low electric rates at night. Electric water heaters are clean, safe, and present no danger of fire or odor.

In the electric water heater, the cold water is connected to the top of the storage tank. From that point, it is delivered to the bottom of the storage tank through a dip tube. The warmer water rises to the top of the tank by circulation and is taken off at that point for distribution to the house. Most water heaters have an outlet in the side of the tank for a temperature/pressure relief valve. If there is no special opening for a relief valve, the valve is placed in a tee close to the hot-water outlet in the tank.

Immersion-type heating elements are shown on the heater in Figure 48–2. This heater

355

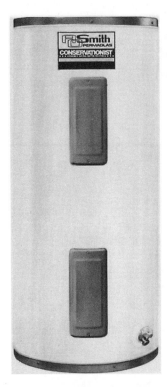

Figure 48-1 *(Courtesy A.O. Smith Water Products Company)*

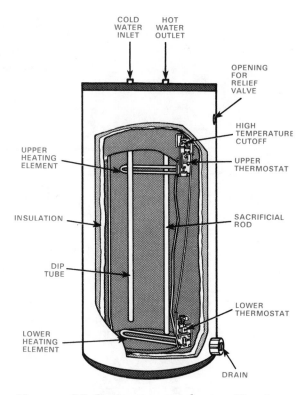

Figure 48-2 Electric water heater *(Courtesy A.O. Smith Water Products Company)*

includes an upper and a lower element. The lower heating element maintains the standby temperature of the tank. The upper element heats a small portion of the water to a higher temperature for immediate use.

Insulation of electric water heaters is very important. A bare tank at an average temperature of 130° will lose 125 Btu per hour.

Water heaters can be obtained using either 120 or 240 volts of electrical current. A 240-volt current is more efficient and economical to use. A magnesium rod is installed to retard corrosion. This is sometimes called a **sacrificial rod.**

Dip Tube

The **dip tube** is made of plastic. It is inserted through the cold water supply in the tank. When the plumber solders the riser tubing to the adapter in the tank top, there is a good possibility that the top of the dip tube will be melted off, and it will drop to the bottom of the tank. The symptom of this mishap is that very little hot water is produced, because as the hot water is drawn off at the top, the cold water rushing in to make up for the lost water short circuits directly over to the hot-water outlet. The cure is to always solder the adapter to the riser copper *before* the adapter is threaded onto the tank inlet nipple.

TEST YOUR KNOWLEDGE

1. What material is the sacrificial rod made from?

2. What is the purpose of a sacrificial rod?

3. Which is more economical, a 120-volt water heater or a 240-volt water heater?

4. Does an electric water heater require a temperature/pressure relief valve?

5. Where is the hottest water in the tank?

THERMOSTATS

OBJECTIVES

After studying this unit, the student should be able to:

❏ Describe the purpose of a thermostat.
❏ Explain the principles governing thermostat controls.
❏ Show the interrelation of gas water heater controls.

The thermostat is a device which will *automatically* take some action that is desired in response to a rise or a fall in *temperature,* which can usually be set with a control on the device.

The thermostat frees us from having to watch a gauge and manually adjust controls in response to changes in temperature and to regulate the performance of a piece of equipment.

Water heaters, space heaters, cooking devices, refrigerators, and air conditioners all use some kind of temperature activated automatic control device commonly called a thermostat.

The thermostat often has two responsibilities: (1) to provide safety from the threat of explosion, fire, health hazards, or simply the destruction of the expensive device which it controls, and (2) to control the environment of the piece of equipment within preselected limits.

Water heaters of all kinds have an automatic temperature/pressure relief valve (TPR valve) in the event that the thermostat fails in its function. Most things expand when heated. For example, gases or liquids contained in a closed vessel like a water heater body can explode when heated. If the water heater is gas fired, this explosion may be followed by a building fire if, as is likely, the fuel gas is then released within the building. If the pilot light on a gas water heater goes out for any reason there must be a device which shuts off the flow of gas to protect the occupants of the building from death by fire or asphyxiation.

PHYSICAL PRINCIPLES

The thermostat always has some way to measure temperature. In response to the results of this measurement the thermostat causes something else to happen. The thermostat on the water heater measures the temperature within the water tank. If the water in the tank is above a preset temperature (no more than 135° F), the thermostat shuts off the flow of gas to the main burner.

In order to measure the temperature the thermostat makes use of one of the following physical facts:

1. Metals and other substances expand when heated and at different rates, depending upon the particular material.

   Example:

   bimetal strips in room thermostats.

2. Substances melt at temperatures unique to that substance.

   Example:

   Safety switch in space heater primary controls; meltable plugs in TPR valves.

3. Dissimilar metals joined at the ends (called a node) produce a small electrical current when heated.

   Example:

   Thermocouples and thermopiles.

4. Some substances have varying electrical resistances depending upon their temperature.

   Example:

   Thermistors, digital control equipment.

5. Some substances have varying electrical resistances when exposed to the light which often accompanies heat.

   Example:

   Cad cells in home heating equipment.

6. Gases expand when heated.

   Example:

   Refrigeration and air-conditioning controls.

PRACTICAL APPLICATIONS

We will cover only the common devices that the plumber might need to repair, replace or make decisions concerning, here.

Bi-Metal Strips

Figure 49–1 shows what happens when we join two dissimilar metal strips together and then

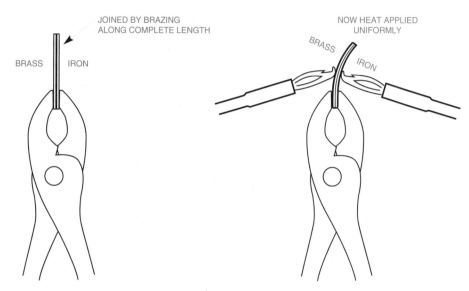

BRASS | IRON

JOINED BY BRAZING
ALONG COMPLETE LENGTH

NOW HEAT APPLIED
UNIFORMLY

BRASS   IRON

Figure 49–1

heat them. Notice that the strip bends away from the side that expands the most. This principle is used to construct a thermostat, Figure 49–2, by joining two very thin long strips together in the same fashion. This is then prebent into a spiral shape so that the whole device can be housed in a small container like a wall thermostat. This spiral is then fastened to a post in the center which may be rotated by the lever which sets the predetermined temperature. The long strip reacts more to the same temperature than the shorter thicker strips shown in Figure 49–1. If electrical contacts are attached to the end of this double strip, we can use the temperature in the space in which this thermostat is contained to close the electrical contacts or to open them. In the case of the room thermostat shown, when the temperature rises above the preset temperature the contacts open. When this happens the heating device shuts off.

Figure 49–2 is a stylized illustration which might actually work in an application for a peri-od of time. In practical applications, however, the electrical arc which would occur each time the contacts closed would erode the contacts and make them inoperable in a short time. This is why two precautions are usually taken with commercial products to ensure their durability: (1) the electrical voltage is reduced to prevent rapid breakdown of the contacts, and (2) the contacts are either made to snap rapidly together from some predetermined distance or the contacts are housed in an airless glass vial and the contacting device is a small quantity of the liquid metal mercury which moves back and forth in the vial as the bi-metal strip coils and uncoils.

Meltable or Fusible Plugs

The temperature pressure relief valve used on most domestic water heaters has a valve which is held closed, trapping the pressure in the water heater tank by a strong spring. When the pressure gets high enough in the water heater, it

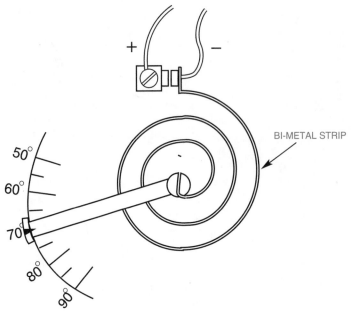

Figure 49-2

compresses the spring and the valve opens, leaving excess pressure out, Figure 49–3. Remember that when the temperature of a liquid or a gas in a closed container goes up, the pressure within that container goes up also. The pressure is relieved within the tank and the temperature goes down. It works both ways. If there is no water being admitted to the water heater for some reason, and the water is all turned into steam, the steam can become superheated, and this will damage the tank and fittings. More common however, is the possibility that the TPR valve becomes corroded shut and inoperable. In this case the fusible plug within the valve melts, and allows the internal heat and pressure to escape. If the fusible plug melts it must be replaced. The usual remedy for this, however, is to replace the old TPR valve with a new one.

In some electrical devices, notably the primary control on heating systems, there is a safe-

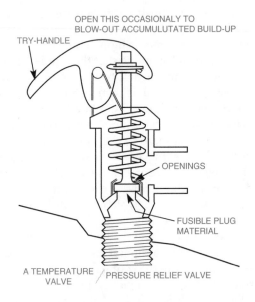

Figure 49-3

ty device in the reset mechanism. This fusible link melts when the system runs, from an electrical standpoint, more than a set period of time without the burner gun firing (60–120 sec). This is arranged so that if given five minutes, the fusible link will re-solidify, rejoining the electrical connection and allowing the system to be reset and another restart attempted.

Thermocouples and Thermopiles

Natural gas heating appliances operate with very low gas pressures, often less than 1 psi. This allows some very lightly structured controls to regulate the flow of gas to the fixture. Gas-fired water heaters do not require electrical power from the building's power supply. Some gas-fired furnaces also do not require electrical power to maintain the pilot flame and the main burner. But they do use a self-contained source of electrical power to maintain the pilot flame. This device is called a thermocouple, Figure 49–4A. The thermocouple uses the principle that two dissimilar metals joined together will produce an electrical current when heated at the connection point of the two metals. The amount of voltage produced is dependent upon the two metals used and the amount of heat applied to the joint. The thermopile produces more electricity than the thermocouple because it is a number of thermocouple joined together in series, Figure 49–4B.

This very small voltage and current is still strong enough to operate some light springs and valves and thereby provide safety and flame control of natural gas-fired heating and refriger-

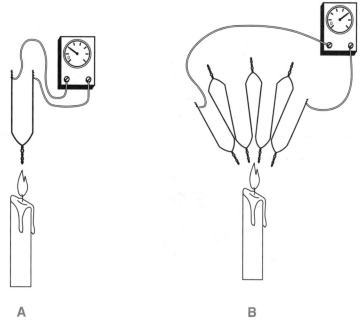

A   B

Figure 49–4

ation devices. Some of the principles explained here will be brought together at the end of this unit to describe the operation of a gas control valve in some detail.

One of the very important devices for automatically controlling flow is the solenoid operated valve, Figure 49–5. This uses another scientific principle, which the student is probably already familiar with: the tendency of coils of wire, when energized with electricity, to become magnetic. This magnetism will deflect the needle of a compass A less known property of this coil of wire is to attempt to center any iron mass within itself. A carpenter's nail, for instance when placed inside of an electrified coil, will tend to position itself in the magnetic field of the coil with as much nail above as is below the coil. This principle is used extensively to control equipment.

The most frequent use is to allow a small amount of electricity to control devices which will then use much higher voltages and current. Another use however, is the one illustrated in Figure 49–5. The device pictured will pull the valve closed when the electricity is attached to the coil, because the iron of the valve attempts to center itself in the coil's magnetic field. With a little rearranging we can use the same device to open a valve when it is energized.

Thermistors and Cad Cells

These are included for completeness and are really not within the scope of a textbook on plumbing. (Suffice it to say that the thermistor, a resistor or transistor which changes its resistance depending upon its temperature, or the cad cell, a device which changes its resistance in

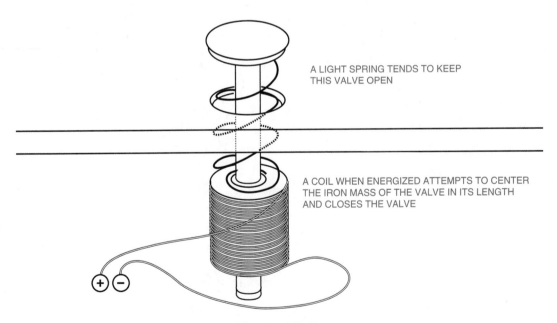

A LIGHT SPRING TENDS TO KEEP THIS VALVE OPEN

A COIL WHEN ENERGIZED ATTEMPTS TO CENTER THE IRON MASS OF THE VALVE IN ITS LENGTH AND CLOSES THE VALVE

Figure 49–5

an electrical circuit depending upon the presence of light, can be used to control devices.) Many commercial and domestic heating devices which light themselves with an electrical arc utilize the cad cell and/or the thermistor to detect if the lighting was successful.

Pressure Actuated Heat Controlling Devices

The expansion of gases and liquids can be used to control devices also. This can be used for heating equipment, but is used more frequently to control refrigeration and air-conditioning installations. Freon, in an enclosed bulb with a tube attached to a bellows, will produce a gas when heated and expand the bellows. The opposite will happen when the bulb of Freon is cooled. When the bellows expands and contracts, it can be used to open and close an electrical circuit. This could turn an electric motor on and off. It could also be used to open and close a valve, Figure 49–6.

A CONTROL DEVICE FOR A GAS FIRED WATER HEATER

If controls for a gas-fired water heater were to be designed, there would first need to be some questions asked as to what the needs are.

Safety Needs

1. The control/s must shut off the flow of gas if the pilot light goes out for any reason.
2. The control/s must not permit the heating device to overheat and destroy itself.

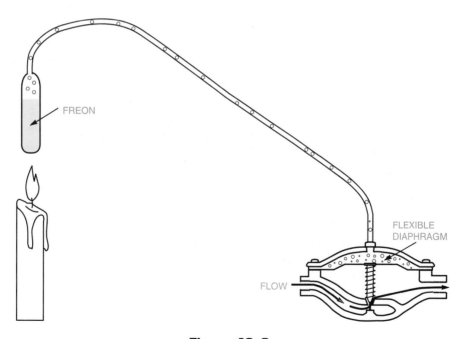

Figure 49-6

Comfort Needs

1. The water in the tank must be maintained within a range of temperatures which are both useful and not dangerous or excessively uncomfortable, 120°F to 135°F.

The temperature pressure relief valve, Figure 49–3, will prevent the water tank from being overheated or over pressurized. This will satisfy safety need number 2.

In order to design some of the other components, we will need to refine some of the earlier descriptions to make their application more useful.

Maintaining Correct Water Temperature Ranges

We will use the different expansion rates of materials when heated to design a valve which closes and opens the gas supply to the main burner when the temperature rises above our preset maximum, or falls below it. Figure 49–7 shows such a valve.

The immersion well of this valve, the main burner supply valve, is installed in the wall of the tank and is surrounded by the water inside of the water heater. Copper, the metal that the well is made of, expands and contracts consid-

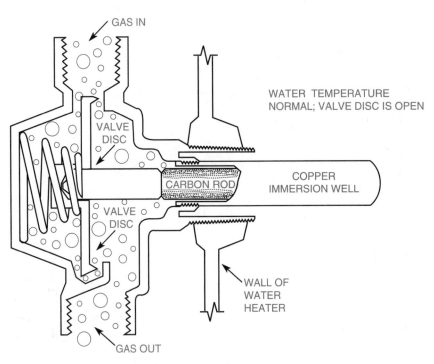

Figure 49-7

erably when heated and cooled. The carbon rod inside of the well, however, expands and contracts hardly at all. The carbon rod is pressed against the closed end of the immersion well. When the copper well expands it reaches farther inside of the tank. The carbon rod does not expand, but is forced by the spring to follow the end of the well. If the well expands enough, the valve disc, which must follow the carbon rod because of the pressure of the spring, will close and the flow of gas must cease, see Figure 49–8.

When the gas valve closes, the fuel to the main burner is cut off and the fire goes out. The pilot flame remains, though waiting to light off the main burner when it is again needed. If someone draws off some hot water from the system, cold water comes into the tank to replace

the hot water that was drawn off. This reduces the temperature in the tank. The copper immersion tube begins to shrink at the lower temperature of the water. When it shrinks, it pushes the carbon rod down toward the spring. If the well shrinks enough, the valve disc will be pushed against the spring and off of its seat. The gas will begin to flow to the main burner. When it gets there, the pilot flame is waiting and the main burner is again lighted.

The drawings for this unit have been simplified to show only the most essential elements of the devices so that they will be easily understood. If the student can visualize an adjustment knob on the left side of the drawings, (Figure 49–7 and Figure 49–8) and the addition of a rod and a spring inside of the disc follower, the device

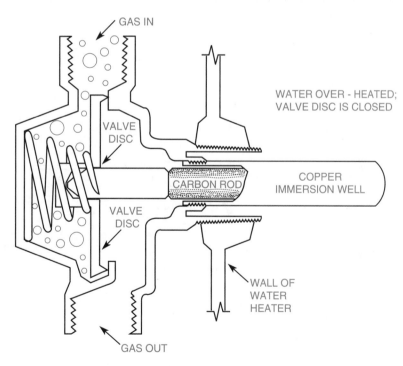

Figure 49-8

would become manually adjustable for range.

The commercial version of our system does just that. It also combines this and the other remaining controls into one neat, compact casting which will be shown schematically later on.

Notice how large the valve disc is compared to other kinds of valves. This is so that a small amount of expansion or contraction on the part of the immersion well will result in a comparatively large opening for gas to flow through.

Ensuring That the Gas IS Cut Off if the Pilot Flame Goes Out

Note the two devices that were explained previously, the thermocouple and the solenoid operated valve, the thing that is most important is that, if the pilot goes out, **all of the gas supply must be cut off.** This includes the gas

to the main burner valve explained in the last segment.

Figure 49–9 schematically demonstrates the interaction between these components. The thermocouple produces the small amount of electricity necessary to hold the main gas valve open as long a the pilot light remains lit. If the pilot flame were to be blown out by a sudden gust of wind, then the main gas valve would close. The pilot light would need to be relit by pressing down on the button, marked pilot, and holding it down while a match is held to the pilot gas jet. Then the button would still have to be held until the heat from the flame would induce the thermocouple to generate enough electricity to hold the main gas valve open by itself.

With the main gas valve operating correctly, some of the gas would continue to supply the pilot flame. The remainder of the gas would

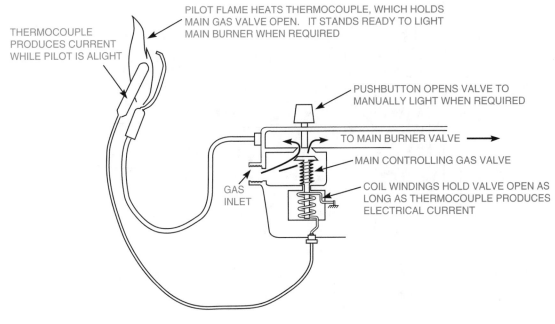

Figure 49-9

then pass through the main burner valve demonstrated in Figures 49–7 and 49–8 and explained in the accompanying text.

SUMMARY

The plumbing student has no doubt noticed that we have included items, relief valves, etc. which are not normally what we might call thermostats, which, after all, is the heading of this unit. The reason is that we have incorporated all of those things which do controlling tasks after sensing changes in temperature. Because temperature and pressure are so closely related, indeed some would even say "pressure is temperature, temperature is pressure," we have to include some mention of pressure related factors also.

The devices explained in this unit will be observed frequently controlling seemingly unrelated devices, household appliances, water supply pumps, warm air heating systems, ventilation equipment, air compressors, and so forth. If the student has a good feel for what was discussed here, many of the control devices encountered will be seen to be simply variations on the themes discussed.

◆ ◆

TEST YOUR KNOWLEDGE

1. A thermostat reacts to changes in temperature. What does it do in response to temperature changes?

2. How does the thermostat benefit its owner?

3. How does the bi-metal strip respond to heat?

4. Where would you find a cad cell in a home?

5. What device changes heat into electricity?

6. Why do room thermostats operate on low voltage?

7. How does the fusible plug operate?

8. How can freon and other gases be used to open or close valves or even switches?

◆ ◆ ◆ ◆ ◆ ◆ ◆ ◆ ◆ ◆ ◆ ◆ ◆ ◆ ◆ ◆ ◆ ◆ ◆

Relief Valves

Key Terms

diaphragm relief valve
lever and weight relief valve
safety or relief valve
spring relief valve

OBJECTIVES

After studying this unit, the student should be able to:

❑ Describe the purposes and uses of relief valves.

RELIEF VALVES

A safety or relief valve is a device that is placed on a closed water, air, steam, or oil system to prevent damage from excessive pressure. A properly placed and adjusted relief valve prevents pressure from exceeding a safe level, thereby preventing explosions.

Small relief valves, 1/2 to 2 inches, are made of brass with threaded connections. Large safety valves, 2 1/2 inches and larger, are made of cast iron or steel with brass seats and working parts. Large valves have flanged ends.

LEVER AND WEIGHT RELIEF VALVE

The **lever and weight relief valve** uses the principle of the lever and weight to balance the pressure within a tank or system, Figure 50–1. In this illustration, F is the fulcrum, W is the weight, S is the stem, L is the lever, and P is the internal pressure. Numbers and notches on the lever show where to place the weight to hold the desired pressure.

This relief valve operates like a third-class lever. When the internal pressure exceeds that

371

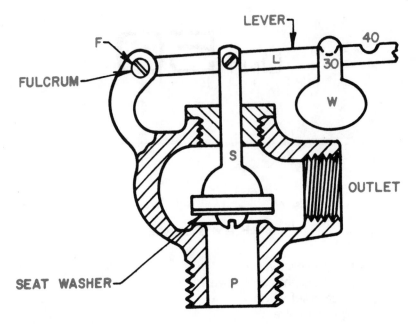

Figure 50–1 Lever and weight relief valve

for which the valve is set, it overcomes the weight and raises the washer from the seat. This relieves the pressure through the outlet and the weight closes the valve.

After the correct position of the weight has been determined, a hole is drilled through the weight and the lever. It may then be fastened in a permanent position with a bolt. The end of the lever beyond the weight should be cut off. This prevents the movement of the weight beyond the safety point. Additional weight on the lever may cause an explosion.

Relief valves are set 25 percent above the water pressure on the system but below the working pressure of the equipment on which the valves are placed.

Relief valves are placed directly into or very close to the tanks or equipment which they pro-

tect. Valves should be opened occasionally to prevent closing by corrosion. The discharge of a relief valve should be run to 2 inches above an open floor drain to prevent contamination of the water supply by back siphonage.

Note:
Check with the local code before using the lever relief valve.

SPRING RELIEF VALVES

The spring relief valve is sometimes known as a *pop valve* because it opens and closes suddenly. These valves use the tension of a coiled

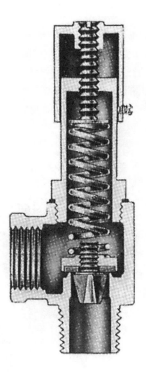

Figure 50–2 Spring relief valve

spring to withstand the internal pressure, Figure 50–2. They are used on water piping, hot-water tanks, pumps, heating boiler, air compressors, and oil lines. They are made in 1/2- and 3/4-inch sizes.

Pressure may be increased by water hammer. Water hammer is caused by the sudden closing of valves by automatic switches failing to operate or by the failure of oil burners or compressors to shut off.

Spring relief valves are set at about 25 percent higher than the pressure on the system. Adjustment is made by turning the adjusting screw in to increase the pressure. Turning it out decreases the discharge pressure.

A spring relief valve should have a large

waterway to prevent clogging by corrosion. Some spring relief valves have a try handle attached that, if operated occasionally, prevents clogging. All parts exposed to water should be made of brass or bronze.

The adjusting screw must not be tightened. This would increase the discharge pressure to a dangerous point.

DIAPHRAGM RELIEF VALVES

The principle upon which the diaphragm relief valve operates is area. This means that pressure exerted equally in all directions upon equal area. The diaphragm is 35 times larger in area than the

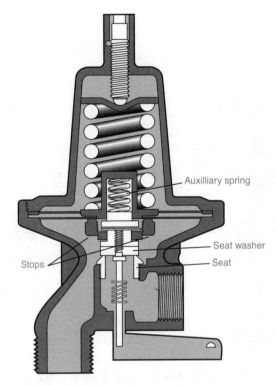

Figure 50-3 Diaphragm relief valve

seat area. The heavy spring, Figure 50–3(A), is necessary to overcome the difference in area.

The diaphragm relief valve is made in 1/2-inch size only. It has a 3/4-inch male thread inlet and a 1/2-inch outlet. The pressure ranges available are 50, 75, 100, 125, and 150 psi. Valves may be obtained that are set at any pressure between 5 to 160 psi.

Diaphragm relief valves are set with the spring chamber up. The valves are located as close to the hot-water tank as possible. The distance from the hot-water tank determines the effectiveness of a relief valve. The size of a relief valve is chosen according to its rated capacity in British thermal units per hour. This capacity must be above the Btu input of the heater.

In the diaphragm relief valve in Figure 50–3, the seat washer does not carry the heavy spring load. It is carried by the stops (B) in the valve body. The auxiliary spring (C) seats the valves with pressure when the excess pressure is relieved from the diaphragm. By using this lightweight spring, a soft seat washer may be used which opens with only a slight increase of pressure.

CHOOSE THE PROPER RELIEF VALVE

Relief valves are seldom needed but must be effective when they are. Spring relief valves depend upon a heavy spring to close the valve against the pressure. After long periods of nonuse, the spring tension tends to groove the seat washer. This fact, and the corrosion that takes place, causes the valve to stick. The pressure required to open it may damage the valve and create a dangerous situation.

Heated water can be as dangerous as dynamite. Therefore, every person installing heating devices should select the proper relief valve and know the best location for it on each particular job.

SUMMARY

Any heating device that operates on a closed system must have a relief valve. Closed systems with automatically controlled pumps installed must also have a relief valve. The relief valve releases internal pressures that rise above a preset upper limit. The considerations for relief valve installations are:

❑ The relief valve should prevent pressures from going above the maximum working pressure of the system.

❑ The relief valve should be in a location that allows a human to get close enough to operate the valve manually from time to time to check for proper operation.

❑ The outlet of the valve should be piped down close to the floor to minimize damage from escaping steam or hot water.

❑ There should be no valves between the relief valve and the heating device, because these could render the relief valve useless.

TEST YOUR KNOWLEDGE

1. On adjustable relief valves, which way would you turn the adjustment screw to increase the pressure at which the valve blows off?

2. What is the purpose of the try handle?

3. If the working pressure of a system is 40 psi and there is no equipment instruction sheets to go by, what would you set the blow-off pressure for?

4. What type of relief valve uses the principle that a small amount of pressure exerted over a large area can overcome a higher pressure exerted over a smaller area?

5. Which relief valve utilizes a third-class lever to overcome the internal pressure?

6. Can water hammer cause a relief valve to blow?

7. Are relief valves ever needed on devices other than heating devices?

8. How would you select a relief valve rated in British thermal units per hour?

THE "SUMMER-WINTER HOOKUP"

OBJECTIVES

After studying this unit, the student should be able to:

❏ Describe the installation of immersion coils.
❏ Explain the principles of the heat exchanger.

SUMMER-WINTER HOOKUPS

The **summer-winter hookup** is a popular method of heating water in residential construction. In this system the hot-water space heating system for the residence is also used to supply the domestic hot water. The student should recall that:

1. **Domestic water** is the water that comes out of the fixtures on kitchen sinks and the shower heads in the bathrooms. This supply system is open. The well or the street main supplies the water and it is used at a rate of about 50 gallons per day on the hot-water side per person.

2. The **circulating hot water** is the water con-

tained in the *closed* home or space heating system. This water is more or less static in quantity. Hot water heating systems are referred to as **hydronic.** While the pressure in the domestic water supply system may be as high as 65 psi, the hydronic system's safety relief valve will "blow off" if the pressure within it gets higher than 30 psi. A positive factor about a closed system is that after the initial fill up water has used all of its free oxygen, by causing an initial amount of corrosion, there is no more free oxygen left to cause corrosion. There is a small amount of **make-up water** admitted to the system, but the quantity is very small. As a result, black steel pipe used for hydronic heating systems shows very little corrosion after years of use. Black steel pipe used in a domestic water supply would corrode shut in very little time depending on minerals and the acidic content of the water.

THE IMMERSION COIL

In use, the immersion coil, Figure 51–1, is bolted into the hydronic boiler. The copper pipe which is attached to the flange as pictured is immersed in the water of the heating system. The temperature of the boiler water is maintained between 180° and 200°F to maintain a constant source of heat, and a healthy heat exchange rate for the domestic hot water system. This is different than if there were no immersion coil being used. In this case the lower boiler temperature could be set at 150°F. Because domestic hot water is required during the summer months as well as the winter ones the boiler must be prepared to run at all times. This could result in unwanted hot water being circulated by natural circulation (*as hotter water rises and cooler water falls*) during the summer months. Therefore this hydronic system must have what is called a *flow control* valve installed over the boiler

to prevent natural circulation.

The student will notice that there are grooves cut into the copper water pipe attached to the immersion coil flange. This increases the surface exposed to the boiler's hot water and enables the many coils to pick up heat, Btus, more readily. In fact the modern immersion coil can pick up heat so quickly that a storage tank to hold the heated water is not necessary. By the time that the water gets through the coil it is up to heat. This means that, unlike other water heating plans, this one should never run out of hot water no matter how much it is used! Because the quantity of water retained in the boiler in the copper coils is so small compared to the rest of the water supply system, this is one water heating system that can get along safely without a temperature/pressure relief valve.

Caution:
Because the water is so hot coming out of the coil, something must be done to prevent scalding. Remember, the boiler has two jobs, to heat the house and to provide domestic hot water. To heat the house efficiently, it must use a water temperature higher than the safe water temperature for domestic use, 135°. Some installations do not concern themselves with this danger and allow 200°F water to arrive at the water outlets throughout the house. Others have a disk with a hole in it, a **flow constrictor,** in the coil outlet adapter. This system slows down the flow through the coil, enabling it to have more time to pick up heat, and also forces the user to add cold water at the outlet to get sufficient flow. The best solution is the **tempering valve,** with perhaps also the flow constrictor, see Figure 51–2. This is a device which adds cold water to the

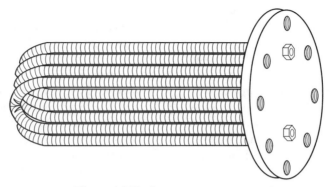

Figure 51-1 An immersion coil

flow of hot water coming out of the immersion coil. It may be preset with a little testing and adjusting to maintain a constant flow of 135° water to the fixtures throughout the house.

than water. It will supply 240 Btu per square foot of radiation opposed to water's 150 Btu. The equipment necessary to heat with steam is too complex and expensive for the average residence however.

HEAT EXCHANGERS

The **immersion or tankless coil** is a heat exchanger. Heat exchangers do exactly what the name implies, they exchange heat. The coils which are immersed do not always pick up heat, however. Sometimes the purpose of the heat exchanger is to lose heat or to give it off. If the liquid vapor, or gas in which the coils are immersed is colder than the liquid, vapor, or gas in the coils, the coils will lose heat. If the situation is reversed the immersed coils will gain heat. Solar water heating requires a heat exchanger to avoid freezing up in winter months.

The student will, in his or her career, see many kinds of heat exchangers especially in commercial buildings. Buildings which use steam for heating will often use heat exchangers in which the domestic water coils run through a tank containing steam to pick up heat. Steam, by the way, is a more efficient heating medium

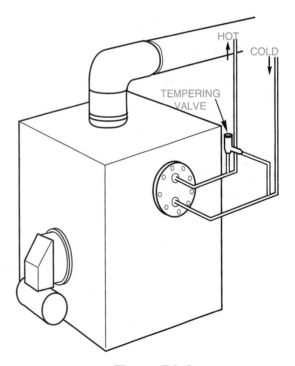

Figure 51-2

The student may also see a process called *staged* water heating in large buildings in which a number of comparatively small boilers or heaters are used to provide heat for a large heat exchanger to supply domestic hot water. This allows the building engineer to use just one or two small boilers in off-peak hours and all of the boilers during peak hours. In this way the system is not shut down if a boiler fails, and boilers, when used, are run in their most efficient way.

SOME "SUMMER-WINTER HOOKUP" CONSIDERATIONS

This system of water heating is very durable and will usually outlast three or four conventional water heaters. By the way, it is wise to avoid calling them "hot water heaters." They don't heat hot water. They are simply "water heaters."

The tankless coil fails in two ways:

1. The pipe in the immersed coil thins with use and ruptures. The customer will complain that the boiler relief valve is constantly dripping or blowing off. The domestic water pressure is escaping into the boiler. The relief valve, which can only handle 30 psi, must get rid of the extra pressure. The immersion coil must be replaced. Watch out for broken studs when removing the flange nuts!

2. The pipe scales up on the inside from mineral deposits. The customer complains that there is not a sufficient supply of hot water any more. If there once was an adequate supply and now there isn't, this may be correctable by de-scaling. One kind of de-scaling apparatus is comprised of a small plastic tank with a pump and plastic inlet and outlet hoses. A commercial acid solution is circulated through the coils until the coils are clear. It is not uncommon, however, for the acid used to eat through the copper coils

and ruin them. Some plumbers and heating mechanics insist on a replacement rather than cleaning for this reason.

Customers often feel that they are having great savings in their water heating costs by having a summer-winter hookup. Their reason is that the hot water is nearly free because the space heater supplies it as a by-product of heating the residence. It is not known if this has ever been costed out in a scientific manner but let's try a quick "rule of thumb" appraisal:

1. An average person's use of hot water is 50 gallons per day. This means that if this is used 350 days a year, 145,950 pounds of water (73 tons) must be brought from an average temperature of 55° (the constant underground temperature in much of the United States), to 135°F. This amounts to 11,676,000 Btu (a unit of heat explained fully in a later unit) in a year. It makes no difference if those Btu are supplied by natural gas, heating oil, or electricity, but physics tells us that those Btu *must be supplied from some source and they are not free.* Solar heat gain is free but negligible in a well-insulated house. The heat pump uses ambient heat in the outside air but has to use electricity to obtain it.

2. The heating boiler must be kept at a higher average temperature than it would have to be if it were just used for space heating. A low temperature of 180° as opposed to 150° for a plain home heating system would have to be maintained. Because of this higher-than-average temperature we have a higher heat loss to, usually, the cellar where it does little good. Cellars and crawl spaces usually have no insulation and so this heat is lost to the surrounding environment.

3. The large house boiler must run even during the summer to supply domestic hot water. Since the boiler is much oversized

An Approximate Comparison of Water Heating Costs				
	Efficiency Btu/Unit		Cost Unit	Costs/Year
Electricity				
Off Peak	100%	3413	.0584/KW	$200
Regular	100%		.1389/KW	$475
Natural Gas	85%	1050	.00574 CU. FT.	$75
#2 Fuel Oil	80%	140,000	.80 GAL	$75

for this, it runs in an inefficient manner. Probably the worst place to install a summer-winter hookup would be in a summer home. But then if the area is subject to freezing temperatures, hydronic heating itself may be a poor choice of heating systems.

4. The cost of heating water must also include the cost of periodic replacement of the equipment. The immersion coil excels here as it should easily outlast at least three or four gas or electric water heaters.

Some installations have what are called tempering tanks, not to be confused with tempering valves, installed just before the cold water enters the coil. This tank is without insulation. The object of the tank is to absorb heat from the surroundings and raise the temperature of the water within the tank. This means that the heater may only have to raise the temperature of the water from 65° to 135° instead of from 55° to 135° if it is in the cellar. The tempering tank would absorb heat from the living space to raise its temperature if it were located in a utility room on the living level of the house. This may not make sense depending upon the insulation and layout of the house.

Another installation has the outlet of the coil entering an electric or gas water heater. During the winter months the water heater simply maintains the water temperature of the water passed to it. In the summer months the house space heater can be turned completely off and the gas or electric water heater can take over.

SUMMARY

The summer-winter hookup utilizes an indirect heater. The term "summer-winter hookup" refers to the fact that the homeowner's house heater is going to run in all seasons. An indirect heater uses steam or hot water to heat water contained within its immersion coil. In practice, the coils are looped many times to get the maximum length of pipe in the smallest space. This ensures that water flowing through the coil will stay in contact with the heat source for as long as possible.

Producing domestic hot water is a common indirect heater task. Because the source hot water must be hot enough whenever the indirect heater is called upon to produce its own hot water, the source heater must have its own steam or hot water up to usable heat at all times. For homes, this means that the heating boiler will be in use all the time rather than in only the cold months. That may or may not be more expensive, depending on the cost of the fuel used by a stand alone water heater.

◆ ◆

TEST YOUR KNOWLEDGE

1. Where does the indirect water heater get its heat to warm domestic water for the home's fixtures?

2. What is the highest temperature of domestic water which is comfortable for people?

3. What does the flow control valve on an hydronic heating system accomplish?

4. What is the purpose of grooving the copper pipe of the immersion coil?

5. The immersion coil is installed into the boiler. How can you detect when the immersion coil has developed a leak?

6. What does a tempering tank do?

7. What does a tempering valve do?

◆ ◆ ◆ ◆ ◆ ◆ ◆ ◆ ◆ ◆ ◆ ◆ ◆ ◆ ◆ ◆ ◆ ◆ ◆

SOLAR WATER HEATING

KEY TERMS

collector (solar)
heat exchanger
temperature differential control box
thermosiphon systems

OBJECTIVES

After studying this unit, the student should be able to:

❑ Discuss how solar water heaters are constructed and used.

SOLAR WATER HEATING

Using the sun to heat water is not new. People in the south have used the sun for domestic water heating for many years. This method usually consists of a coil of copper soldered to a galvanized sheet and covered with glass, Figure 52–1. The assembly is painted black and placed on the roof. The sun's rays penetrate the glass and warm the galvanized sheet and, along with it, the coiled copper tubing.

The heat from the sun's ray then accumulates under the glass cover. This causes a kind of *greenhouse effect*. If a tank is placed in the peak of the roof, a natural circulation is set up between the collector and the tank, Figure 52–2. The tank becomes warm and supplies hot water to the house. Systems which depend on natural circulation are called **thermosiphon systems**. For this to work correctly the *collector* must be lower than the holding tank, because hotter water tends to rise and colder water tends to fall. The water in the solar collector will be hotter than the water in the tank and will tend to, or "want" to, rise. If the tank is located above this level, this will take place easily. If the tank is lower than the collector, as it is in most cases, the thermosiphon system is inapplicable and a small fractional horsepower pump is necessary to move the water where it is desired to be.

383

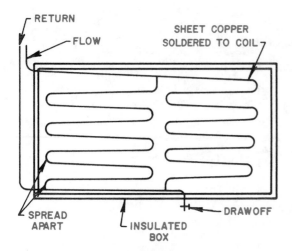

Figure 52–1 Solar coil

When the holding tank or *heat exchanger* is lower than the collector the natural circulation tendency is in the wrong direction. As a matter of fact, if precautions are not taken, in the cool evening hours the warm water in the holding tank will attempt to rise into the darkened, sunless, and cooler solar collector above. As it does this the process will be *reversed* and the heat collected will be released back into the environment! Obviously, this cannot be allowed to happen.

It is also a difficult system to use where the temperature drops below freezing. However, solar systems are being tested in the north. Collectors are drained down during cold nights, and shutters are installed over the collector. Antifreeze solutions are used along with a heat exchanger. Most systems being manufactured use a pump and a heat exchanger to eliminate the problems of the solar water heating system.

COLLECTORS

The collector is the device that is exposed to the sun to collect the heat or Btu, Figure 52–3. There are many different kinds of collectors. It may be a copper coil soldered to a copper sheet or aluminum sheets welded together with passages for liquid between them. It could be corrugated material where water trickles down the corrugations to pick up heat.

Because collectors are subjected to high heat, chemical reactions take place at a rapid rate. If these chemical reactions are not controlled, the collector will eventually develop leaks. Tap water contains percentages of free oxygen and minerals, and it often has a low pH (high acid) content. Therefore, additives are nec-

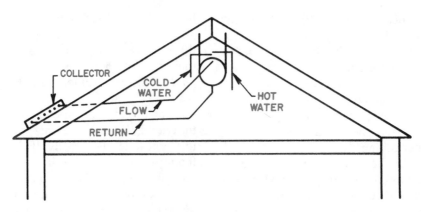

Figure 52–2 Attic tank

Figure 52–3 A solar collector

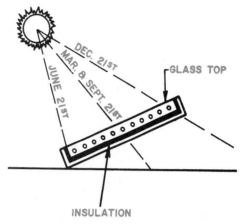

Figure 52–4 Position of coil

essary to achieve a satisfactory life span of the unit. If antifreeze and anticorrosion chemicals are added to the collector water, a heat-exchanger unit must be used.

Collectors should face toward the south, if possible, and never deviate more than 30 degrees from due south. The angle of the collector from the horizontal should be the latitude plus 10 degrees in sunny areas of the country, Figure 52–4.

HEAT EXCHANGERS

If a coil of aluminum or copper tubing is immersed in a tub of cold water, and hot water is passed through the coil, the tub of water will become warmer. This is the principle of the heat exchanger. The chemically treated water from the collector is passed through the coil in the heat exchanger. The tap water in the exchanger tank is, in turn, heated by the coil.

There are, as in the case of the collector,

many different heat-exchanger designs. Most manufacturers are combining the functions of the heat exchanger with that of the storage tank, Figure 52–5.

STORAGE TANK

Because the sun does not shine during cloudy weather or at night, hot water must be stored for sunless periods. The size of the storage tank depends on the size of the collector. If the storage tank is too large, the collector may not have the capacity to bring the temperature up to a usable temperature (120° to 160°).

A storage tank large enough to supply hot water under all possible conditions would have to be very large and very expensive. Because of this, solar-heating units are constructed to handle only ordinary weather conditions, and a conventional backup unit is added. Often, an electrical heating element is added to the storage tank. This heating element supplements the heat during long, sunless periods.

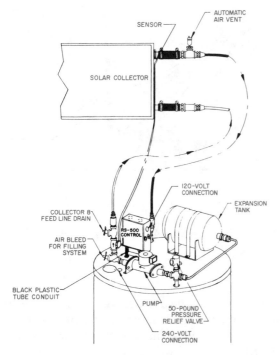

Figure 52–5 Heat exchanger and storage *(Courtesy Mor-Flo Industries Inc.)*

COMPONENT ARRANGEMENT

Because we are depending upon naturally occurring phenomena, the sun's rays, to provide the heat to accomplish the task which we have set, we must pay special attention to the following considerations.

1. **The sun doesn't shine at night.**

   The problems of a reverse circulation have been addressed above. With the addition of a fractional horsepower electric pump a *flow control* valve can be installed which prevents natural circulation. The vanes of the pump at rest also prevent circulation. Check valves in a natural circulation system to prevent backflow would prevent all circulation because the forces involved are not strong enough to consistently go through the check valve even in the correct direction.

2. **The sun doesn't provide nearly as many Btu of heat when it is cloudy.**

   Some areas of the country have more cloudy, rainy weather than others. Although there is solar heating even on cloudy days, the size of the system must be adjusted for this. In these areas a larger, and more expensive, system is needed when it is cloudy. It is not uncommon for the cloudy weather to last a week or more. Should systems be built which will store enough hot water to supply a week's worth of hot water? No, of course not.

   The answer is a conventional back-up system. All solar systems need a back-up system. This is sometimes incorporated into the heat exchanger. A better solution is to install a gas or electrical water heating system along with a solar water heating system. This will provide an independent heat source when the solar system breaks down. The complete solar water heating system is not, however, inexpensive.

3. **The sun provides more heat in the summer than it does in the winter.**

   The angle and placement of the solar collector is critical. Each solar collector is covered with glass or plastic. The cover must allow infrared light, heat, to pass through and then to hold that heat in by preventing air currents from immediately carrying it away. Commercial heat collectors are much more efficient than home-built collectors because of the great attention to detail involved in their manufacture.

   If the rays strike the glass at a 90-degree angle, most of the light will be absorbed by the collector. However, a the light ray varies from that 90 degree angle, more and more of the heat is *reflected*. In Figure 52–4 the light ray labeled December 21st would have most of its heat energy reflected back into space. Most of the heat from the June 21st ray, on the other hand, would be absorbed by the collector and converted to hot water. Since the days of the summer represented by June 21st in the figure are much longer than the days represented by December 21st, yielding a much longer number of heating hours during the day, do you think an angular adjustment of the collector would be wise? How would you adjust it for your area?

Different glass or plastic compositions have different refraction factors. That means that some covers will allow a greater deviation from the 90-degree optimum angle and still resist reflecting the light ray back into space. The absorption material under the cover is coated a flat black because black absorbs heat better than any other color. This absorption material is also convoluted to resist reflecting the ray right back out through the cover glass.

Wind is a cooling factor. If the collector can be sheltered from the wind while receiving the direct 90-degree impact of the sun on the shortest day of the year, it will probably produce the most heat when it is needed the most.

ACCESSORIES

When the storage tank is below the level of the collector, the collector fluid must be pumped up from the tank. A small-capacity pump must be part of the solar-heating equipment.

A thermostatically controlled device controls the pump. It regulates the pump so it will only operate when the water in the collector is warmer than the water in the storage tank. Because the control must measure temperatures from two different points, it is often called a temperature differential control box.

The water within the storage tank can become so hot that a tempering value must be installed. This prevents the occupants from being scalded. The tempering valve mixes some cold water with the outgoing hot water when it is necessary.

The collector and heat-exchange coils form a closed system. Since water expands when heated, an expansion tank must be installed. This provides a place for the expanded water to go. All water-heating systems must have a temperature / pressure relief valve also.

SUMMARY

The rising cost of fuel oil for generating electricity has influenced many people to consider solar water heating. In general, gas- and oil-fired water heaters are more economical at the present time. A customer who uses an electrical water heater, however, could very well benefit from solar heating. With the rapidly increasing cost of heating fuels, solar water heating will become more economical in all cases in the future.

Even though sunlight is free, solar heating equipment is expensive. In addition, colder areas and areas where the sun does not shine all the time still require a backup system. This is usually a conventional water heating system that comes on as needed.

Solar water-heating equipment for a home can cost $3600. If the life expectancy of the equipment is 20 years, the system must show a savings of 10 dollars a month to simply pay for itself. In general, if water heating is done by electricity, the solar system will pay for itself in 6 to 7 years. At present prices, if water heating is done by gas or oil, it may take as long as 30 years.

◆ ◆

TEST YOUR KNOWLEDGE

1. What is the most important factor influencing the efficiency of solar collectors?

2. What is the task of the solar heat exchanger?

3. In which direction should the collector face?

4. Why is a back-up system necessary?

5. What distinguishes a thermosiphon solar water heating system?

6. Would you need a relief valve on the collector side of a heat-exchanger solar system?

7. What kind of a control must be used to prevent cool collector fluid from chilling the domestic water tank?

SECTION 6

SPECIAL CASES

◆ ◆ ◆ ◆ ◆ ◆ ◆ ◆ ◆ ◆ ◆ ◆ ◆ ◆ ◆ ◆ ◆ ◆

SOVENT DRAINAGE SYSTEMS

KEY TERMS

aerator
deaerator
SOVENT drainage system

OBJECTIVES

After studying this unit, the student should be able to:

❑ Explain how a SOVENT drainage system works.
❑ Discuss why a SOVENT drainage system costs less than a two-pipe system.

DRAINAGE SYSTEMS

In drainage systems, traps prevent sewer gas from entering a building. Because traps must be simple, absolutely reliable, and have no moving parts, the water-filled trap is used.

The seal in a water-filled trap is quite fragile. A small pressure from above (siphonage) or from below (back pressure) will upset the trap seal. The design of all drainage systems must, therefore, ensure that pressure within the system is balanced with atmospheric pressure.

In conventional two-pipe systems, balance is maintained by separate systems for the movement of air and waste. The SOVENT drainage system, however, accomplishes this by using only one stack in a self-venting system.

SOVENT DRAINAGE SYSTEM

The SOVENT drainage system is designed to simplify drainage, waste, and vent piping in multistory buildings. It was developed in Switzerland

391

as a less expensive alternative to the two-pipe system commonly used in the United States.

The SOVENT system used one pipe and specialized fittings to prevent the stack from becoming completely filled at any point, Figure 53–1. If the cross section of the stack is never completely filled, then air is free to move within the same stack as the waste. This will permit the air to equalize pressures within the stack.

The SOVENT design:

❑ Slows down the speed of the flow at each floor.

❑ Causes a turbulence at each floor that introduces air into the waste.

❑ Removes the air as the waste reaches the base of the stack.

❑ Prevents positive pressures from developing.

The result is a single stack that is self-venting with the fittings balancing pressures throughout the system. SOil stack and VENT combine into a single SOVENT stack.

The SOVENT design utilizes two unique fittings: the *aerator* and the *deaerator*. These special fittings are the basis for the self-venting features of the SOVENT.

AERATOR

The aerator fitting has a built-in offset, which serves to slow down the flow, Figure 53–2. It has a mixing chamber, one or more branch inlets, one or more waste inlets for connection of smaller waste branches, a baffle in the center of the chamber, and the stack outlet at the bottom of the fitting.

The aerator provides a chamber where the flow of soil and waste from the horizontal branches can unite smoothly with the air and liquid already flowing in the stack. The aerator fitting does this so efficiently that the stack cannot become completely filled and cause pressure changes.

Aerator fittings are placed in the stack at every floor where there is a substantial branch. Where there is no substantial branch line, a double in-line offset is installed in its place, see Figure 53–2. The purpose of the offset is to slow down the flow, just as the aerator does.

DEAERATOR

The deaerator is placed near the base of the stack. It prevents waste from slowing down at the base and forming a slug of water that would cause pressure changes. For the same reason, a deaerator is placed at every offset in the stack, except where there is a double in-line offset, Figure 53–3.

The deaerator functions along with the aerator above it to make a single stack self-venting. It consists of an air separation chamber with an internal projection, a stack inlet, a pressure relief outlet at the top and a stack outlet at the bottom, Figure 53–4. In practice, some of the air in the descending slug is ejected into the relief line. This is led back into the house drain at a point at least 4 feet from the base of the stack, Figure 53–5.

BRANCH LINES

The SOVENT design does not require back venting until the horizontal run exceeds 27 feet. The long horizontal distances without a separate vent permitted by the SOVENT design are achieved by oversizing the branches. This prevents them from ever being completely filled.

PLUMBING CODE REGULATIONS

In many places, the plumber must get special permission from the sewer authority to use the SOVENT system. This system has, however, been installed in more than one hundred apartment

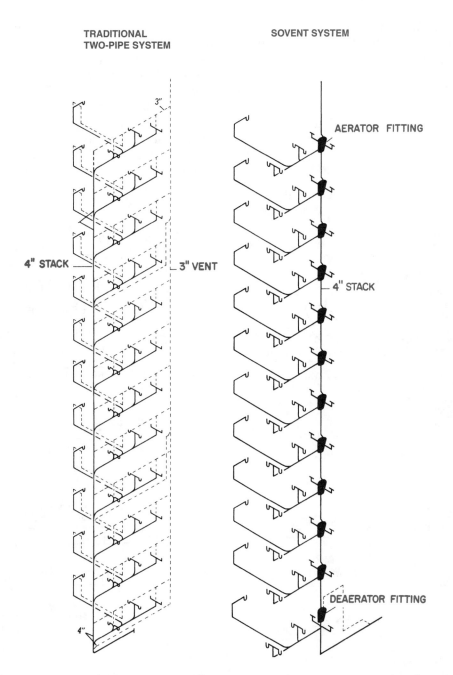

TRADITIONAL
TWO-PIPE SYSTEM

SOVENT SYSTEM

AERATOR FITTING

4" STACK

3" VENT

4" STACK

DEAERATOR FITTING

Figure 53–1 Compare the two-pipe and SOVENT drainage systems. Note that the SOVENT system requires less material—a cost-saving feature. *(Courtesy Copper Development Association Inc.)*

and office buildings in the United States alone.

Additional information about SOVENT plumbing may be obtained from

Copper Development Association, Inc.
405 Lexington Avenue
New York, New York 10017

SUMMARY

Plumbing codes have been worked out through many years of trial, error, and some studies, which have resulted in a workable system. The working plumber has been handed a book, the locality's plumbing code, with instructions, which if followed carefully will produce good results. So why change? Those good results have been hard won.

The SOVENT system has been engineered from its onset and then tested; it works also. It should be cautioned that the SOVENT system is an integrated system. To use it, you must use all

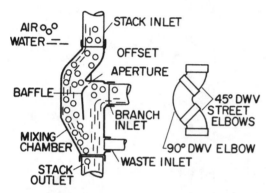

Figure 53–2 SOVENT aerator *(Courtesy Copper Development Association Inc.)*

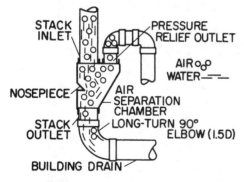

Figure 53–3 SOVENT deaerator *(Courtesy Copper Development Association Inc.)*

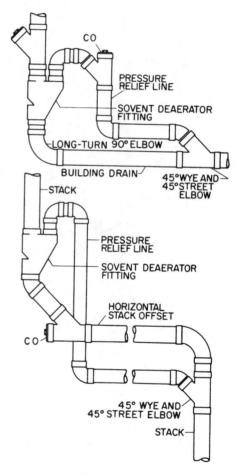

Figure 53–4 *(Courtesy Copper Development Association Inc.)*

of it. And, of course, if it is different from the local plumbing code, permission must be obtained.

Notice also that this system has been especially designed for high-rise buildings. The advantage that SOVENT has over conventional plumbing is cost in materials and in labor. It is presented here only to demonstrate that there are other stack and venting systems that work well also.

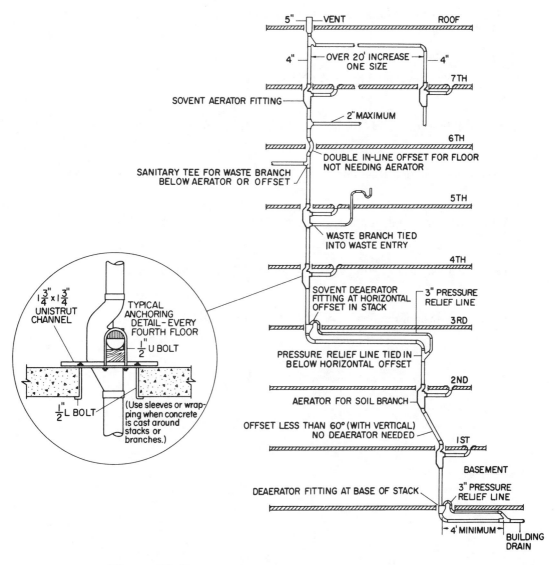

Figure 53–5 *(Courtesy Copper Development Association Inc.)*

◆ ◆

TEST YOUR KNOWLEDGE

1. What is the major difference between conventional plumbing practice and the SOVENT system?

2. What country developed the SOVENT system?

3. What is it that introduces the air into the waste as it falls?

4. What is the task of the aerator?

5. What two words make up the acronym SOVENT?

6. How are the long trap arm lengths possible with the SOVENT system?

7. What is the task of the deaerator at the base of the stack?

UNIT 54

CROSS CONNECTIONS

OBJECTIVES

After studying this unit, the student should be able to:

❏ Describe how and where cross connections are used.
❏ Discuss the causes and dangers of back siphonage through water pipes.

CROSS CONNECTIONS

Cross connections are direct connections between a water supply and a system carrying unsafe water. Preventative measures may now be taken to ward off this dangerous situation.

Back siphonage through water pipes from drains is caused by failure or a drastic reduction of the water pressure. Pressure reduction may be caused by the sudden flushing of street mains, fire engines pumping, breaks in mains, or improper closing of street valves. Inside buildings, it is caused by meters, strainers, pressure regulators, or undersized piping. Adding fixtures to existing pipelines results in an excessive pressure drop.

When any one of these conditions occurs while a fixture is clogged, the water falling back in a water riser creates a partial vacuum, which draws sewage into the water system, Figure 54–1.

In tall buildings, siphonage may be severe enough to draw water from a lavatory into a faucet even though the faucet nozzle is above the water surface. This may be corrected by elevating the nozzle to at least 1 1/2 inches above the rim of the fixture, Figure 54–2.

In Figure 54–3, a reliable vacuum breaker (A) installed between the flushometer valves (B) and

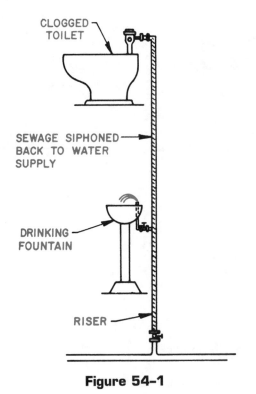

Figure 54–1

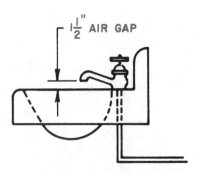

Figure 54–2

bowls usually prevents this condition in toilets.

In some large buildings, untreated river water may be used to flush toilets, test tanks, and other fixtures. A separate system of pipe is required, but the two systems are cross-connected in an emergency such as a fire, Figure 54–4. The danger occurs when someone unknowingly opens the valve between the pure and the unsafe water supplies. If this occurs, many people may become dangerously ill.

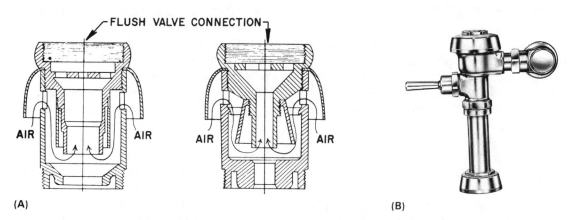

Figure 54–3 *(Courtesy Sloan Valve Co.)*

Note:
Check with your local code. Using untreated river water is strictly forbidden in some areas of the United States.

All pipelines containing unsafe water are painted red. The valve must be labeled and back-flow preventers installed. This prevents unsafe water from entering the safe water supply.

Overhead drain lines should be inspected periodically to prevent sewage water from leaking into tanks containing food. Since fixtures below sewer level are subject to stoppage and flooding, backwater valves should be installed. When water supply systems are designed, the mains must be large enough to maintain 15 psi of pressure at the highest fixture in addition to the probable flow.

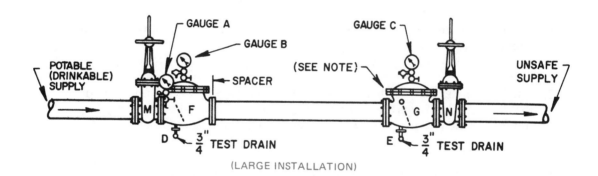

(LARGE INSTALLATION)

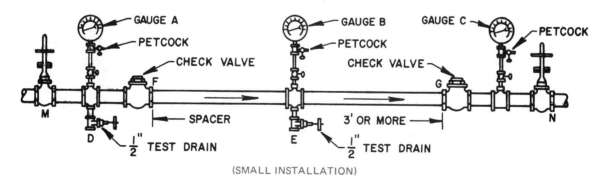

(SMALL INSTALLATION)

Figure 54-4

◆ ◆

TEST YOUR KNOWLEDGE

1. Is a direct connection between a fire main and a drinking water supply pipe a cross connection?

2. What is the greatest danger with cross connections?

3. A hose has been attached to the faucet and it hangs down into the sink or tray. What do you see wrong with this common situation in many commercial kitchens?

4. What is the term applied to the situation in which bad water is drawn into the supply system?

5. The flushometer valve on many commercial toilets exits below the flood rim of that fixture. What keeps the water from a stopped and flooded bowl from being drawn into the water supply system if the water pressure should fail?

6. What is the purpose of the air gap between the opening of the faucet and the height of the floor rim of the fixture it serves?

◆ ◆ ◆ ◆ ◆ ◆ ◆ ◆ ◆ ◆ ◆ ◆ ◆ ◆ ◆ ◆ ◆ ◆

SUMP PUMPS AND CELLAR DRAINERS

KEY TERMS

sump
sump pump

OBJECTIVES

After studying this unit, the student should be able to:

❏ Explain how water may be removed from a cellar.
❏ Describe the installation and operation of a sump pump.

SUMP PUMPS AND CELLAR DRAINS

In localities where the groundwater is close to the surface, basements are likely to be wet, especially in rainy weather. This can be a health problem.

The depression in a basement floor where water collects is known as a sump. In new construction, this condition can be remedied by laying a system of field tile or perforated drainpipe around the outside of the basement wall. The tile is then connected to a sump pump in the basement and below the floor, Figure 55–1. Elevator pits are drained in this manner. The field tile is covered with broken stone to ensure better drainage.

To remedy such a condition in an existing building:

1. Build a sump under the basement floor and place holes in the sides of it.
2. Extend field tile around the outside of the cellar wall and connect it to the sump.
3. Install an automatic cellar drainer to pump

401

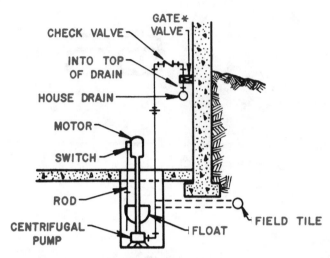

*When connected to a house drain, some codes require both the check valve and gate valve.

Figure 55–1

the water up to the drain if it will not run out by gravity.

There are several types of cellar drainers. Older, now obsolete types used water, air, or steam pressure through a jet to siphon the water out. This type of siphon raised water 1 foot for every 4 psi of pressure. It wasted considerable water. It also posed a constant threat of forcing contaminated water back into the potable (drinkable) water lines when connected to it.

One of the newer cellar drainers consists of an electrically driven, vertical centrifugal pump. It is controlled by a float and is more efficient than older styles. The motor is installed above the highest possible water level so that it will remain dry. As the water rises in the sump, the float pushes the rod up and turns on the switch, starting the motor. As the water recedes, the float pulls the rod down and shuts off the motor.

Another still newer type of cellar drainer is a completely submersible unit that operates underwater, Figures 55–2 and 55–3. The unit, therefore, is never flooded out and operates until the current source is cut off or the desired water level is reached. It can be lowered into a flooded cellar or pit to pump out water through a flexible pipe or hose. The switch, which is a sealed unit, is more effective than the rod-and-float type.

Care must be taken to avoid electrical shock when working around cellar drainers since they are always located in damp areas. Be certain that the unit is properly grounded.

Unions should be placed on all pipes that lead to the pump. This makes removal of the pump for repairs when the sump is full of water easier. The discharge from the cellar drainer should extend to a safe terminal, such as a storm drain or suitable dry well. It should also be protected by a suitable check valve to prevent the backflow of surface water into the sump. Check the plumbing code before joining the discharge of a pump to the house drain.

Figure 55–2 (Courtesy American Kenco® Inc.)

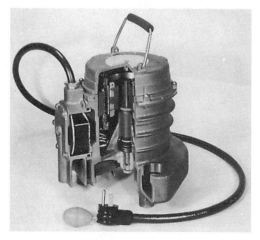

Figure 55–3 (Courtesy American Kenco® Inc.)

◆ ◆

TEST YOUR KNOWLEDGE

1. What is a sump?

2. There are two kinds of electrically operated cellar drainers. Can you describe their configuration?

3. What is the major safety consideration when working around cellar drainer pumps?

4. Is use of a three-pronged plug a way of grounding the pump?

5. Why is a union in the discharge pipe of a cellar drainer important?

6. Why is the union between the pump and the check valve in Figure 55–1?

WELDED PIPING

OBJECTIVES

After studying this unit, the student should be able to:

❏ Describe welding processes and types of welds.
❏ Prepare pipe ends for welded fittings.
❏ Properly align pipes for welding.

WELDING PROCESSES

There are two welding processes for welded fittings: gas welding and electric arc welding.

Gas Welding

Gas welding mixes acetylene and oxygen to create temperatures high enough to melt steel. The gas welding setup is similar to the soldering operation, except there are two tanks and two hoses leading to the torch tip. The pure oxygen supplied by one tank permits much higher temperatures than if acetylene gas alone were used.

In operation, the flame melts the base metal together; that is, it melts the pipe metal to the fitting metal. Some filler metal is added with a steel welding rod much the same as soldering. The gas welding apparatus may be fitted with a cutting attachment. The cutting torch is used to

405

raise the temperature until the pipe melts. Then a stream of pure oxygen is directed into the molten puddle. The steel actually burns. If the operator moves the torch around the pipe circumference, it cuts through the pipe.

Electric Arc Welding

Arc welding uses electric current and is less expensive than gas welding. The electricity for arc welding may be either direct current or alternating current. If it is direct current, it may have a positive or negative ground.

The type of welding electrode used for filler metal must be matched to the machine, the type of electricity used, and the base metal to be welded. Manufacturers' booklets show the proper rods and machine settings to use for a variety of conditions.

TYPES OF WELDS

A properly welded joint is leakproof, vibration proof, and stronger than the pipe itself. Many underground pipelines have welded joints. Hydraulic and steam lines are welded with increasing frequency. Almost any kind of metal can be joined by welding.

There are two basic types of weld forms: the butt weld and the fillet weld, Figure 56–1. The butt weld is used to join metal that is laid edge to edge. The fillet weld is used to join metal with faces that form an angle.

Fittings are either fillet-welded or butt-welded. A fillet weld-type fitting is similar to a copper sweat fitting. The sweat fitting is soldered so that the solder flows to the bottom of the socket. The fillet weld fitting is joined only at the face of the fitting and to the pipe.

More and more copper-to-brass fittings are being brazed. In brazing, the joint is welded with a brass rod rather than soldered. Brass flows like solder to the depth of the socket. While it is possi-

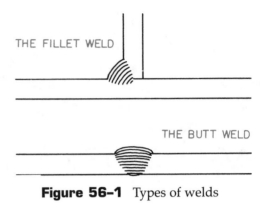

Figure 56–1 Types of welds

ble to weld cast iron and malleable iron by welding with iron rods, brazing these materials seems to give a more satisfactory job.

Fillet Weld End Preparation

A fillet weld is used with a socket fitting. A socket fitting is self-aligning. The end of the pipe is cut square and reamed. The pipe is inserted into the fitting and marked at the face of the fitting. The pipe is then withdrawn 1/16 inch from the fitting. The 1/16 inch allows the heated pipe to expand into the fitting without disturbing the joint as it is being welded, Figure 56–2. The fitting allowance for a socket type, 90-degree fitting equals the pipe size.

Butt Weld End Preparation

The butt weld fitting has the same inside and outside diameter as the pipe to which it is joined, Figure 56–3. Long-radius and short-radius patterns are available for 90-degree elbows. This is similar to the long sweep and the short sweep encountered in soil fittings.

The allowance for the short-radius elbow is equal to the pipe size. The long-radius allowance is equal to 1 1/2 times the pipe size. The allowance for a 45-degree elbow is equal to 5/8

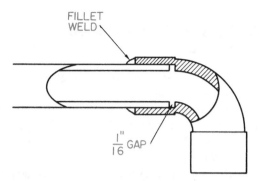

Figure 56-2 Fillet weld on socket fitting

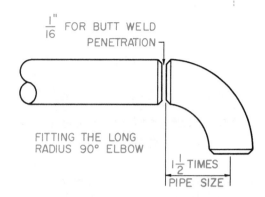

Figure 56-3 Butt weld

times the pipe size. An additional allowance must be made at the joint to allow the weld to penetrate to the inside diameter of the pipe. A 1/16-inch separation is used for standard-weight pipe.

Before the joint is set up for tacking, the pipe is beveled to the same angle as the fitting, Figure 56–3. A tack is a short weld that holds the fitting in line, while the weld is being made. When tacking fittings to pipe, remember that the joint will pull or shrink toward the weld as the weld cools, Figure 56–4.

PIPE ALIGNMENT

Proper alignment of piping is very important when welding. If done correctly, welding is easier to do and the piping system will function properly.

There are many ways to align piping. Most are done using framing squares, levels, and rules. The following procedures suggest methods of obtaining a good alignment quickly.

Pipe-to-Pipe Alignment

Move the pipe lengths together until the bevels nearly abut or touch, allowing a gap for the weld (1/16 inch), Figure 56–5. Center squares on top of both pipes and move the pipe up and down until

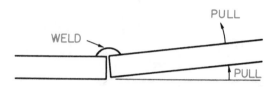

Figure 56-4 When the weld cools, the metal attempts to pull toward the welded side.

the squares are aligned. When the pipes are aligned, the numbers on the two squares will match up. Tack weld the top and bottom. Repeat the procedure by placing the squares on the side of the pipe. Correct the alignment by moving the pipe left or right. Tack weld each side.

90-Degree Elbow-to-Pipe Alignment

Place the fitting's bevel so it nearly touches the bevel of the pipe, Figure 56–6. Allow a gap for welding. Tack weld on top. Center one square on top of the pipe. Center the second square on the elbow's alternate face. Move the elbow until the squares are aligned.

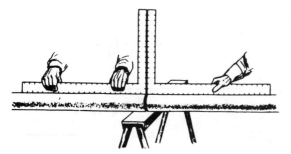

Figure 56-5 Pipe-to-pipe alignment
(Courtesy Tube Turns)

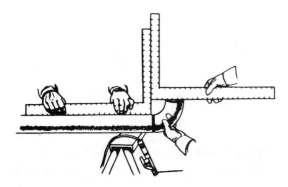

Figure 56-6 90° elbow-to-pipe alignment
(Courtesy Tube Turns)

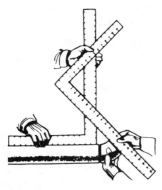

Figure 56-7 45° elbow-to-pipe alignment
(Courtesy Tube Turns)

45-Degree Elbow-to-Pipe Alignment

The procedure for aligning a 45-degree elbow to pipe is the same as aligning a 90-degree elbow to pipe except the squares are crossed, Figure 56–7. To obtain a correct 45-degree angle, move the squares until the same numbers on the inside scales are aligned.

Alternate method. Abut the fitting and the pipe allowing a gap for the weld. Center a spirit level on the pipe, Figure 56–8. Center a 45-degree spirit level on the face of the elbow. Move the elbow until the 45-degree bubble is centered.

Tee-to-Pipe Alignment

Abut the bevels, allowing a gap for the weld. Tack weld on top. Center a square on top of the pipe, Figure 56–9. Place the second square on the center of the branch outlet. Move the tee until the squares are aligned.

Alternate method. Abut the bevels. Place a square on the tee, Figure 56–10. Center a rule on top of the pipe. The blade of the square should be parallel with the pipe. Check this by measuring with the rule at several points along the pipe.

Flange-to-Pipe Alignment

Abut the flange to the pipe. Align the top two holes of the flange with a spirit level, Figure 56–11(A). Move the flange until the bubble is centered. Tack weld on top. Center the square on the face of the flange, Figure 56–11(B). Center

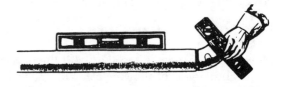

Figure 56-8 45° elbow-to-pipe alternate method *(Courtesy Tube Turns)*

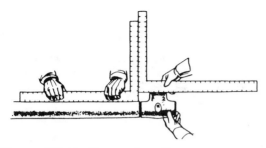

Figure 56–9 Tee-to-pipe alignment
(Courtesy Tube Turns)

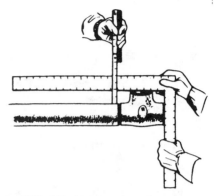

Figure 56–10 Tee-to-pipe alternate method
(Courtesy Tube Turns)

the rule on top of the pipe. Move the flange until the square and pipe are parallel. Tack weld on bottom. Center the square on the face of the flange. Center a rule on the side of the pipe and align as before. Tack both sides.

SUMMARY

Many plumbing apprenticeships have pipe welding as part of their curriculum now. With pipefitters, welding instruction is a must. The plumber will use welding skills to fabricate pipe and equipment hangers, to extend well casings above grade, and to fabricate fire, brine, steam, and water supply lines on larger projects.

There are hazards to the welding process. One is the possibility of receiving a "flash" from arc welding. This is similar to looking at the sun without any eye protection. The symptoms are burning eyes and the feeling that there is sand in the eye. This should be treated by a doctor. The prevention for this malady is to use the protective shield over the entire face when arc welding and to turn the face and eyes in another direction when someone nearby is arc welding. In gas welding, the danger to the eyes is less, but still is there. Gas welding goggles must be worn. Also when welding on galvanized metal, the zinc coating should be com-

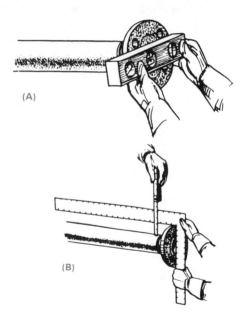

Figure 56–11 Flange-to-pipe alignment
(Courtesy Tube Turns)

pletely removed with a surface grinder beforehand. Burning zinc gives off smoke that can make you ill. After the welding is done, there are a number of protective coatings that can be applied to prevent rusting.

◆ ◆

TEST YOUR KNOWLEDGE

1. What is the common fuel gas used in gas welding?

2. Name the two basic joint types.

3. What is the most serious hazard when arc welding?

4. How could the welded joint be stronger than the pipe itself?

5. What is meant by the joint pulling when it cools?

6. Why must butt-welded pipe joints be beveled?

◆ ◆ ◆ ◆ ◆ ◆ ◆ ◆ ◆ ◆ ◆ ◆ ◆ ◆ ◆ ◆ ◆ ◆ ◆

Valves

KEY TERMS

backwater valve
butterfly valve
check valve
flush valve

gate valve
globe valve
plug cock
vacuum breaker

OBJECTIVES

After studying this unit, the student should be able to:

❏ Identify the kinds and sizes of globe, gate and check valves.
❏ Explain how these valves operate and where and how they are used.
❏ Describe the types and purpose of backwater valves.

Many types of valves are made for various purposes. They are made for water, steam, oil, gasoline, and other types of pipelines. They are also made to withstand certain temperatures and pressures.

GLOBE VALVES

Globe valves are used on water, air, gas, oil, or steam lines. Essentially, they are control valves.

They may be either fully opened or partially closed to regulate the flow.

Globe valves are made of brass, bronze, steel, or cast iron. Brass and bronze valves are made in 1/8- to 3-inch sizes. They can withstand pressures of 125, 150, 240, and 300 psi. Steel and cast-iron valves are made in 2 1/2- to 12-inch sizes and can withstand pressures of 125 psi. The 2 1/2- to 6-inch sizes can also withstand pressures of 1500 psi. Brass-trimmed iron body valves are made in 2- to 16-inch sizes.

411

A disc stops the flow in a globe valve. There are four types, three of the more common types are shown in Figure 57–1:

❏ The *conventional disc* closes against a beveled seat.

❏ The *composition disc* has a wide choice of disc material for hot or cold water, air, oil, or steam.

❏ The *needle valve type* is used for fine throttling control. It is usually placed on gasoline or oil lines.

❏ The *plug-type valve* has a broad contact between the plug and the seat. The plug and seat are made of a special alloy steel and may be used where conditions are severe, such as throttling steam.

The better valves have renewable seats, Figure 57–2. All these valves operate on the principle of the screw to raise or lower the disc.

Globe valves are made in straight, angle, wye, and radiator patterns. The bonnet screws into the body on some. Others have a yoke under the body or a loose nut. Larger valves are bolted. The friction in globe valves is about 60 times greater than in gate valves.

Globe valves may be joined to pipe by screwed, welded, sweated, or flanged joints. When making a screwed joint, the wrench is placed on the valve end into which the pipe is inserted. This prevents strain on the valve. On a welded or sweated joint, the stem is removed, or at least backed off, to avoid damage to the disc.

On water systems, the valve is installed so that the incoming pressure is under the seat. This allows the stem packing to be repaired without shutting down the entire system. A mechanic attempting to replace the packing could be burned if the valve is not installed this way.

On some steam systems, the valves are installed so the pressure is over the seat. This keeps the stem from contracting as it cools and prevents the valve from opening.

On systems where it may be necessary to completely drain the lines, valves on horizontal lines must be installed with the stems horizontal. On vertical globe valves, the seat forms a dam that may only allow the line to be partially drained.

Most globe valves may be repacked at the stem. This is done by removing the packing nut, gland, and old packing and then reassembling it with new packing. The nut should not be tightened too much as the stem may be scarred.

CONVENTIONAL

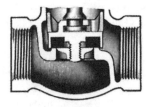
COMPOSITION

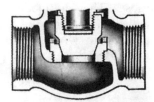
PLUG-TYPE

Figure 57–1 Globe valves

GATE VALVES

The gate valve, also called a *stop valve,* operates on the screw principle. The wedge-shaped gate moves up and down at right angles to the path of flow between two perpendicular rings. When seated against these rings, it shuts off the flow.

The gate valve causes no more friction than a straight pipe. It is used for liquid, pump, and main lines where maximum flow is required. Gate valves should not be used for throttling as they will chatter and vibrate. When throttled, the seat may be cut by wire drawing.

Gate valves are made with a rising or non-rising stem, Figure 57–3. The nonrising stem type has a left-hand thread upon which only the disc is raised or lowered.

Gate valves are also made with three types of gates. The solid-wedge gate with angle seat is recommended for steam, water, oil, air, or gas lines. It may be installed in any position, but must not be strained. Strain may change the fixed faces of the seat and cause leaks. The split-wedge and the double-disc gate valves must be installed in a vertical position. Otherwise, sediment may interfere with their operation. They may chatter if placed on lines having a high velocity.

Three types of bonnets are used on gate

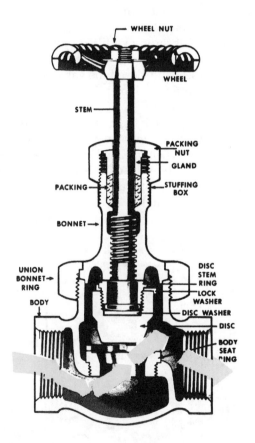

Figure 57–2 Plug-type disc globe valve

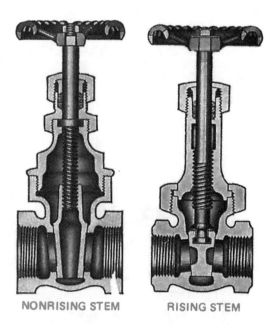

Figure 57–3 Types of gate valves

valves. The inside screw bonnet is used for low pressure. The union ring bonnet, Figure 57–4, reinforces to the body and is ideal for frequent inspection. The bolted bonnet is used on cast-iron and steel valves for high pressure.

Valves may be attached to pipe by screwed, sweated, flanged, or welded joints. Unions or flanges should be placed between the valves and equipment. The pipe should be well supported and allow for expansion to prevent sagging or strains.

Gate valves are made in bronze, ranging in sizes from 1/4 to 3 inches. Standard bronze valves are made to withstand 125, 150, 175, or 250 psi of steam pressure. However, the pressure recom-

mended varies inversely as the temperature rises. Larger sizes are made of cast iron and steel.

CHECK VALVES

Check valves are installed in pipelines to prevent fluids from flowing in the wrong direction. They are made in two basic types, swing check and lift check, each having several different patterns. Check valves are made of brass in the smaller sizes (1/8 to 2 inches) and cast iron in the larger sizes (3 to 6 inches).

In the *swing check valve*, Figure 57–5, the flow is straight through a tilted seat. When the flow stops or reverses, the swing disc falls against the seat and closed the valve. This valve causes very little friction and is used for low to moderate pressures.

Swing check valves are used on return-circulating systems, pumps, waterlines, and main house drains. When installed on return-circulating lines, they prevent cold water from being drawn from the last fixture.

The flow through a *lift check valve* is similar to that of a globe valve, Figure 57–6. It causes

Figure 57–4 Union ring bonnet

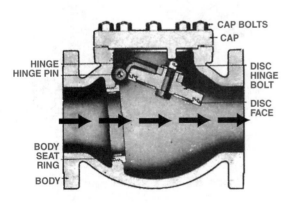

Figure 57–5 Swing check valve

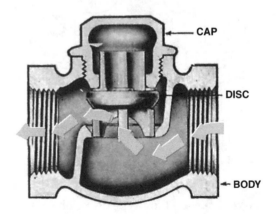

Figure 57–6 Lift check valve

more friction than the swing check valve. It is generally used on the same systems as the globe valves. It holds tighter and can withstand higher pressure. It is used for water, steam, air, gas, and vapor systems. Lift check valves are made in the same sizes as swing check valves.

Lift check valves are made in globe, angle, vertical, and horizontal patterns. Other patterns include the ball disc (a stainless steel ball) and the cushioned disc (a composition disc).

Lift check valves are used on mixing valves to prevent either hot or cold water from passing to the opposite line whenever there is a drop of pressure in one of the lines. They are also used to prevent polluted water from being siphoned back into the water system whenever there is a pressure failure.

Care must be taken when installing swing and lift check valves. They operate by gravity and will not work properly if installed at the wrong angle.

Other Check Valves

This is only a partial list of the types of check valves. Two other commonly used types are the *ball check valve* and the *vertical check valve*. Both are used on domestic water systems where a well is the water source.

OTHER VALVES

Cocks

Valves come in many kinds and varieties, but there are only a very few methodologies used to stop or control the flow of liquids or gases. The gate-type valve and the globe or compression-type valve have been shown. After the cock-type valve and a few of its variations are explained, the student should have a good

understanding of the way that valves work.

One of the earliest words for valve was *cock*. This term was used generically to describe valves. Another old term for valve is *bibb*. Some still call the faucet on the laundry tray, *laundry tray bibbs*. The current generic term is *stop* for valve. The most common design for a valve was the **plug cock**, Figure 57–7. It is easy to see that the design is very simple. This can actually be whittled out of wood and used to stopper water, wine, and whiskey casks, and often was. The design is so simple, sturdy, and effective that it remains with us to this day. The example shown is the common gas cock. Another familiar term is *petcock*. This is a cock-type valve which is applied to small or slow draining jobs. The inside design is much the same. Unfortunately when we try to get nice neat naming conventions for designs, we nearly always begin to run into exceptions in the plumbing and pipe fitting trades. The *ball cock* is an example. This is the valve which refills the water closet tank after it is flushed. It is actually not a cock.

The tapered / cylindrical plug is slipped into a matching hole. Since there is a hole drilled through the plug all that one has to do to turn on

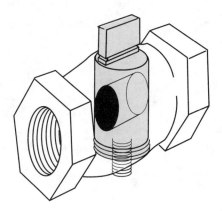

Figure 57–7 A plug cock

the flow is to turn this plug in the valve body until the hole in the plug lines up with the holes in the valve casting. This amounts to a 90-degree turn of the plug in either direction from full on to full off. If the old wooden cock started to leak, it was easily fixed by tapping the tapered plug more deeply into it tapered socket. With its modern counterpart, the nut on the bottom, Figure 57–7, can be tightened to accomplish the same effect.

The ball valve is a modern design using the same principle. In this case, it is a ball with a hole drilled through it rather than a tapered plug with a hole drilled through it. See Figure 57–8. These valves are used extensively in new water supply work, and rightfully so. These valves restrict the flow of water about 1/10 as much as the old compression valve used in the past.

The butterfly valve uses a flat disk rather than a ball or a plug to accomplish the same task. This valve, Figure 57–9, has a locking mechanism in the handle so that the flow within the valve will not cause it to swing shut or open when that is not the intention of the operator.

Figure 57–8 A ball valve (*Courtesy Tylok International – C.B. Crawford, Inventor/Founder*)

Flush Valves

The flush valve or "flushometer" is designed to replace the flush tank in urinals and toilets in commercial applications. In use, it forces 4 gallons of water into the fixture bowl under pressure. This has the advantage of aiding the self-cleaning features of the fixture which it serves. Since there is no tank to fill up before the fixture can be reflushed, it is ready instantly to be flushed again. It would work in residences also but few residences have the amount of water necessary for the flush valve to be effective. See Figure 57–10A and Figure 57–10B.

This valve is mounted directly to the bowl of the fixture it serves. If the fixture becomes blocked the contents of the bowl would cover and possibly enter the outlet pipe to this device. If precautions are not taken, a reduction in supply pressure could result in waste water being drawn up into the potable water supply system. For this reason a vacuum breaker is installed at the outlet of the flush valve to prevent *back siphonage*. The vacuum breaker has a rubber device which will allow the passage of water into the fixture but which will close up and refuse to allow water to run back into the valve.

Flush valves are remarkably trouble free, but occasionally flush valves refuse to shut off. Often a sharp rap with a plastic mallet will solve the problem. Shutting the control valve off (shown on the right side of the cutaway diagram in Figure 57–10), rapping the valve as described, and then turning the control valve back on will often work with stubborn cases.

BACKWATER VALVES

Backwater valves are check valves placed in drain-pipes to prevent sewage from flowing

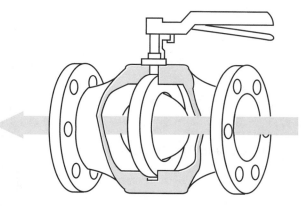

Figure 57–9 A "butterfly" valve

backwards into buildings. Backwater valves must be installed where fixtures are placed in a basement, where sewage is pumped up to a house drain, or where buildings are near rivers.

The two types of backwater valves are the swing and the balance type.

The *swing-type backwater valve,* Figure 57–11, has a light, brass disc that is hinged at the top and closes against the brass seat. The disadvantage of the swing type is that it is closed, except when water is flowing in the right direction. When it is closed for these long periods, the drainage system is not being ventilated.

The *balance-type backwater valve,* Figure 57–12, has a disc on one end of an arm that is balanced in the center. An adjustable weight is placed on the opposite end to keep the valve open under normal use. Check with the local code before installing this valve.

In case of a stoppage in the drain or a flood, the water backing up closes the valve. Both valves are equipped with a large cleanout. A gate valve may be used on the drain as an emergency shutoff.

SUMMARY

Valves are devices that stop or control the flow. Valves have been around in many forms as long as plumbing has. They are known by many names, including stops, ball cocks, plug cocks, petcocks, bibbs or bibs, paves, bungs and, of course, valves.

The simplest valve was probably a small bank of soil that a primitive farmer used to control the flow of water between the rows of vegetables. The farmer used a shovel to open a channel and then used the same tool to close it. Next probably came the bung, which is no more than a tapered wooden plug driven into a hole in a wine barrel top. In time, someone figured out that less wine would be lost if a hole from end to end was drilled through the bung and another, smaller bung with a hole from side to side was inserted through that, thus producing the first plug cock.

There are valves that control or stop the flow automatically. These range from the simple swing check, which allows flow to travel in one

Figure 57–10A *(Courtesy Sloan Valve Company)*

DIAPHRAGM TYPE

Sloan ROYAL Quiet - Flush II Flush Valve

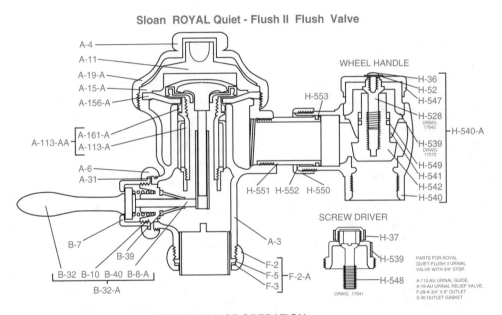

DESCRIPTION OF OPERATION

The Segment Diaphragm (A-156-A) separates the valve into an upper and lower chamber with the pressure the same on both sides of the diaphragm, equalized through the bypass in the diaphragm. The greater pressure area on top of the diaphragm holds the valve close on the seat. The slightest touch of handle grip (B-32) in any direction pushes the plunger (B-8-A) inward, which tilts relief valve (A-19-A), releasing the pressure in the upper chamber. Then the pressure below raises the diaphragm and working parts as a unit, allowing the water which flushes the bowl to go down through the barrel of the valve. While this is occurring, a small amount travels up through the bypass, gradually filling the upper chamber to close the valve.

Figure 57–10B Sloan ROYAL Quiet-Flush II Flush Valve *(Courtesy Sloan Valve Company)*

direction only, to huge dam gate valves using electric motors to turn a threaded valve stem, to the solenoid valve, which is controlled electronically by a thermostat.

There are some general usage rules to the installation of valves. Check- and globe-type valves can be installed backwards (incorrectly). Always look for the arrow on the valve casting before installing. The arrow points in the direction of the flow. Swing checks must be installed right-side up, or the gate, which depends on gravity, will not operate.

The brass-bodied gate valve must be used in an application requiring either a fully open or a fully closed valve. The gate wobbles in the partially open or closed positions and it will rattle against the seat. This ruins both the seat and the gate. Some control valves will be ruined by the smallest amounts of grit in the flow and must be installed with a strainer upstream.

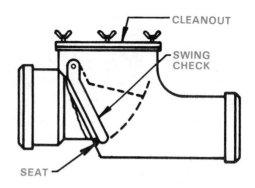

Figure 57-11 Swing-type backwater valve

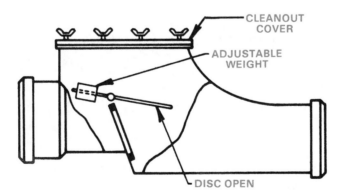

Figure 57-12 Balance-type backwater valve

◆ ◆

TEST YOUR KNOWLEDGE

1. What is the name of the globe valve part that closes off the flow?

2. What is the purpose of a check valve?

3. What operates the swing and lift check valves?

4. Which valve must be used in the fully open or fully closed position?

5. What kinds of valves present the least resistance to flow?

6. What is the purpose of a vacuum breaker?

7. Where would you find a flush valve?

8. What is a backwater valve and where would it be installed?

BRITISH THERMAL UNIT AND THE EXPANSION OF WATER

OBJECTIVES

After studying this unit, the student should be able to:

❑ Describe the British thermal unit and how it relates to plumbing.
❑ Discus the expansion of water and the pressures it produces

THE BRITISH THERMAL UNIT

Heat may be measured in degrees or in British thermal units. A **British thermal unit (Btu)** is the quantity of heat necessary to raise the temperature of 1 pound of water 1 degree. A pint of water is approximately 1 pound. Therefore, 1 Btu will increase a pint of water from 54° to 55°F. Likewise, 10 Btu will raise 1 pound of water 10 degrees and 25 Btu will raise 25 pounds of water 1 degree.

To estimate the quantity of heat in Btu required to raise water a certain number of degrees, multiply the number of pounds of water by the number of degrees it is to be raised.

Example:

Find the number of Btu necessary to heat 100 pounds of water 40 degrees.

Solution:

$$100 \text{ lb} \times 40° = 4000 \text{ Btu}$$

421

If the quantity of water is given in gallons, multiply by 8.33 pounds (weight of a gallon). If the quantity is given in cubic feet, multiply by 62.5 pounds (weight of a cubic foot). If the two temperatures are given, find the temperature difference by subtracting.

Example:

Find the number of Btu necessary to raise 30 gallons of water from 40 degrees to 160 degrees.

Solution:

8.33 lb × 30 gal = 249.9 lb
160° − 40° = 120° temperature difference
249.9 lb × 120° = 29,988 Btu

EXPANSION OF WATER

The expansion of water differs from that of metals. Metals expand as long as heat is applied and contract as long as heat is extracted. Water, however, has a point of maximum density. At 39.2°F, water reaches a point where it can no longer contract. Water expands when heated above this temperature. It also expands when cooled below 39.2°F. Water expands about 1/26 of its volume when heated to the boiling point, Figure 58–1.

One cubic inch of water expanded to steam will fill a cubic foot of space, expanding 1728 times its original volume. Water converted to steam will drive a locomotive. Ice exerts a pressure of 33,000 pounds per square inch. Water converted to ice will most likely break a pipe.

The weight of a cubic foot of water becomes lighter as it is heated or cooled below 39.2°F, Table 58–1.

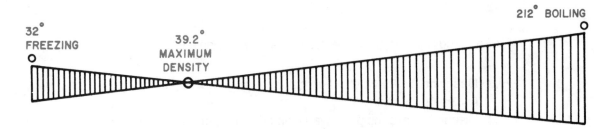

Figure 58–1

Ice	at	32°	weighs	62.418	lb.	per	cubic	foot
Water	at	39.2°	weighs	62.425	lb.	per	cubic	foot
Water	at	60°	weighs	62.372	lb.	per	cubic	foot
Water	at	160°	weighs	60.991	lb.	per	cubic	foot
Water	at	212°	weighs	59.760	lb.	per	cubic	foot

Table 58–1

Note:
Realizing the tremendous pressure exerted by expanding water, reliable safety valves must be placed on all water heating apparatus.

CHANGES OF STATE

Most materials exist in the physical world in three states. The three states are: solid, liquid, and gas. At the temperatures found normally on the earth's surface, that is between –80° to +135°F, most materials remain in one state. Metal remains a solid. Gasoline remains a liquid. Air and natural gas remain gaseous. Water, however, is a compound that can exist in two states at these temperatures. Because the plumber must deal with water in all of its states, the plumber must be familiar with the changes that take place in water. Water, when it is heated above 212°F, boils and becomes a gas. The factor that causes these changes in all materials is heat.

As has already been shown, it takes 1 Btu to increase the temperature of 1 pound of water 1°F. To raise the temperature of 1 pound of water from 200° to 212° would require only 12 Btu. But to change the temperature of 1 pound of liquid water at 212° to 1 pound of gaseous water at 213°, a change of only 1 degree, would take 971 Btu. This is because 970 Btu are used up simply to change the state of the water from a liquid to a gas. In the same fashion, it takes the loss of 144 Btu to change 1 pound of water at 32° to 1 pound of ice at 32°.

When making steam in a boiler for space heating, this must be taken into account. The heat that is absorbed or given off when a compound changes its state is used in many heating and refrigerating applications.

COMMON USES OF THE BRITISH THERMAL UNIT

The Btu calculation is used when the size of any heat-using equipment needs to be determined. This includes refrigeration and air-conditioning equipment, as well as strictly heating appliances. The following examples may give the student a better "feel" for the use of the British thermal unit.

Example:

A freezer has a heat loss of 12,000 Btuh (Btu per hour).

"Freezer" means: the equipment must cause a temperature *drop* or *difference* within the space.
"Btuh" means: the Btus removed must be removed on an hourly basis.
"Has a heat loss of" means: the amount of Btus have already been calculated, allowing for the (1) thickness and the thermal transmission rating of the insulating materials in the walls, ceilings, floors, doors, and windows, (2) the temperature differences: the differences between the inside and outside temperatures at all of the preceding points, and (3) the number given, 12,000 Btu, represents the amount of heat which has to be removed *each hour* to maintain the inside temperature of the freezer, perhaps 20°F.

Considerations:

If the equipment supplied for this job is rated at 12,000 Btuh, will the equipment have to run continuously to maintain an inside temperature of 20 degrees?

This needs to be checked by reading the documentation for the proposed unit. Is this the maximum output of the equipment considered or is it the output which can be expected when the equipment runs X percent of the time? It may be that this is the maximum capability of the equipment. If this is the case and the equipment needs to also bring down the temperatures of, let's say, meats and produce that the customer may place in it from time to time, one may wish to buy equipment with double the heat removing capability, 24,000 Btuh. Now it can handle new additions of meats and produce, and when everything is down to designed temperature, the freezer compressor will run only 50 percent of the time. A much healthier *duty cycle* for the compressor.

Example:

A brand of 80-gallon gas-fired water heater has an advertised *recovery rate* of one hour. The customer wants to use it for her restaurant to wash dishes. She must wash dishes every two hours at the minimum, and this will use all of the 80 gallons of hot water each time. Will this appliance do the job?

"Recovery rate of one hour" means: if the hot water has all been used and is now running cold from the water heater, in one hour the water in the heater will once again be at 130°F, heated from 55° ground water temperature. This is normal user temperature.

"Restaurant to wash dishes" means: the water delivered to some commercial dishwashers must be to 190° not 130°. Assume that this dishwasher has this requirement.

Solution:

Since the only information that is known about this water heater is that it has an 80-gallon capacity and that its recovery rate is one hour, this information will be used to calculate its Btuh, (Btu per hour), firing rate. From this it can also be determined if it can recover to 190° in two hours.

Pounds: The British thermal unit is the heat required to raise one pound of water one degree Fahrenheit. Units of gallons, not pounds, are needed, so convert gallons to pounds. One gallon of water weighs 8.34 pounds. So 80 gallons would weigh 80×8.34 or 667.2 pounds.

Heat: The amount in degrees of temperature which the appliance is raising the 80 gallons to is: 130° user temperature, minus 55°, ground water temperature, is equal to 75°, i.e. $130 - 55 = 75$.

Btuh: The advertised recovery rate is accomplished in one hour and the number of Btus desired is rated for one hour so no conversion is needed here. So 667.2 pounds, the weight of the water, times 75 degrees, the temperature rise, is equal to 50,040 Btu, i.e. $667.2 \times 75 = 50,040$. The main burner of the water heater will deliver 50,040 Btu per hour.

Will the water heater raise the temperature to 190° in two hours?: The heater will consume 50,040 Btu in one hour, therefore it will use $2 \times 50,040 = 100,080$ Btu in two hours. The weight of the water remains the same, 667.2 pounds, but the temperature difference is not the same. It was $130 - 55 = 75$; now it will be $190 - 55 = 135°$. The question that needs to be answered so that 100,080 can be compared with the answer is: How many Btus are required to raise 667.2 pounds of water 135 degrees? $667.2 \times 135 = 90,072$. The answer is *yes*. This water heater needs 90,072 Btus to do the job and has 100,082 available and is capable of doing the task set for it.

SUMMARY

Water is an interesting substance that defines some generalities of science. One of those is that the colder an object becomes, the smaller it gets. And, if you heat something, it swells as heat is added. With water it depends. Water expands when cooled below 39.2°F until it becomes ice at 32°. When cooled below that temperature, it behaves like other substances—it shrinks. Water exists naturally in three states: gas, water, and liquid.

The British thermal unit (Btu) is an invention that represents a quantity of heat. The degree of heat as in "it's 90° out there," is a measure of the intensity of that heat. A burning match may produce a flame with a temperature of 2000°. If you had a container with water in it and a thermometer, and they were all at room temperature and all together weighed one pound, you may not be able to bring the temperature on the thermometer up one degree if you held a match under it. This means, that a match burning at 2000° did not apply enough of the quantity of heat necessary to add up to 1 Btu. If you continued lighting matches and holding them beneath the apparatus, perhaps the temperature would go up one degree. When that happens, you have applied 1 Btu. Instead of using those very hot matches you might soak a towel in hot tap water and wrap the test apparatus with it. After some time, the temperature on the thermometer will rise one degree. If you take the wet towel away, you have added 1 Btu with a towel, which was originally at only perhaps 120°F. The British thermal unit is a measurement of quantity, not of intensity. We start the experiment with everything at room temperature so that the air in the room itself has no effect on our experiment.

There is a certain number of Btu that must be added to or subtracted from a substance to change its state. If ice becomes water, that is a change of state and 970 Btu will have been absorbed to accomplish this. This is a useful fact of physics that makes efficient refrigeration and air-conditioning possible.

◆ ◆

TEST YOUR KNOWLEDGE

1. The British thermal unit is a measurement of heat, but what precisely is it measuring?

2. If one Btu is applied to an object and its temperature rises by 1°F, how much does it weigh?

3. If you have a one-pound can of water with a thermometer in it, and you place a burning candle under the can, is it possible to rate the candle's heat production in Btuh (Btu per hour)?

4. What will happen if a steel pipe is filled with water and then frozen solid?

5. At what temperature is a full cup of water the heaviest?

6. If contained, how much pressure is exerted by freezing water?

CONDUCTION, CONVECTION, AND RADIATION

◆ ◆

KEY TERMS

conduction
convection
radiation

OBJECTIVES

After studying this unit, the student should be able to:

❏ Discuss conduction, convection, and radiation and their importance in the plumbing trade.

CONDUCTION

Conduction is the passage or transfer of heat from molecule to molecule, from the hottest to the coldest region of a substance. Every substance is made up of molecules that are in constant motion. When heat is added, molecules move at a faster rate. When a bar of iron is heated at one end, Figure 59–1, the vibrations of the molecules increase. These vibrations are passed from molecule to molecule along the bar until the opposite end is hot.

The amount of heat lost is computed by the conductivity of the material. Figure 59–2 shows how heat in a room is lost through the ceiling, windows, and walls. Figure 59–3 shows how many Btu may be lost through a brick wall.

The rate at which a material conducts heat is expressed in the amount of British thermal units passing through 1 square foot of material 1 inch thick in 1 hour, with 1 degree temperature difference. Metal and glass have the highest ratings. Concrete, brick, and stone have a medium rating. Wood and the light, fluffy substances

427

Figure 59–1 Conduction through bar

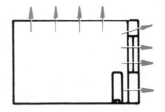

Figure 59–2 Heat loss in a room

used for insulation have a low rating.

Copper and brass are used for heating coils because of their high rating, 240 Btu.

CONVECTION

If part of a fluid body (liquid or gas) is heated, it expands and becomes lighter than the other parts. Since the molecules are free to move, the cooler, heavier part of the substance sinks to the bottom and pushes the hotter part of the substance to the top. Thus, ascending and descending currents are set in motion whenever any part of the liquid or gas at the bottom is heated to a hotter temperature than the rest. Convection, or circulation, is the circular movement caused by a difference in weight as a result of a temperature change.

If a full pan of cold water is heated, some of the water will run over the sides due to expansion. The pan will still be full of hot water, but lighter in weight. A cubic foot of water at 39.2° weighs 62.425 pounds, while at 160° it weighs 60.9 pounds, or 1.5 pounds less.

If a cubic foot of water at these temperatures is placed on opposite sides of a balance scale, the cold water will sink and raise the cubic foot of hot water. This takes place in all bodies of water when heat is applied.

Circulation between a heater and hot water tank takes place in the circulating pipes. The hot water travels up the flow (top pipe), and the cold water travels down the return pipe (bottom), Figure 59–4. All circulating pipes should

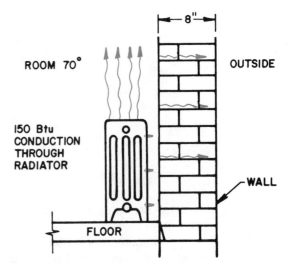

Figure 59–3 Heat loss through brick wall (3 Btu)

pitch in the direction that the water naturally travels to assist the flow.

Convection currents are also set in motion in air. In a fireplace, for instance, the air passing through the fire is heated and rises up the chimney, carrying the smoke and creating a draft. Convection can apply to a circular movement caused by differences in temperature in a body of either water or air.

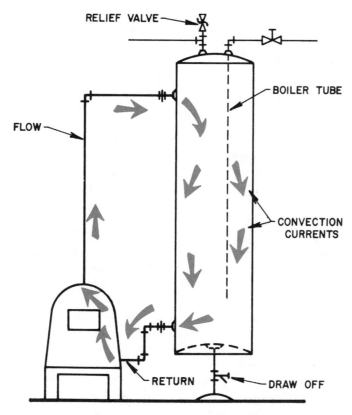

Figure 59–4 Convection currents in an old-style water heater

RADIATION

Radiation is the passage of heat through space. The distance may be great or small. The heat absorbed is inversely proportional to the distance between the two objects. The heat absorbed from a radiator 5 feet away is $5 \times 5 = 25$, or 1/25 of the heat given off.

Heat from the sun, 93 million miles away, is absorbed by the Earth. This heat travels with the infrared light in the atmosphere, not by the air itself. A 3/4-inch space of still air is a good insulator. Heat can travel through a vacuum; otherwise, we would receive no heat from the sun.

Heat loss by radiation is continually taking place wherever different temperatures exist. The sun heats the Earth by day, and the Earth's heat is given up to the air and objects at night. Heat produced in buildings in winter is lost to the outside air by conduction through the walls, glass, and roof. A dull, rough surface absorbs and radiates heat to a greater extent than a bright and smooth surface. This is because bright, shiny surfaces reflect heat. Aluminum foil is used for insulation because it reflects, or turns back, heat.

If a black cloth and a white cloth are placed on snow, the heat of the sun is absorbed to a greater extent by the black cloth and causes the snow to melt faster in this place.

SUMMARY

Any system tends to have heat equilibrium. The universe seeks heat equilibrium. Our solar system seeks heat equilibrium. Earth does also and so does every system on it. A building is a system. This means that, if it were possible for heat to have a desire, heat would want everything to be the same temperature. In order to do this, heat travels to colder areas in the system in an attempt to equalize all temperature. If the inside of a home is warmer than the outside, the heat travels through the walls to the outside. If the inside is cooler, then the outside heat tries to warm the interior of the home.

In order for heat to travel toward the colder area, it uses one or more of three modes of heat transference: conduction, convection, or radiation. The purpose of insulation in homes, refrigerators, and vacuum bottles is to reduce the rate at which heat flows.

Heat transference by conduction is the major method in a solid, but to some extent conduction occurs in liquids and gases as well. Think of conduction as the transference of heat by objects in contact. In fluids, like liquids and gases, the objects in contact are molecules. Notice that air and other gases are also fluids.

When fluids are heated, the individual molecules tend to move. The warmer and therefore less dense molecules tend to rise, and the cooler molecules must travel to fill in the space vacated by the warmer molecules. This is convection—the second method of heat transference. But notice that convection is a kind of auxiliary method, in that the movement of the molecules aids heat transference by conduction—our first method.

All substances emit light when heated. Light is another form of energy. The light produced by a cast-iron radiator is invisible and in the infrared color range, which is invisible to humans. The light energy travels at 186,000 miles per second until it strikes something solid to absorb its energy and convert it back into heat. As long as it strikes nothing, it produces no heat. This is a useful property. In meat refrigerators an infrared heater can be used to keep a worker warm without appreciably raising the temperature of the refrigerated room. Some air molecules are struck by the light and heated in the process, but in the world of atoms there is a surprising amount of space between objects for the light to pass through without touching anything but the worker who may be seated at a desk keeping records.

TEST YOUR KNOWLEDGE

1. In what direction does heat travel?

2. Within an iron bar that is being heated what is the means by which the heat travels to the cooler end?

3. Why is the coolest water near the bottom of a pond in the summer?

4. On a frosty still day, you may feel the sun's rays against your face. How is this heat being transferred to your body?

5. Does heat travel between the rooms of a building?

6. How is the transference of heat recorded?

INDIRECT HEATERS AND STEAM BOILERS

KEY TERM

indirect heater

OBJECTIVES

After studying this unit, the student should be able to:

❑ Describe the indirect heater and steam boiler.

INDIRECT HEATERS AND STEAM BOILERS

Another method used to heat water is the indirect heater. This appliance consists of a copper coil within a cast-iron jacket. It is piped so that it is below the waterline of the steam boiler. In this way, the hot water in the steam boiler circulates through the indirect heater and around the copper coil, Figure 60–1.

A storage tank is used with the boiler and the indirect heater, Figure 60–2. The flow and return from the copper coil is run to the storage tank. The water from the storage tank flows up through the coil and back to the tank. This is similar to the summer-winter hookup.

In the indirect heater, the outlet is at the bottom of the steam boiler. The inlet is run to the boiler outlets 2 inches below the waterline. In larger systems, all boiler sections are yoked together as in Figure 60–2. A single connection may be made for the return. In Figure 60–2, it should be noted that the four 1-inch connections are equal to one 2-inch connection. It is advisable to place gate valves between the heater and the boiler so that repairs may be made.

The bottom of the storage tank should be at least 12 inches above the heater coil for gravity circulation. The upper connection on the tank hooks to the upper connection of the heater coil.

433

The lower outlet of the tank hooks to the lower connection of the heater coil.

All water-heating equipment must have a safety valve. Valves should never be placed between the heater and the tank.

The indirect heater may also be used with direct steam, Figure 60–3. The indirect heater, in this case, is placed above the waterline of the boiler. The steam inlet is at the top; the return is taken from the bottom. A dirt pocket prevents dirt from entering the steam trap. Steam traps should be located at least 5 feet away to allow the steam condensation to cool before entering the trap. The trap outlet is connected to a vacuum return.

INDIRECT HEATERS AND HOT-WATER BOILERS

Indirect heaters may also be used with hot-water heating systems. While steam systems transfer heat with more efficiency than hot water, steam boilers are more elaborate and costly. For residential use, the hot-water boiler is used much more frequently. Figure 60–4 shows a modern indirect heater used with a hot-water boiler. To assist the flow of hot water from a boiler to a floor-standing indirect heater, a circulating pump is used, Figure 60–5.

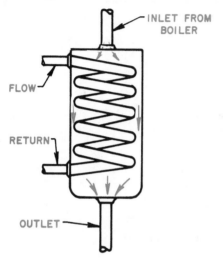

Figure 60–1 Indirect heater

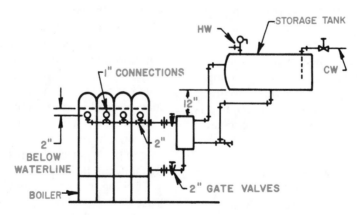

Figure 60–2 Connections to steam boiler

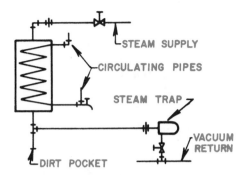

Figure 60-3 Direct steam

SUMMARY

Indirect heaters use the heating medium from a primary heating device to heat their own contents. A summer-winter hookup is an example of an indirect heater. In order to allow the heating coils to be filled with antifreezing solution, a solar water heater may use indirect heating methods. In some applications, coils may be placed in or around flue stacks to use waste heat to heat domestic water. Steam boilers are a common source of heat for indirect heaters.

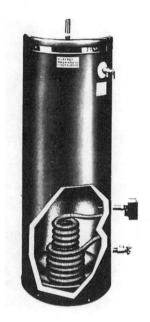

Figure 60-4 Indirect heater

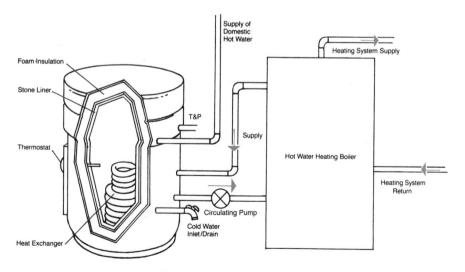

Figure 60-5 Using a circulating pump

TEST YOUR KNOWLEDGE

1. What features or characteristics give an indirect heater its name?

2. What is the usual first destination for the hot water of an indirect heater?

3. How should the storage tank be placed to ensure natural circulation?

4. Why should the steam trap be at least 5 feet away from the indirect heater jacket?

SECTION 7

(Courtesy Kohler Company)

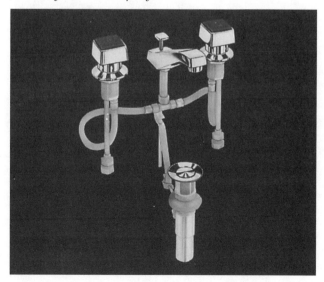

◆ ◆ ◆ ◆ ◆ ◆ ◆ ◆ ◆ ◆ ◆ ◆ ◆ ◆ ◆ ◆ ◆ ◆

FIXTURE
INSTALLATION

KEY TERMS

centersets
stub pieces
trip lever assembly
waste and overflow

OBJECTIVES

After studying this unit, the student should be able to:

❑ Rough-in and install fixtures following the manufacturer's instructions.

FIXTURE INSTALLATION

An important part of many plumbing jobs is the installation of the terminal equipment. These fixtures include toilets, lavatories, sinks, bathtubs, water softeners, air-conditioning units, water pumps, or heating equipment.

Fixtures are constantly changing in design. The plumber often has to rough-in and install equipment that is unfamiliar. The plumber can do the job properly by following the manufacturer's rough-in information provided with each fixture.

A neat and clean fixture installation depends on an accurate rough-in. This involves locating and installing all pipes connected to the fixtures, Figure 61–1. A careful examination of the manufacturer's information sheet will save time and material during rough-in and installation.

439

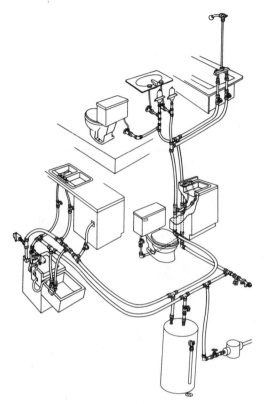

Figure 61-1 *(Courtesy Genova Products Inc.)*

Fixtures and other equipment are often delicate and easily damaged. They should be protected after the installation is complete. There are a number of preparations on the market designed to protect fixture surfaces. Equipment can often be protected with its own packing material.

Pipe openings must be sealed after rough-in to prevent the entry of building residue. Pipe sticking out from walls and floors should be capped or plugged. Strainers should be removed and a piece of paper placed under them to prevent the entry of plaster and other building materials.

ROUGH-IN

The plumber must know wall and floor thicknesses before the fixtures are roughed in. Much of the information given on rough-in sheets is from the finished wall or floor. The plumber must allow for this distance. Backing boards and pipe support must be provided before subfloors and wall coverings are in place. **Stub pieces,** or capped pipe nipples, should penetrate the walls and floors. These help locate supply and drain lines after the floor and wall coverings are in place, Figure 61–2.

Often, the plumber is just given approximate locations for fixtures. A good deal of judgment must be exercised in this case. Fixtures along a wall look best if the spaces between them are equal. However, allowable trap-arm lengths, an existing medicine cabinet, windows, and the swing of the door must be considered. There must be access to the bathtub plumbing and enough clearance to make the fixtures usable.

It is very important that wall thickness and precise fixture locations be known with some precision before the roughing-in is started. Figure 61–3 gives the *minimum* allowable clearances for some common bathroom fixtures.

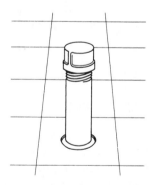

Figure 61-2 A stub piece

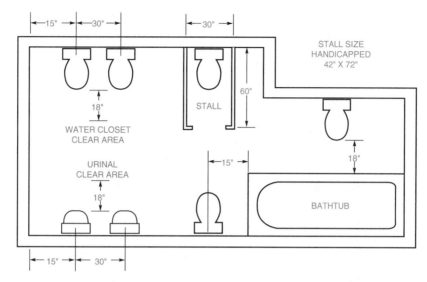

Figure 61–3 Minimum fixture clearances (*Courtesy Universal Rundle Corp.*)

The student should take note that:

1. If this is a commercial building, one of the toilet enclosures will most likely use the measurements based on handicapped usage.
2. The side clearance for a toilet when it is located next to a wall or a low fixture like a bathtub may be closer (15") than when it is next to a urinal or another toilet (30").
3. Adequate frontal clearance for knee room and standing room must be provided for toilets and urinals.

TOILETS

The toilet or water closet comes in a variety of styles. Wall-hung types require a special carrier behind the wall covering to hold the fixture weight. Some flush-tank toilets attach to the wall. These require careful alignment between the tank and the bowl in order to avoid leaks in the connecting flush ell.

There are also two-piece and one-piece close-coupled toilers. Floor-mounted toilets are roughed-in 10, 12, or 14 inches from the finished wall, Figure 61–4. The most common rough-in measurement is 12 inches. Assembly of the tank to the bowl must be done carefully to avoid breaking the fixture. Toilets are set on wax to the closet flange. Putty is not acceptable for setting the toilet bowl as it will dry out and shrink in time.

LAVATORIES

Lavatories or basins must be level and the proper floor-rim height above the floor, Figure 61–5. A good quality, properly installed backing board or wall carrier is important.

Consult the manufacturer's rough-in sheet to determine the distance from the floor to the centerline of the screws in the hanger bracket, Figure 61–6. Add the thickness of the finished floor. Using a 2-foot level, draw a horizontal line on the tile. Hold the hanger in place. Mark the screw holes. Drill the holes and set the hanger carefully. If a P trap is used the centerline of the basin should be directly over the drain in the wall, Figure 61–7.

Lavatories may also be of the vanette or "vanity" type. These are bowls which are set into a cabinet top. The installation of the fixture varies little from the wall hung lavatory with

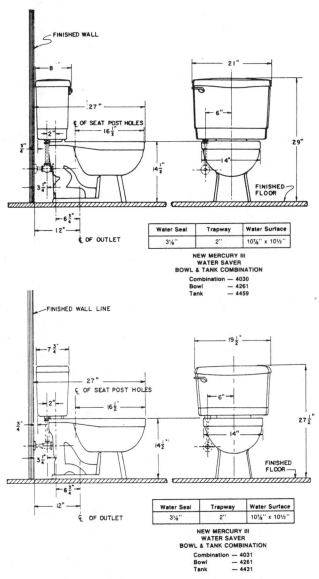

Figure 61–4 Roughing-in dimensions

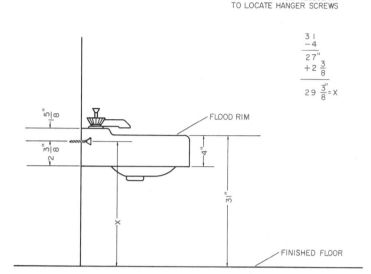

Figure 61-5 Placing the lavatory

the exception that a wall hanger is not installed. The vanette cabinet is open at the back which provides for easy installation of the P trap and wall supplies. A board runs across the back of the cabinet. This should be fastened securely with wood screws to the underlying studs in the partition. It is tempting to depend on the weight of the cabinet and pipe connections to hold the cabinet in place, but this is a mistake and will eventually result in leaks from frequently disturbed pipe connections.

The vanette has a plastic "marble" or a composite top. The bowl for the lavatory may be cast right into this top. In these cases the holes for the trim and faucets are pre-cut. Other tops, especially with custom cabinetry, will have to have the holes cut into them for the bowl and also the chromed trim. The bowls purchased to go into this top will be either rimmed or rimless. With the rimmed bowl, the hole cut must be larger than the maximum dimensions of the bowl. A plated or stainless rim is then fastened into the hole and the basin is mounted to this. With the rimless bowl, the hole cut through the top is smaller than the maximum dimensions of the bowl. The bowl in these cases is puttied into place and fastened from the bottom.

The vanette top is very expensive. Great care should be exercised cot cut this hole accurately. The bowl must be on hand to be carefully measured before the hole is cut. If there is one on the job, it is better to have a finish carpenter or cabinet maker cut the hole. Remember that the blade of a hand jig saw cuts on the *up* stroke. If a coarse blade is used, or the blade is forced through the work, the thin lamination of the top surface may pull loose from the flakeboard base. Also, if too much downward pressure is put on the saw as it is moved around the cut, the worker may be dismayed to discover scratches in the counter top outside of the area to be covered by the basin rim. It is best to tape the top surface of the counter top, cut right through the masking tape, and the remove the tape. To cut the faucet and trim holes, use a hole saw rather than a spade type bit, for the same reasons.

BATHTUBS

Bathtubs come in a variety of shapes and sizes,

Figure 61–8 and Figure 61–9. They are made of cast iron, steel, fiberglass, or even wood. In most cases bathtubs are built in. They are installed against the studs in the wall rather than the finished wall.

The centerline of the drain hole is not usually in the center of the width of the tub, Figure 61–10.

A left-hand tub has the waste opening on

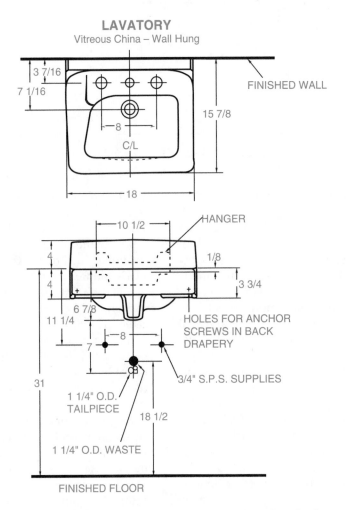

Figure 61–6 Lavatory. (*Courtesy American Standard*)

the left side when the viewer is standing at the skirted, long side of the tub. The drain line from the tub never comes up under the hole in the

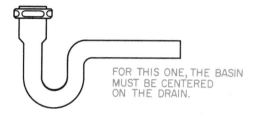

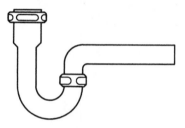

FOR THIS ONE, THE BASIN MUST BE CENTERED ON THE DRAIN.

Figure 61-7 Two kinds of P traps

bottom of the bathtub. Use the rough-in sheet to locate the drain line.

The assembly that goes directly against the bathtub is called the waste and overflow. It may also be called a trip lever assembly. This assembly must be obtained separately, Figure 61–11.

FAUCETS

Sink and basin faucets are often called centersets, Figure 61–12. When installing, seal the faucet to the fixture with putty if a rubber gasket is not included. Sinks and lavatories (basins) installed in countertops must also be sealed to prevent water from leaking under the fixture, Figure 61–12 and Figure 61–13.

SUMMARY

The installation of fixtures is determined by the rough-in. The plumber plans the roughing-in with the finished installation firmly in mind.

Figure 61-8 *(Courtesy Universal Rundle Corp.)*

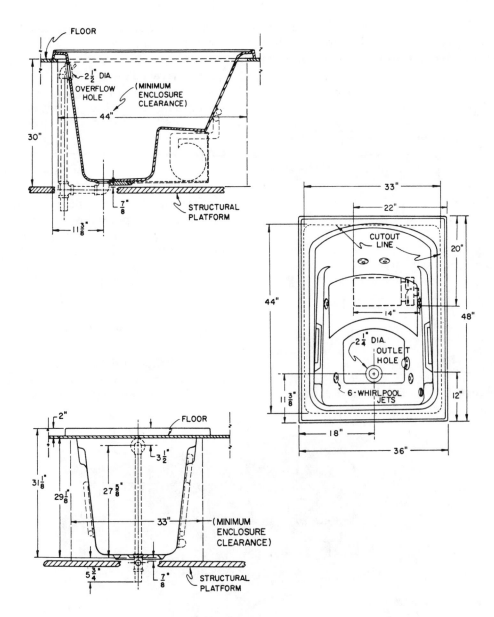

Figure 61–9

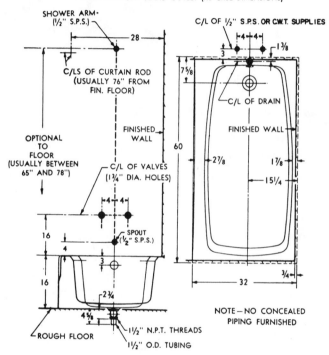

SPECTRA BATH

ENAMELED CAST IRON — RECESS
SHOWN WITH
1303. SER. B/S SUPPLY FITTING &
1560. THRU 1562. SER. C.D.&O.

2605.SER.
2607.SER.

2607. SER. LEFT HAND OUTLET (SHOWN)
2605. SER. RIGHT HAND OUTLET (REVERSE DIMENSIONS)

NOTE — NO CONCEALED
PIPING FURNISHED

NOTE: FITTINGS NOT-INCLUDED WITH FIXTURE AND MUST BE ORDERED SEPARATELY.

PLUMBER NOTE — Provide suitable reinforcement for all wall supports.
IMPORTANT: Dimensions of fixtures are nominal and may vary within the range of tolerances established by ANSI Standards A112.19.1.
These measurements are subject to change or cancellation. No responsibility is assumed for use of superseded or voided leaflets.

AMERICAN STANDARD

Figure 61-10 Bathtub *(Courtesy American Standard)*

Figure 61–11 Trip lever assembly *(Courtesy Kohler Company)*

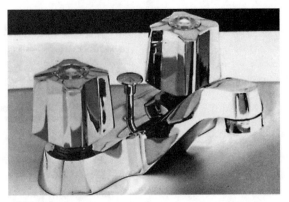

Figure 61–12 Centerset *(Courtesy Kohler Company)*

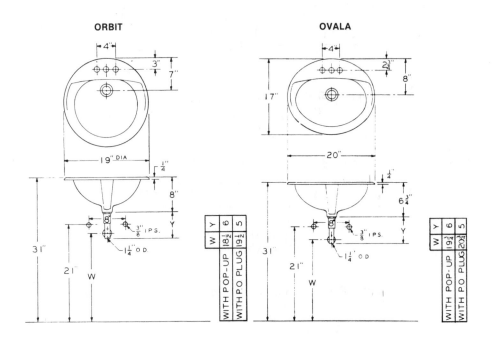

Figure 61–13 *(Courtesy Universal Rundle Corp.)*

The thickness of the finished wall and floor coverings will have an effect with the finished installation, therefore the plumber must be alert for charges from the original specifications. Architectural plans, more often than not, do not contain the center or side measurements to the individual fixtures. These are left to the plumber's discretion. The local plumbing code will have a chart or a drawing in it showing the various minimum side and front clearances for the individual fixtures. After that is taken into account, then consideration should be given to providing equal spaces between fixtures. This is a publicly visual representation of the entire plumbing system's quality. A wall with one lavatory crowded to the left and the other having ample room on both sides will forever remind the occupants of their plumbing contractor. Fixtures with cold water on the left and hot on the right are only slightly worse than misplaced ones.

After the fixtures are installed, the plumber shouldprotect them from damage. Other tradespeople are not as concerned as is the plumber over the high luster of the fixtures or their fittings. Building paper taped over the delicate surfaces will help somewhat. If the water is left off until the final stages of construction, the paper is less likely to be removed to obtain water to mix mortar or to clean paint brushes.

♦ ♦

TEST YOUR KNOWLEDGE

1. What is the purpose of a stub piece?

2. What is meant by the term roughing-in?

3. Where would you find a trip lever assembly?

4. What is a centerset?

5. What is the plumber's name for the bathroom basin?

6. Where would you find the roughing-in dimensions for a fixture?

7. How much clearance should be provided in front of a water closet?

8. How large should a toilet stall be for handicapped access?

◆ ◆ ◆ ◆ ◆ ◆ ◆ ◆ ◆ ◆ ◆ ◆ ◆ ◆ ◆ ◆ ◆ ◆

FAUCETS

OBJECTIVES

After studying this unit, the student should be able to:

❏ List several types of faucets.
❏ Explain how different faucets control the flow of water.

FAUCETS

Faucets are used to control the water at plumbing fixtures, Figure 62–1. They are made of brass and have a nickel-plated, or chrome-plated finish. The chrome-plated finish looks best.

Most faucets are made in combination pat-terns because this permits mixing or tempering the hot and cold water. The better quality faucets have removable seats that, if worn or cut, may be removed and replaced.

Sink and laundry tray faucets have *swing spouts*. The types of handles on faucets are the tee, cross, and lever. Each of these is initialed H or C, indicating hot and cold water.

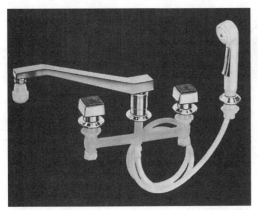

Figure 62–1 *(Courtesy Kohler Company)*

Draw-off and lawn faucets are equipped with hose threads. The latter have loose key or wheel handles.

TYPES OF FAUCETS

In the compression faucet, the principle of the screw is used to raise or lower the stem. By turning the handle clockwise, the stem is lowered until the washer is pressed against the seat, thereby closing the faucet. Turning the handle counterclockwise opens it, Figure 62–2.

The quick-pression, or quick opening, faucet, operates on the screw principle, but the threads have more pitch. This enables the faucet to be fully opened with a quarter turn of the handle, Figure 62–3.

In the self-closing, or quick closing, faucet, the principle of the inclined plane or the eccentric is used to raise the stem. It is closed by a spring when released. This type is used in public buildings to conserve water. They sometimes cause water hammer, the sharp sound of moving water against the sides of a containing pipe, since they close quickly, Figure 62–4.

Faucets are attached to pipes with male or female American Standard threads or solder fittings. They have solid or adjustable flanges and some have ground joint couplings.

Faucets have either composition rubber or fiber seat washers and graphite packing at the stem. Some have a stuffing box, an enclosure around a faucet's valve stem.

Facing a fixture, the hot faucet should always be placed on the left and the cold on the right. Faucet spouts should be at least 1 1/2 inches above the rim of the fixture to prevent waste water from being siphoned back into the water pipe that would pollute the water supply, Figure 62–5.

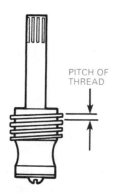

Figure 62–2 Stem of a compression faucet

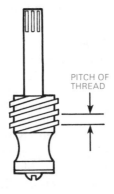

Figure 62–3 Stem of a quick-pression faucet

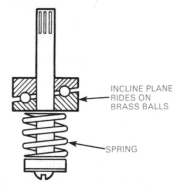

Figure 62–4 Stem of a self-closing faucet

SINGLE-LEVER AND WASHER-LESS FAUCETS

Single lever faucets are one control handle to control the amount and also the temperature of the water. The design of these faucets is radically different from the older stem and disk type. These faucets do not have valve stems and washers to stop and control the flow but rather O-rings and the *plug cock*-type valve action. The handle rotates a brass cylinder or plug within a hole in the valve body. Some designs use a spherical plug, Figure 62–6. When the hole in the plug lines up with the hole in the valve body, liquid will flow. Of course, to mix hot and cold water at variable temperatures requires well-engineered hole placement. The method of controlling the amount of water flow varies, but often when the cylinder is raised by the handle a greater flow occurs because more opening to the spout is exposed by this action.

O-rings, which are made of neoprene, have been found to be a very reliable seal against leakage. Many single-lever or single-control, and also *dripless*, faucets make use of the O-ring to prevent leakage, Figure 62–7.

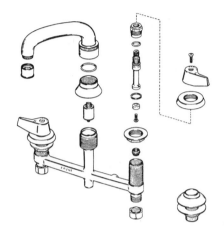

Figure 62–5

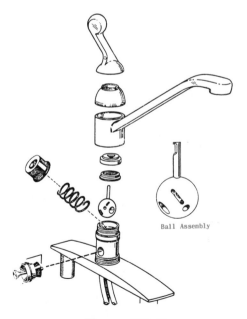

Ball Assembly

Figure 62–6

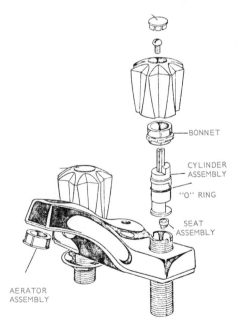

BONNET

CYLINDER ASSEMBLY

"O" RING

SEAT ASSEMBLY

AERATOR ASSEMBLY

Figure 62–7

TEST YOUR KNOWLEDGE

1. On which side of the fixture should the hot water faucet be placed?

2. What is the purpose of the coarse thread on the faucet stem?

3. What is the purpose of a swing spout?

4. Where would you expect to find self-closing faucets and what is their purpose?

◆ ◆ ◆ ◆ ◆ ◆ ◆ ◆ ◆ ◆ ◆ ◆ ◆ ◆ ◆ ◆ ◆ ◆ ◆

Connected Waste and Overflow Outlets

Key Term

connected waste and overflow

OBJECTIVES

After studying this unit, the student should be able to:

❑ Describe how bath wastes and overflow outlets are designed.

❑ State how bath wastes are installed.

Connected waste and overflows are made of chrome-plated brass. They are used for the bathtub waste and to prevent the bathtub from overflowing in case the faucet is left running.

Connected waste and overflows consist of a slip-joint tee and two special elbows with brass tubing between, Figure 63–1. The waste shoe fits against the bottom outlet of the tub, into which the waste plug is screwed.

The top elbow fits against the outside of the overflow outlet. A screw passes through the strainer and holds the elbow and a washer in place, Figure 63–2.

The outlet is made tight by putting putty on the inside and a washer on the outside of the tub. The waste plug is screwed into the waste elbow to draw up the elbow. The putty and the washer provide a watertight joint.

The ends of the tubing are made watertight by washers. The short tubing in the bottom of the tee is connected to the tub trap tailpiece with a slip coupling.

Connected waste and overflows are made in 1 1/2-inch and 1 1/4-inch sizes. Most Bureau of Health Codes require the larger size.

A pop-up waste similar to those used for lavatories is made for bathtubs. It may be removed for cleaning.

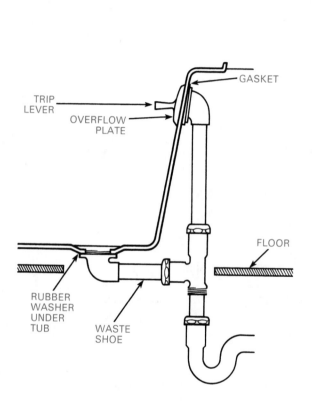

Figure 63-1 Connected waste and overflow outlets

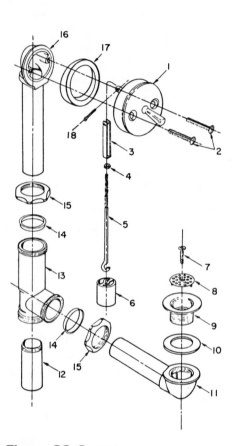

Figure 63-2 *(Courtesy Universal Rundle Corp.)*

◆ ◆

TEST YOUR KNOWLEDGE

1. What is the purpose of the connected waste and overflow fitting?

2. Where would you look for the waste shoe of a connected waste and overflow fitting?

3. Regarding the installation of the waste shoe, on what side of the tub would the putty go, and on what side would the rubber washer go?

4. Where is the trap for the tub located in relation to the connected waste and overflow?

♦ ♦ ♦ ♦ ♦ ♦ ♦ ♦ ♦ ♦ ♦ ♦ ♦ ♦ ♦ ♦ ♦ ♦ ♦

PATENT OVERFLOWS

KEY TERMS

patent overflow or P.O. plug

OBJECTIVES

After studying this unit, the student should be able to:

❑ State how overflows are designed to operate.
❑ List and describe the parts of a patent overflow plug.
❑ Install a patent overflow plug.

The patent overflow, or P.O. plug, Figure 64–1, is the waste connection used in a lavatory having an integral overflow. It consists of two parts: the plug that passes through the bottom of the lavatory, and the bottom section that acts as a locknut. It is 1 1/4-inches in diameter and has a plug and chain, Figure 64–2.

The plug section has two holes in the side that allow the overflow to run down the drain, even though the plug is in place. Placing one of these holes in line with the overflow outlet makes it eas-

ier to clean. The bottom of the plug has a long thread on which the locknut section is screwed.

To install a P.O. plug, paint the bottom of the flange on the plug with pipe dope and apply a ring of putty around it. Put it through the lavatory and slip the gasket in place, Figure 64–2.

The locknut is put in place next and tightened. This piece is thin and may be crushed if it is tightened too much. The patent overflow and the plug and chain are generally being replaced by the pop-up waste.

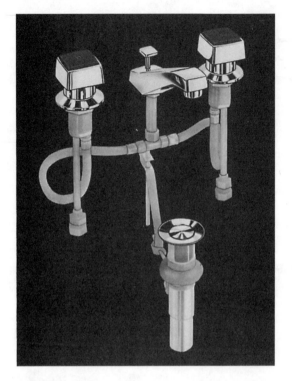

Figure 64–1 *(Courtesy Kohler Company)*

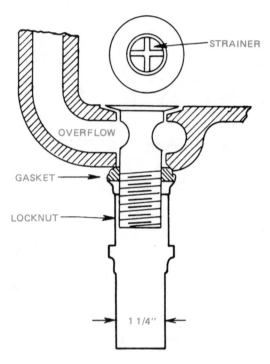

Figure 64–2 P.O. plug

Pop-Up Waste

The *pop-up waste,* Figure 64–3, is made for lavatories and bathtubs. It consists of a brass plug machined to fit the waste fitting. This plug is raised or lowered by a rod attached to a lever and extended up to the slab of the lavatory.

A ball joint is located where the lever passes through the waste fitting, with packing and a spring inside a locknut to prevent leakage. The tension of the spring maintains the plug.

The overflow feature is the same as that for the P.O. plug. The parts inside the waste pipe do become clogged with soap, hair, and other foreign matter, and should be cleaned out periodically.

When used on a bathtub, it is operated by a handle and an eccentric through the wall or overflow. The pop-up waste is more sanitary than the plug and chain.

SUMMARY

The patent overflow or P.O. plug as it is called in the trade, allows the fixture bowl to be plugged by the use of a lever or a lift rod. Another important feature of the P.O. plug is that it allows the overflow water from the bowl to pass through it

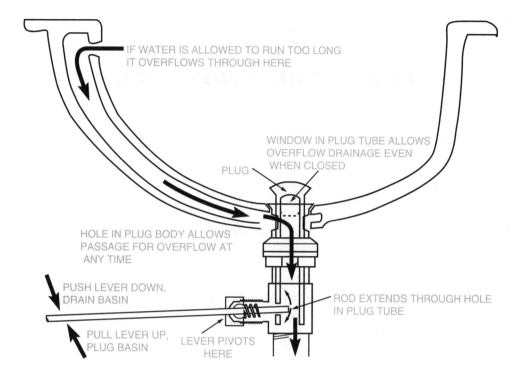

IF WATER IS ALLOWED TO RUN TOO LONG
IT OVERFLOWS THROUGH HERE

WINDOW IN PLUG TUBE ALLOWS
OVERFLOW DRAINAGE EVEN
WHEN CLOSED

PLUG

HOLE IN PLUG BODY ALLOWS
PASSAGE FOR OVERFLOW AT
ANY TIME

PUSH LEVER DOWN,
DRAIN BASIN

ROD EXTENDS THROUGH HOLE
IN PLUG TUBE

PULL LEVER UP,
PLUG BASIN

LEVER PIVOTS
HERE

Figure 64–3

and then go down the drain rather than onto the bathroom floor. Note that plumber's putty is placed on the rim of the strainer insert, which compresses the putty on the the inside of the fixture. The rubber gasket goes against the bottom of the bowl with the more narrow end up. A thin thrust washer is then put between the locknut and the rubber gasket. The thrust washer allows the locknut to turn freely without dragging on the rubber gasket and twisting it out of shape. Avoid overtightening any of the parts of the patent overflow assembly.

◆ ◆

TEST YOUR KNOWLEDGE

1. What does P.O. stand for?

2. What is the purpose of the P.O. plug?

3. What keeps the plug raised so that water entering the bowl can continually run down the drain when this is desired?

4. What part of the assembly is responsible for raising and lowering the plug?

DUPLEX STRAINERS

OBJECTIVES

After studying this unit, the student should be able to:

❑ Describe the construction of sink strainers.
❑ Explain how sink strainers are installed.

A duplex strainer is used in an enameled cast-iron, steel, or vitreous chain sink, Figure 65–1. It is made of chrome-plated brass, Figure 65–2. The strainer is put together in the order in which it is arranged in Figure 65–2. There is a crossbar strainer inside the plug in addition to the round perforated strainer shown at the top.

A ring of soft putty is applied on the underside of the flange of the plug and placed in the waste hole of the sink. Then, the gasket is painted with pipe dope and placed on the plug outside the sink and the locknut is tightened on the plug.

The coupling washer is then placed between the plug and the tailpiece and is tightened by the coupling nut. Be careful not to tighten it too much or the washer will be squeezed inside the tailpiece.

An improved strainer is the drain plug with crumb cup and stopper. This has a 4-inch removable strainer that may be lifted out. This collects all food particles. The strainer has a ground joint stopper that allows water to be kept in the sink for dish washing, Figure 65–2. It is commonly referred to as a *basket strainer*.

SUMMARY

The sink strainer provides threads for the rest of the plumbing to be attached to a sink bowl. It has cross bars in its outlet to keep objects too large for a sink waste from passing through. The duplex part is a chrome or stainless steel basket that provides an easily removable device called a crumb cup for removing vegetable peelings and other washing and cooking debris, not to mention rings and other jewelry, from going down the drain, Figure 65–3.

Figure 65–1 *(Courtesy Kohler Company)*

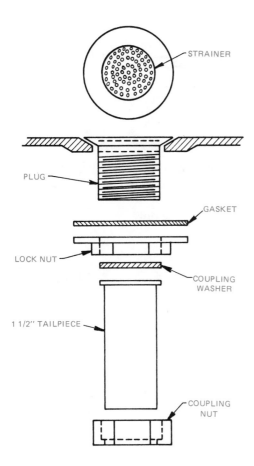

Figure 65–2 Duplex sink strainer

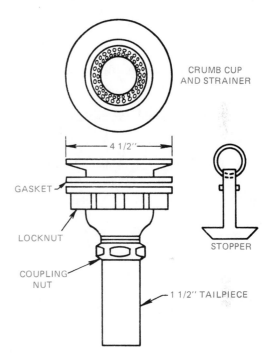

Figure 65–3 Drain plug with crumb cup and stopper

TEST YOUR KNOWLEDGE

1. What is a sink strainer?

2. What provides the basic water seal between the strainer and the sink bowl?

3. Why should you avoid overtightening the tailpiece coupling nut?

4. What is another name for the inner basket?

TRAP AND FAUCET CONNECTIONS

◆ ◆ ◆ ◆ ◆ ◆ ◆ ◆ ◆ ◆ ◆ ◆ ◆ ◆ ◆ ◆ ◆

KEY TERMS

faucet shank
ground joint
slip joint
washer joint

OBJECTIVES

After studying this unit, the student should be able to:

❏ List some of the trap and faucet connections.
❏ Properly use waste and slip joints.

The washer joint is used to connect supply pipes to flush tanks and brass tailpieces to fixture waste. A disadvantage of this type of joint is that if the water pressure is great enough, the joint will be pushed apart.

The slip joint is used on brass tubing traps, connected waste and overflows, and flush elbows.

As shown in Figure 66–1, the tubing is slipped into the fitting on top of the trap.

Packing or a rubber washer is placed around the slip ring tubing and a coupling nut is tightened over it. This connection is not recommended for pressure pipes since it may be pushed apart. An example of the type of joint in use is the lavatory waste, Figure 66–2.

The best connection between brass supply pipes and faucets is the ground joint. It consists of parts with surfaces machined on an angle to a perfect fit, Figure 66–3, Figure 66–4,

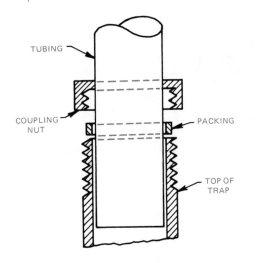

Figure 66-1 Slip joint

and Figure 66–5.

The **faucet shank** is machined on the inside and threaded on the outside. The coupling is machined on the outside and has a nut by which both parts are drawn together, Figure 66–4. This provides a watertight joint. Pipe dope, packing, and washers are not required for this joint. Care must be taken not to scratch the two machined surfaces or leaks will occur.

The *speedway flexible lavatory and tank supplies* use the same principle. They have the additional feature of being soft, and can be adjusted to any offset, Figure 66–5. These tubes are made in 5", 8", 12", 30", and 36" lengths. Elbows and adaptors are also available.

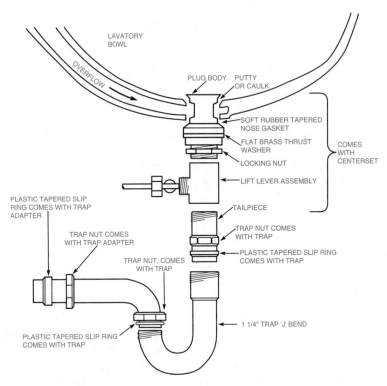

Figure 66-2 Lavatory waste connections

Tank supply pipes, constructed of soft copper, are also made with a flat top for those ball cocks that require a washer, Figure 66–6. The connection made at the valve in Figure 66–7 is called a compression joint. The part that is called a tapered ring, on this figure, is also called a compression ferrule. Compression ferrules come in many different shapes, including cylindrical, ovoid or egg-shaped, and spherical.

This particular kind of joint is especially strong and is capable of withstanding several hundred pounds per square inch of internal pressure. When the joint is tightened initially the ferrule is compressed by the internal ramps inside of the valve outlet and inside of the compression nut. Figure 66–7 shows a compression nut just started. The ferrule is not compressed yet. When the joint is made the first time firm pressure should be used *using two wrenches*. From then on you will discover that the ferrule

is fastened permanently to the tubing. The detail, detail "A," shows the forces at work here. The dark arrows demonstrate the direction of the crushing forces as the wrenches are applied.

If the tubing is to be replaced, a new compression ferrule will have to be installed.

SUMMARY

The terminal connections at the fixtures are easily removable. These are union-type joints, which make whatever is being connected portable. One of the drawbacks of union-type joints is that they are vulnerable to leaks caused by vibration and should always be in a visible location. Vibration can be caused by traffic on nearby thoroughfares or even earth tremors, which are not uncommon, or by the rough usage of humans. For example, cleaning people

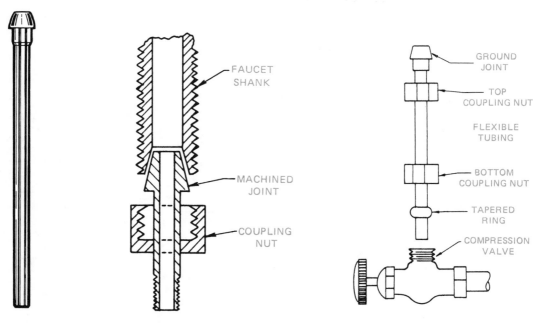

Figure 66–3   **Figure 66–4** Faucet shank   **Figure 66–5** Speedway lavatory supply

LAVATORY TUBE
With Metal Nosepiece

One-piece construction with insert for added strength

CLOSET TUBE
with Plastic Insert

One-piece construction

Figure 66–6

can thump and bump fixtures, people step on flush valve handles rather than use their hands, and people sit on lavatories. All of these things cause stress and strain and, occasionally, leaks.

Because of their exposed locations, these fittings are chrome plated. The plumber must use smooth-jawed wrenches or strap wrenches to leave the surfaces pristine and unmarred. Slip lock pliers are generally not smooth-jawed. There are not that many different sizes of the smaller chromed hex nuts. A good quality open-end wrench set is inexpensive and will save rounded corners on hex nuts and roughed-up chrome. Here, again, the danger is in overtightening joints rather than undertightening them.

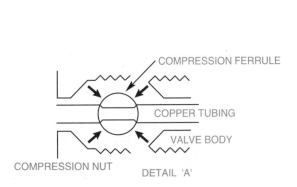

COMPRESSION FERRULE

COPPER TUBING

VALVE BODY

COMPRESSION NUT

DETAIL 'A'

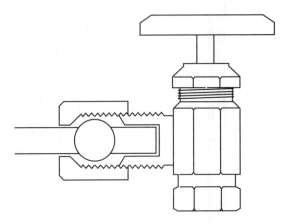

Figure 66–7

TEST YOUR KNOWLEDGE

1. What is the disadvantage of a washer-type joint?

2. What makes a ground joint superior for pressure connections on fixtures?

3. Where would you find the faucet shank?

4. Why are soft fixture supply tubes an advantage?

5. How much pressure can a compression joint withstand?

6. What is the difference between a lavatory supply tube and a closet supply tube?

7. What is the valve on a Speedway supply called?

PVC and CPVC Pressure Fittings

PVC & CPVC Schedule 80 Pressure Fittings

NOTE: All weights are for PVC I Gray and are listed in pounds each.
To obtain CPVC weights, multiply by 1.116.

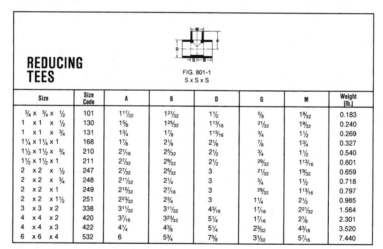

TEES

FIG. 801
S x S x S

FIG. 805
T x T x T

Size	Size Code	A Slip	A Thread	M	G	Weight (lb.) 801	Weight (lb.) 805
1/4	002	27/32	27/32	27/32	1/4	0.040	0.045
3/8	003	1 1/32	1 1/32	1	1/2	0.062	0.064
1/2	005	1 7/16	1 19/32	1 3/16	9/16	0.132	0.137
3/4	007	1 11/16	1 13/32	1 13/32	21/32	0.210	0.189
1	010	1 7/8	1 27/32	1 13/16	3/4	0.318	0.319
1 1/4	012	2 1/8	1 31/32	2 3/32	23/32	0.515	0.440
1 1/2	015	2 3/8	2 1/8	2 3/8	1 25/32	0.719	0.575
2	020	2 13/16	2 3/8	2 7/8	1 5/16	1.064	0.837
2 1/2	025	3 23/32	3 5/32	3 9/16	1 19/32	1.676	1.508
3	030	3 3/16	3 9/16	4 7/32	1 13/16	2.634	2.321
4	040	4 5/8	4 1/4	5 7/16	2 11/32	4.008	3.986
6	060	6 3/4	—	7 5/8	3 23/32	9.550	—
8	080	8 5/8	—	9 3/4	4 9/16	17.450	—

REDUCING TEES

FIG. 801-1
S x S x S

Size	Size Code	A	B	D	G	M	Weight (lb.)
3/4 x 3/4 x 1/2	101	1 17/32	1 21/32	1 1/2	5/8	1 9/32	0.183
1 x 1 x 1/2	130	1 5/8	1 25/32	1 13/16	21/32	1 9/32	0.240
1 x 1 x 3/4	131	1 3/4	1 7/8	1 13/16	3/4	1 1/2	0.269
1 1/4 x 1 1/4 x 1	168	1 7/8	2 1/8	2 1/8	7/8	1 3/4	0.327
1 1/2 x 1 1/2 x 3/4	210	2 1/16	2 5/32	2 1/2	3/4	1 1/2	0.540
1 1/2 x 1 1/2 x 1	211	2 7/32	2 9/32	2 1/2	29/32	1 13/16	0.601
2 x 2 x 1/2	247	2 7/32	2 5/32	3	21/32	1 9/32	0.659
2 x 2 x 3/4	248	2 11/32	2 1/4	3	3/4	1 1/2	0.718
2 x 2 x 1	249	2 15/32	2 7/16	3	29/32	1 13/16	0.797
2 x 2 x 1 1/2	251	2 23/32	2 3/4	3	1 1/4	2 1/2	0.985
3 x 3 x 2	338	3 11/32	3 11/32	4 3/16	1 7/16	2 21/32	1.564
4 x 4 x 2	420	3 7/16	3 23/32	5 1/4	1 7/16	2 7/8	2.301
4 x 4 x 3	422	4 1/4	4 3/8	5 1/4	2 3/32	4 3/16	3.520
6 x 6 x 4	532	6	5 3/4	7 5/8	3 1/32	5 7/16	7.440

TEES

FIG. 802
S x S x T

Size	Size Code	M	A	G	C	Weight (lb.)
1/2	005	1 9/32	1 3/8	15/32	1 3/8	0.135
3/4	007	1 1/2	1 11/16	1 17/32	1 11/16	0.192
1	010	1 13/16	1 27/32	7/16	1 27/32	0.295

30° ELLS

FIG. 815
S x S

Size	Size Code	G	H	M	Weight (lb.)
6	060	2 11/16	5 5/8	7 21/32	5.54

(Courtesy Spears Manufacturing)

90° ELLS

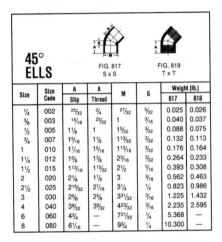

FIG. 806 S x S   FIG. 807 S x T   FIG. 808 T x T

Size	Size Code	A Slip	A Thread	M	G	Weight (lb.) 806	807	808
1/4	002	1	15/16	13/16	11/32	0.028	—	0.032
3/8	003	1 7/32	1 1/32	31/32	7/16	0.043	—	0.046
1/2	005	1 3/8	1 3/8	1 3/16	1/2	0.072	0.099	0.089
3/4	007	1 9/16	1 9/16	1 13/16	1/2	0.101	0.139	0.128
1	010	1 13/16	1 13/16	1 13/16	11/16	0.205	0.218	0.209
1 1/4	012	2 7/32	1 15/16	2 7/32	15/16	0.324	0.311	0.321
1 1/2	015	2 15/32	2 3/32	2 1/2	1 1/16	0.458	0.409	0.425
2	020	2 3/4	2 3/8	2 29/32	1 1/4	0.573	—	0.607
2 1/2	025	3 3/8	3 5/32	3 9/16	1 9/16	1.111	—	1.162
3	030	3 23/32	3 9/16	4 3/16	1 13/16	1.488	—	1.789
4	040	4 25/32	4 1/4	5 7/16	2 1/2	1.882	—	1.952
6	060	6 9/16	—	7 5/8	3 17/32	7.420	—	—
8	080	8 5/8	—	9 3/4	4 9/16	14.060	—	—

45° ELLS

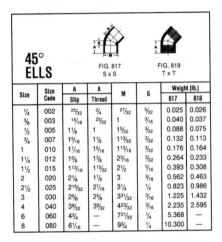

FIG. 817 S x S   FIG. 819 T x T

Size	Size Code	A Slip	A Thread	M	G	Weight (lb.) 817	818
1/4	002	25/32	3/4	27/32	5/32	0.025	0.026
3/8	003	15/16	25/32	1	3/16	0.040	0.037
1/2	005	1 1/8	1	1 5/32	3/32	0.088	0.075
3/4	007	1 5/16	1 1/8	1 13/16	3/32	0.132	0.113
1	010	1 7/16	1 5/16	1 13/16	3/32	0.176	0.164
1 1/4	012	1 5/8	1 3/8	2 3/16	3/32	0.264	0.233
1 1/2	015	1 13/16	1 15/32	2 1/2	3/16	0.393	0.308
2	020	2 1/8	1 7/8	3	3/16	0.562	0.463
2 1/2	025	2 15/32	2 7/16	3 1/4	1/4	0.823	0.986
3	030	2 5/8	2 5/8	3 31/32	3/16	1.225	1.432
4	040	3 9/32	3 9/32	4 23/32	3/16	2.235	2.595
6	060	4 3/4	—	7 21/32	1/4	5.368	—
8	080	6 1/16	—	9 3/4	1/4	10.300	—

COUPLINGS

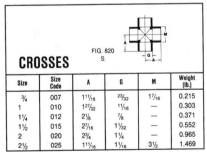

FIG. 829 S x S   FIG. 830 T x T

Size	Size Code	L Slip	L Thread	M	N	Weight (lb.) 829	830
1/4	002	1 3/8	1 15/16	27/32	1/8	0.022	0.023
3/8	003	1 5/8	1 15/16	1	1/8	0.033	0.035
1/2	005	1 27/32	1 9/32	1 7/32	3/32	0.047	0.065
3/4	007	2 1/8	1 23/32	1 1/2	3/32	0.096	0.086
1	010	2 7/8	2 1/16	1 13/16	3/32	0.150	0.145
1 1/4	012	2 19/32	2 7/32	2 7/32	3/32	0.227	0.218
1 1/2	015	2 7/8	2 1/2	2 1/2	3/16	0.303	0.261
2	020	3 1/8	3	3	3/16	0.423	0.342
2 1/2	025	3 13/16	3 11/32	3 3/4	1/4	0.641	0.675
3	030	4 7/32	3 15/32	3 21/32	1/4	0.829	0.958
4	040	5 5/16	3 21/32	4 23/32	1/4	1.416	1.468
6	060	8 1/4	—	7 21/32	1/4	3.412	—
8	080	8 5/8	—	9 3/4	1/4	6.640	—

CROSSES

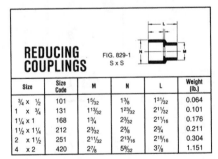

FIG. 820 S

Size	Size Code	A	G	M	Weight (lb.)
3/4	007	1 11/16	23/32	1 7/16	0.215
1	010	1 27/32	11/16	—	0.303
1 1/4	012	2 1/8	7/8	—	0.371
1 1/2	015	2 7/16	1 1/32	—	0.552
2	020	2 3/4	1 1/4	—	0.965
2 1/2	025	1 11/16	1 1/16	3 1/2	1.469

REDUCING COUPLINGS

FIG. 829-1 S x S

Size	Size Code	M	N	L	Weight (lb.)
3/4 x 1/2	101	1 5/32	1 3/8	1 31/32	0.064
1 x 3/4	131	1 13/32	1 23/32	2 1/32	0.101
1 1/4 x 1	168	1 3/4	2 3/32	2 11/16	0.176
1 1/2 x 1 1/4	212	2 3/32	2 3/8	2 3/4	0.211
2 x 1 1/2	251	2 11/32	2 13/16	2 15/16	0.304
4 x 2	420	2 7/8	5 5/32	3 7/8	1.151

(Courtesy Spears Manufacturing)

PVC & CPVC Schedule 80 Pressure Fittings (continued)

FEMALE ADAPTERS

FIG. 835 S x T

Size	Size Code	M	N	L	Weight (lb.)
1/4	002	27/32	1/8	15/16	0.022
3/8	003	1	1/8	1 15/32	0.033
1/2	005	1 9/32	3/32	1 3/4	0.073
3/4	007	1 1/2	3/32	1 29/32	0.098
1	010	1 13/16	3/32	2 7/32	0.153
1 1/4	012	2 7/32	3/16	2 3/8	0.216
1 1/2	015	2 1/2	3/16	2 1/2	0.287
2	020	3	3/16	2 3/4	0.384
2 1/2	025	3 17/32	1/4	3 1/32	0.658
3	030	4 1/4	1/4	3 25/32	0.995
4	040	5 3/8	1/4	4 1/4	1.599

MALE ADAPTERS

FIG. 836 M x S

Size	Size Code	L	M	C	Weight (lb.)
1/2	005	1 27/32	1 3/16	7/8	0.041
3/4	007	2	1 1/2	1	0.074
1	010	2 11/32	1 13/16	1 1/8	0.116
1 1/4	012	2 17/32	2 3/16	1 1/4	0.177
1 1/2	015	2 23/32	2 3/8	1 3/8	0.251
2	020	2 7/8	2 7/8	1 1/2	0.326
2 1/2	025	3 23/32	3	1 3/4	0.545
3	030	3 15/16	4 5/16	1 7/8	0.797
4	040	4 23/32	5 7/16	2 1/4	1.115

PLUGS

FIG. 849 Spigot / FIG. 850 Thread

Size	Size Code	L	E	Weight (lb.)
1/4	002	9/16	3/16	0.008
3/8	003	13/16	3/16	0.012
1/2	005	15/16	3/16	0.018
3/4	007	1 1/64	7/32	0.029
1	010	1 1/4	1/4	0.046
1 1/4	012	1 9/32	9/32	0.077
1 1/2	015	1 15/32	5/16	0.108
2	020	1 9/16	1/2	0.176
2 1/2	025	1 15/16	3/8	0.311
3	030	2	3/8	0.452
4	040	2 1/8	3/8	0.716

CAPS

FIG. 847 Soc / FIG. 848 Thread

Size	Size Code	M	Fig. 847		Fig. 848		Weight (lb.) Fig. 847	Weight (lb.) Fig. 848
			L	C	L	C		
1/4	002	13/16	1	5/8	11/16	19/32	0.013	0.018
3/8	003	1	1 1/2	3/4	15/16	5/8	0.022	0.024
1/2	005	1 3/16	1 1/4	7/8	1 3/32	3/4	0.045	0.046
3/4	007	1 13/32	1 7/16	1 1/4	1 15/32	25/32	0.062	0.056
1	010	1 23/32	1 5/8	1 1/8	1 7/16	1	0.078	0.101
1 1/4	012	2 7/32	1 27/32	1 1/4	1 1/2	1	0.159	0.141
1 1/2	015	2 1/2	2 1/16	1 3/8	1 19/32	1 1/32	0.211	0.187
2	020	3	2 11/32	1 1/2	1 3/4	1 1/16	0.327	0.283
2 1/2	025	3 9/16	2 11/16	1 3/4	2 3/16	1 5/8	0.485	0.508
3	030	4 5/16	3	1 7/8	2 21/32	1 5/8	0.780	0.828
4	040	5 7/16	3 11/16	2 1/4	3	1 3/4	1.387	1.338
6	060	7 7/8	4 15/16	3 1/4	—	—	2.966	

FLANGES

FIG. 851 Soc / FIG. 852 Thread / FIG. 853 Blind

Size	Size Code	M	R	A	N	Bolt Dia.	No. of Bolts	L Fig. 851	C Fig. 851	L Fig. 852	Weight (lb.) Fig. 851	Weight (lb.) Fig. 852	Weight (lb.) Fig. 853
1/2	005	3 1/2	1/2	2 3/8	9/16	1/2	4	7/8	27/32	15/16	0.192	0.197	0.017
3/4	007	3 7/8	1/2	2 3/4	9/16	1/2	4	1 1/8	1 1/8	15/16	0.247	0.247	0.232
1	010	4 1/4	5/8	3 1/8	9/16	1/2	4	1 1/4	15/32	15/32	0.323	0.330	0.323
1 1/4	012	4 5/8	5/8	3 1/2	9/16	1/2	4	1 3/8	1 9/32	15/32	0.414	0.405	0.436
1 1/2	015	5	23/32	3 7/8	5/8	1/2	4	1 5/8	1 7/16	1 1/4	0.492	0.481	0.500
2	020	6 1/16	3/4	4 3/4	3/4	5/8	4	1 23/32	1 1/2	1 3/16	0.742	0.719	0.797
2 1/2	025	7	15/16	5 1/2	3/4	5/8	4	2 7/32	2 3/32	1 3/4	1.271	2.568	0.366
3	030	7 15/16	1	6	3/4	5/8	4	2 5/16	2 1/8	2 5/16	1.601	1.684	1.392
4	040	9 1/16	1 5/32	7 1/2	3/4	5/8	8	2 1/2	2 1/4	2	2.544	2.469	2.338
5	050	10	1 31/32	8 1/2	7/8	3/4	8	3 1/4	3 1/16	—	2.877	—	—
6	060	11	1 1/4	9 1/2	7/8	3/4	8	3 1/4	3	—	4.106	—	3.854
8	080	13 1/2	1 13/32	11 3/4	7/8	3/4	8	4 1/2	4 1/4	—	6.850	—	6.850

(Courtesy Spears Manufacturing)

REDUCER BUSHINGS

FIG. 837 S x S FIG. 838 S x T FIG. 839 T x T

Size	Size Code	M	Fig. 837 L	Fig. 837 C	Fig. 838 D	Fig. 838 L	Fig. 839 L	Weight (lb.) Fig. 837	Weight (lb.) Fig. 838	Weight (lb.) Fig. 839
3/8 x 1/4	052	13/16	31/32	3/4	3/4	15/16	13/16	0.011	0.012	0.011
1/2 x 1/8	071	31/32	—	—	3/4	15/16	—	—	0.020	—
1/2 x 1/4	072	31/32	17/32	1	9/16	3/4	1	0.019	0.020	0.021
1/2 x 3/8	073	31/32	17/32	1	9/16	3/4	15/16	0.013	0.017	0.015
3/4 x 1/4	098	1 3/16	—	—	1	1 1/4	1	—	0.035	0.037
3/4 x 3/8	099	1 3/16	—	—	—	—	1	—	—	0.030
3/4 x 1/2	101	1 3/16	1 7/32	3/4	1	1 17/32	1	0.026	0.035	0.024
1 x 1/4	128	—	—	—	—	—	—	—	—	0.057
1 x 3/8	129	1 1/2	—	—	1 5/32	1 13/32	1 1/4	—	0.065	0.053
1 x 1/2	130	1 1/2	1 3/8	3/4	1 5/32	1 13/32	1 1/4	0.066	0.062	0.056
1 x 3/4	131	1 1/2	1 3/8	1	1 1/8	1 3/8	1 1/4	0.042	0.062	0.043
1 1/4 x 1/2	166	1 27/32	1 3/4	3/4	1 1/4	1 17/32	1 9/32	0.085	0.109	0.062
1 1/4 x 3/4	167	1 27/32	1 7/32	1	1 1/4	1 17/32	1 9/32	0.113	0.099	0.076
1 1/4 x 1	168	1 27/32	1 21/32	1 1/8	1 1/4	1 17/32	1 9/32	0.079	0.080	0.063
1 1/2 x 1/2	209	2 3/32	1 9/16	3/4	1 5/16	1 5/8	1 11/32	0.138	0.135	0.112
1 1/2 x 3/4	210	2 3/32	1 9/16	1	1 5/16	1 5/8	1 11/32	0.151	0.128	0.123
1 1/2 x 1	211	2 3/32	1 9/16	1 1/8	1 5/16	1 5/8	1 11/32	0.141	0.132	0.102
1 1/2 x 1 1/4	212	2 3/32	1 9/16	1 1/4	1 5/16	1 5/8	1 11/32	0.087	0.087	0.063
2 x 1/2	247	2 9/16	1 3/4	3/4	1 3/8	1 3/4	1 7/16	0.204	0.203	0.176
2 x 3/4	248	2 9/16	1 3/4	1	1 3/8	1 3/4	1 7/16	0.218	0.263	0.170
2 x 1	249	2 9/16	1 3/4	1 1/8	1 3/8	1 3/4	1 7/16	0.208	0.200	0.191
2 x 1 1/4	250	2 9/16	1 3/4	1 1/8	1 3/8	1 3/4	1 7/16	0.228	0.189	0.165
2 x 1 1/2	251	2 9/16	1 29/32	1 3/8	1 3/8	1 3/4	1 7/16	0.167	0.176	0.138
2 1/2 x 1/2	287	3 1/8	—	—	2	2 3/8	—	—	0.412	—
2 1/2 x 3/4	288	3 1/8	—	—	2	2 3/8	—	—	0.390	—
2 1/2 x 1	289	3 1/8	—	—	2	2 3/8	—	—	0.446	—
2 1/2 x 1 1/4	290	3 1/8	—	—	2	2 3/8	—	—	0.403	—
2 1/2 x 1 1/2	291	3 1/8	—	—	2	2 3/8	—	—	0.413	—
2 1/2 x 2	292	3 1/8	2 3/8	1 3/8	2	2 3/8	1 5/16	0.247	0.289	0.259
3 x 3/4	334	3 3/4	—	—	2	2 3/8	—	—	0.604	—
3 x 1	335	3 3/4	—	—	2	2 3/8	—	—	0.564	—
3 x 1 1/4	336	3 3/4	—	—	2	2 3/8	—	—	0.604	—
3 x 1 1/2	337	3 3/4	—	—	2	2 3/8	2 1/64	—	0.574	0.562
3 x 2	338	3 3/4	2 9/32	1 1/2	2	2 3/8	2 1/64	0.597	0.548	0.456
3 x 2 1/2	339	3 3/4	2 1/4	2	2	2 1/4	2 1/64	0.383	0.465	0.383
4 x 2	420	4 3/8	2 11/16	1 1/2	2	2 3/8	2 1/8	1.194	0.756	0.836
4 x 2 1/2	421	4 1/2	—	—	2	2 3/8	2 1/8	—	0.992	—
4 x 3	422	4 1/2	2 11/16	2	2	2 11/16	—	0.738	0.787	0.697
6 x 2	528	6 7/8	4	1 1/2	1 1/2	—	—	2.568	2.445	—
6 x 3	530	6 7/8	4	2	—	2	—	2.738	2.804	—
6 x 4	532	6 1/2	4	1 3/4	3 1/2	4	—	2.681	2.705	—
8 x 6	585	8 23/32	4 15/16	3 1/2	—	—	—	5.694	—	—

WYE

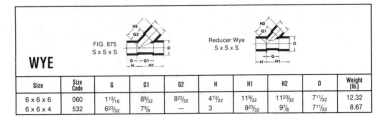

FIG. 875 S x S x S   Reducer Wye S x S x S

Size	Size Code	G	G1	G2	H	H1	H2	D	Weight (lb.)
6 x 6 x 6	060	1 13/16	8 9/32	8 23/32	4 13/32	11 9/32	11 23/32	7 11/32	12.32
6 x 6 x 4	532	6 23/32	7 5/8	—	3	9 23/32	9 7/8	7 11/32	8.67

(Courtesy Spears Manufacturing)

Appendix B

Useful Information

WATER PRESSURE TO FEET HEAD

POUNDS PER SQUARE INCH	FEET HEAD	POUNDS PER SQUARE INCH	FEET HEAD
1	2.31	100	230.90
2	4.62	110	253.98
3	6.93	120	277.07
4	9.24	130	300.16
5	11.54	140	323.25
6	13.85	150	346.34
7	16.16	160	369.43
8	18.47	170	392.52
9	20.78	180	415.61
10	23.09	200	461.78
15	34.63	250	577.24
20	46.18	300	692.69
25	57.72	350	808.13
30	69.27	400	922.58
40	92.36	500	1154.48
50	115.45	600	1385.39
60	138.54	700	1616.30
70	161.63	800	1847.20
80	184.72	900	2078.10
90	207.81	1000	2309.00

NOTE: One pound of pressure per square inch of water equals 2.309 feet of water at 62° Fahrenheit. Therefore, to find the feet head of water for any pressure not given in the table above, multiply the pressure pounds per square inch by 2.309.

FEET HEAD OF WATER TO PSI

FEET HEAD	POUNDS PER SQUARE INCH	FEET HEAD	POUNDS PER SQUARE INCH
1	.43	100	43.31
2	.87	110	47.64
3	1.30	120	51.97
4	1.73	130	56.30
5	2.17	140	60.63
6	2.60	150	64.96
7	3.03	160	69.29
8	3.46	170	73.63
9	3.90	180	77.96
10	4.33	200	86.62
15	6.50	250	108.27
20	8.66	300	129.93
25	10.83	350	151.58
30	12.99	400	173.24
40	17.32	500	216.55
50	21.65	600	259.85
60	25.99	700	303.16
70	30.32	800	346.47
80	34.65	900	389.78
90	38.98	1000	433.00

NOTE: One foot of water at 62° Fahrenheit equals .433 pound pressure per square inch. To find the pressure per square inch for any feet head not given in the table above, multiply the feet head by .433.

**BOILING POINTS OF WATER
AT VARIOUS PRESSURES**

VACUUM, IN INCHES OF MERCURY	BOILING POINT	VACUUM, IN INCHES OF MERCURY	BOILING POINT
29	76.62	7	198.87
28	99.93	6	200.96
27	114.22	5	202.25
26	124.77	4	204.85
25	133.22	3	206.70
24	140.31	2	208.50
23	146.45	1	210.25
22	151.87	Gauge Lbs.	
21	156.75	0	212.
20	161.19	1	215.6
19	165.24	2	218.5
18	169.00	4	224.4
17	172.51	6	229.8
16	175.80	8	234.8
15	178.91	10	239.4
14	181.82	15	249.8
13	184.61	25	266.8
12	187.21	50	297.7
11	189.75	75	320.1
10	192.19	100	337.9
9	194.50	125	352.9
8	196.73	200	387.9

TAP AND DRILL SIZES
(American Standard Coarse)

SIZE OF DRILL	SIZE OF TAP	THREADS PER INCH	SIZE OF DRILL	SIZE OF TAP	THREADS PER INCH
7	1/4	20	49/64	7/8	9
F	5/16	18	53/64	15/16	9
5/16	3/8	16	7/8	1	8
U	7/16	14	63/64	1 1/8	7
27/64	1/2	13	1 7/64	1 1/4	7
31/64	9/16	12	1 13/64	1 3/8	6
17/32	5/8	11	1 11/32	1 1/2	6
19/32	11/16	11	1 29/64	1 5/8	5 1/2
21/32	3/4	10	1 9/16	1 3/4	5
23/32	13/16	10	1 11/16	1 7/8	5
			1 25/32	2	4 1/2

MINUTES CONVERTED TO DECIMALS OF A DEGREE

MIN.	DEG.	MIN.	DEG.	MIN.	DEG.	MIN.	DEG.	MIN.	DEG.	MIN.	DEG.
1	.0166	11	.1833	21	.3500	31	.5166	41	.6833	51	.8500
2	.0333	12	.2000	22	.3666	32	.5333	42	.7000	52	.8666
3	.0500	13	.2166	23	.3833	33	.5500	43	.7166	53	.8833
4	.0666	14	.2333	24	.4000	34	.5666	44	.7333	54	.9000
5	.0833	15	.2500	25	.4166	35	.5833	45	.7500	55	.9166
6	.1000	16	.2666	26	.4333	36	.6000	46	.7666	56	.9333
7	.1166	17	.2833	27	.4500	37	.6166	47	.7833	57	.9500
8	.1333	18	.3000	28	.4666	38	.6333	48	.8000	58	.9666
9	.1500	19	.3166	29	.4833	39	.6500	49	.8166	59	.9833
10	.1666	20	.3333	30	.5000	40	.6666	50	.8333	60	1.0000

DECIMAL EQUIVALENTS OF FRACTIONS

INCHES	DECIMAL OF AN INCH	INCHES	DECIMAL OF AN INCH
1/64	.015625	7/16	.4375
1/32	.03125	29/64	.453125
3/64	.046875	15/32	.46875
1/20	.05	31/64	.484375
1/16	.0625	1/2	.5
1/13	.0769	33/64	.515625
5/64	.078125	17/32	.53125
1/12	.0833	35/64	.546875
1/11	.0909	9/16	.5625
3/32	.09375	37/64	.578125
1/10	.10	19/32	.59375
7/64	.109375	39/64	.609375
1/9	.111	5/8	.625
1/8	.125	41/64	.640625
9/64	.140625	21/32	.65625
1/7	.1429	43/64	.671875
5/32	.15625	11/16	.6875
1/6	.1667	45/64	.703125
11/64	.171875	23/32	.71875
3/16	.1875	47/64	.734375
1/5	.2	3/4	.75
13/64	.203125	49/64	.765625
7/32	.21875	25/32	.78125
15/64	.234375	51/64	.796875
1/4	.25	13/16	.8125
17/64	.265625	53/64	.828125
9/32	.28125	27/32	.84375
19/64	.296875	55/64	.859375
5/16	.3125	7/8	.875
21/64	.328125	57/64	.890625
1/3	.333	29/32	.90625
11/32	.34375	59/64	.921875
23/64	.359375	15/16	.9375
3/8	.375	61/64	.953125
25/64	.390625	31/32	.96875
13/32	.40625	63/64	.984375
27/64	.421875	1	1.

STANDARD PIPE DATA

NOMINAL PIPE DIAM. IN INCHES	ACTUAL INSIDE DIAM. IN INCHES	ACTUAL OUTSIDE DIAM. IN INCHES	WEIGHT PER FOOT POUNDS	LENGTH IN FEET CONTAINING ONE CUBIC FOOT	GALLONS IN ONE LINEAL FOOT
1/8	.269	.405	.244	2526.000	.0030
1/4	.364	.540	.424	1383.800	.0054
3/8	.493	.675	.567	754.360	.0099
1/2	.622	.840	.850	473.910	.0158
3/4	.824	1.050	1.130	270.030	.0277
1	1.049	1.315	1.678	166.620	.0449
1 1/4	1.380	1.660	2.272	96.275	.0777
1 1/2	1.610	1.900	2.717	70.733	.1058
2	2.067	2.375	3.652	49.913	.1743
2 1/2	2.469	2.875	5.793	30.077	.2487
3	3.068	3.500	7.575	19.479	.3840
3 1/2	3.548	4.000	9.109	14.565	.5136
4	4.026	4.500	10.790	11.312	.6613
4 1/2	4.560	5.000	12.538	9.030	.8284
5	5.047	5.563	14.617	7.198	1.0393
6	6.065	6.625	18.974	4.984	1.5008
8	7.981	8.625	28.554	2.878	2.5988
10	10.020	10.750	40.483	1.826	4.0963

CONVERSION FACTORS

PRESSURE

1 lb. per sq. in.	=	2.31 ft. water at 60°F
	=	2.04 in. hg at 60°F
1 ft. water at 60°F	=	0.433 lb. per sq. in.
	=	0.884 in. hg at 60°F
1 in. Hg at 60°F	=	0.49 lb. per sq. in.
	=	1.13 ft. water at 60°F
lb. per sq. in. Absolute (psia)	=	lb. per sq. in. gauge (psig) + 14.7

TEMPERATURE

°C	=	(°F − 32) × 5/9

WEIGHT OF LIQUID

1 gal. (U.S.)	=	8.34 lb. × sp. gr.
1 cu. ft.	=	62.4 lb. × sp. gr.
1 lb.	=	0.12 U.S. gal. ÷ sp. gr.
	=	0.016 cu. ft. ÷ sp. gr.

FLOW

1 gpm	=	0.134 cu. ft. per min.
	=	500 lb. per hr. × sp. gr.
500 lb. per hr.	=	1 gpm ÷ sp. gr.
1 cu. ft. per min. (cfm)	=	448.8 gal. per hr. (gph)

WORK

1 Btu (mean)	=	778 ft. lb.
	=	0.293 watt hr.
	=	1/180 of heat required to change temp of 1 lb. water from 32°F to 212°F
1 hp-hr	=	2545 Btu (mean)
	=	0.746 Kwhr
1 Kwhr	=	3413 Btu (mean)
	=	1.34 hp-hr

POWER

1 Btu per hr.	=	0.293 watt
	=	12.96 ft. lb. per min.
	=	0.00039 hp
1 ton refrigeration (U.S.)	=	288,000 Btu per 24 hr.
	=	12,000 Btu per hr.
	=	200 Btu per min.
	=	83.33 lb. ice melted per hr. from and at $32°$F.
	=	2000 lb. ice melted per 24 hr. from and at $32°$F.
1 hp	=	550 ft. lb. per sec.
	=	746 watt
	=	2545 Btu per hr.
1 boiler hp	=	33,480 Btu per hr.
	=	34.5 lb. water evap. per hr. from and at $212°$F.
	=	9.8 kw.
1 kw.	=	3413 Btu per hr.

MASS

1 lb. (avoir.)	=	16 oz. (avoir.)
	=	7000 grain
1 ton (short)	=	2000 lb.
1 ton (long)	=	2240 lb.

VOLUME

1 gal. (U.S.)	=	128 fl. oz. (U.S.)
	=	231 cu. in.
	=	0.833 gal. (Brit.)
1 cu. ft.	=	7.48 gal. (U.S.)

WEIGHT OF WATER

1 cu. ft. at $50°$F. weighs 62.41 lb.

1 gal. at $50°$F. weighs 8.34 lb.

1 cu. ft. of ice weighs 57.2 lb.

Water is at its greatest density at $39.2°$F.

1 cu. ft. at $39.2°$F. weighs 62.43 lb.

CONVERSION CONSTANTS

TO CHANGE	TO	MULTIPLY BY
Inches	Feet	0.0833
Inches	Millimeters	25.4
Feet	Inches	12
Feet	Yards	0.3333
Yards	Feet	3
Square inches	Square feet	0.00694
Square feet	Square inches	144
Square feet	Square yards	0.11111
Square yards	Square feet	9
Cubic inches	Cubic feet	0.00058
Cubic feet	Cubic inches	1728
Cubic feet	Cubic yards	0.03703
Cubic yards	Cubic feet	27
Cubic inches	Gallons	0.00433
Cubic feet	Gallons	7.48
Gallons	Cubic inches	231
Gallons	Cubic feet	0.1337
Gallons	Pounds of water	8.33
Pounds of water	Gallons	0.12004
Ounces	Pounds	0.0625
Pounds	Ounces	16
Inches of water	Pounds per square inch	0.0361
Inches of water	Inches of mercury	0.0735
Inches of water	Ounces per square inch	0.578
Inches of water	Pounds per square foot	5.2
Inches of mercury	Inches of water	13.6
Inches of mercury	Feet of water	1.1333
Inches of mercury	Pounds per square inch	0.4914
Ounces per square inch	Inches of mercury	0.127
Ounces per square inch	Inches of water	1.733
Pounds per square inch	Inches of water	27.72
Pounds per square inch	Feet of water	2.310
Pounds per square inch	Inches of mercury	2.04
Pounds per square inch	Atmospheres	0.0681
Feet of water	Pounds per square inch	0.434
Feet of water	Pounds per square foot	62.5
Feet of water	Inches of mercury	0.8824
Atmospheres	Pounds per square inch	14.696
Atmospheres	Inches of mercury	29.92
Atmospheres	Feet of water	34
Long tons	Pounds	2240
Short tons	Pounds	2000
Short tons	Long tons	0.89285

Glossary

acrylonitrile butadiene styrene (ABS) - piping used extensively for drainage, waste, and vents

adapter - fitting that changes from one type of pipe material to another

aeration tank - a tank that uses a small electrically operated motor to stir its contents occasionally and to pump air below the surface

aerator - a fitting that has a built-in offset, which serves to slow down the flow

aerobic - oxygen consuming

air chamber - used to relieve the shock resulting from water hammer

air control - a device used to maintain air cushion

air-pressure drop test - when air pressure of 5 pounds per square inch is maintained for 10 minutes

annealing - a process of heating and cooling that is used to avoid brittle or cracked joints

a pump's prime - when a pump is full of water and has no air in it

aquifer - the water-bearing layer of loose material

architect's level - an optical instrument that is set up well to one side of the proposed ditch; also known as builder's level

arc welding - uses electric current and is less expensive than gas welding

area drains - drains that are placed in driveways and other paved surfaces

backfilling - the replacement of soil that was removed during excavation

backing boards - devices for supporting and securing pipe, fixtures, and equipment to walls, ceilings, floors, or other structural members; also called supports and hangers

back pressure - occurs when water slugs which form in soil or waste stacks push air before them

back vent - the portion of the stack that continues upward and eventually joins back into the main stack above the highest fixture; also known as continuous vent

backwater valves - placed in drainpipes to prevent sewage from flowing backwards into buildings

balloon construction - a derogatory term used to describe the speed with which a house of frame-type construction can be built

beam clamp - used to fasten a hanger to a steel beam

bell and spigot - the joint style of the most common type of connection

bending spring - a long closely wound spring

bends - another term for elbows

bid - a proposed price

boxed in - reinforced with wooden cross-members

branched fittings - used to bring two pipelines together into a single line or to install a branch line into an existing main line

branch soil pipe - a pipe that receives the discharge of toilets and urinals

branch waste pipe - a horizontal or vertical pipe that receives the discharge of small fixtures and conveys it to the stack

brazing - when the joint is welded with a brass rod rather than soldered

British thermal unit (Btu) - the quantity of heat necessary to raise the temperature of 1 pound of water 1 degree

Btu - (British thermal unit) a unit of heat measurement

Btuh - British thermal unit per hour

builder's level - an optical instrument that is set up well to one side of the proposed ditch; also known as architect's level

building or house drain - the lowest horizontal drainpipe within a building

building or house sewer - the part of the drainage system that extends from the end of the building drain and carries its discharge to a public or private sewer, an individual sewage disposal system, or some other approved point of disposal

building trap - a trap placed in the main house sewer

bullhead tee - a tee having a branch that is larger than the run

bursting pressure - the pressure at which a pipe and fitting may be expected to fail

bushings - used to reduce the outlets of boilers or tanks and have inside and outside threads

butterfly valve - uses a flat disk rather than a ball or plug; this valve has a locking mechanism in the handle so that the flow within the valve will not cause it to swing shut or open when this is not the intention of the operator

butt weld - used to join metal that is laid edge to edge

bypass - permits water or steam to be used temporarily when the reducing or thermostatic valves are out of order

cased - when a large-diameter steel pipe is lowered into the well hole

cast copper - copper that has been shaped with the use of a mold

centersets - what sink and basin faucets are called by plumbers

center-to-center - the measurement to the center of the fitting

chain tongs - also referred to as chain wrenches

chain wrenches - used for working on larger pipe sizes in even smaller work areas; also referred to as chain tongs

chalk line - used to make a straight line on a vertical or horizontal surface

check valve - installed in pipelines to prevent fluids from flowing in the wrong direction

chlorinated polyvinyl chloride (CPVC) - plastic pipe material used for higher-pressure applications

circuit vent - much the same as a loop vent except the vent portion connects to a vent stack and not a stack vent

circulating hot water - water contained in the hydronic heating system

circulating pipes - pipes that carry water from heaters to hot-water storage tanks

circulation or convection - when water is heated and the molecules expand, setting the currents in motion

cistern - a holding tank or vessel for holding rainwater

cleanout fitting - an access to the pipe in order to clean out obstructions

cleats - short pieces of wood that support the ends of the backing board

clevis hanger - used for large pipes

close nipple - has two standard pipe threads; also called all-thread nipple

coil hanger - supports ceiling coils or a number of parallel pipes

collector (solar) - a device that is exposed to the sun to collect the heat or Btu

combination soil fitting - a fitting that does the work of two separate fittings

combination wye and eighth bend - a branch that has two 45-degree turns

combined sewers - sewers that receive both storm water and sewage

common nails - the most frequently used nails in building construction

compression faucet - a type of faucet that uses the principle of the screw to raise or lower the stem

compression fittings - a type of fitting for use with copper tubing

compression joint - a fitting used to join tubing by means of pressure or friction; also a union-type fitting

conduction - that passage or transfer of heat from molecule to molecule, from the hottest to the coldest region of a substance

connected waste and overflow - used for the bathtub waste and to prevent the bathtub from overflowing in case the faucet is left running

continuous vent - the portion of the stack that continues upward and eventually joins back into the main stack above the highest fixture; also known as back vent

convection - the circular movement caused by a difference in weight as a result of a temperature change

coppering - the installation of copper water pipe in the basement

corporation ferrule - a device inserted into the water main with a threaded connection into the wall of the water main; also called corporation stop

corporation stop - a device inserted into the water main with a threaded connection into the wall of the water main; also called corporation ferrule

couplings - straight fittings designed to join two pieces of pipe of the same size end to end

CPVC - (chlorinated polyvinyl chloride) plastic pipe material used for higher-pressure applications

cross-connection - a direct connection between a water supply and a system carrying unsafe water

curb key - used to reach shutoff valves in the water service pipe

curb stop or cock - a valve installed just inside the property line

curtain rod - another name for shower curtain rod; installed if the bathroom is not designed to have a tub enclosure

cut list - contains the end-to-end measurements of pipes

deaerator - a fitting that prevents waste from slowing down at the base and forming a slug of water that would cause pressure changes

deep seal trap - a trap with a 6-inch seal that prevents the seal in area drains from evaporating in periods of drought

deep-well pump - pushes the water up the well pipe to the tank

developed length - the distance measured along the centerline of the pipe, from the trap to the vent

diaphragm relief valve - operates upon the principle of area

die - another term for the thread-cutting part of a pipe threader

dimensioning - refers to the placement of measurements on the drawing

dip tube - a tube through which cold water is delivered to the bottom of the storage tank

distribution box - may be a pipe with some tees in the middle and elbows on the end whose purpose is to effectively use all of the pipe in the distribution field

domestic - household or family

domestic water - water that comes out of the fixtures on kitchen sinks and the shower heads in bathrooms

double wye - a wye used where branches are to be made on both sides of the pipe at the same point

downstream - toward the lower end or the same direction the contents are flowing

draft - the tendency of rising warm air to form a column

draft hood - a sheet metal device that is placed on the appliance to which the vent or smoke pipe is connected

drainage field - a system of horizontal pipes buried outside of a building that causes the effluent from the septic tank to be distributed slowly into the soil; also a distribution field

drainage fittings - fittings that are made of galvanized cast iron to prevent rusting

drainage fixture unit - a method used to determine drainage system sizing, based on the gallon per minute discharge of the smallest common fixture, the lavatory and all other fixtures are compared to that

drilled wells - the deepest, most dependable, and pollution-free; made with massive machinery under the supervision of a well drilling specialist, the wells are drilled down into the water-bearing strata

drip leg - what is formed when the bottom of the tee is closed with a short pipe nipple and a cap

driven wells - used in loose, sandy, or gravelly soils; established by driving a length of pipe, onto which a well point has been screwed, down into the earth with hammer blows

drop-eared elbow - has two lugs for fastening; used most frequently in the installation of faucets for automatic washers and the riser pipe terminal for shower arms; also known as lug elbow, eared elbow, or drop elbow

drop elbow - has two lugs for fastening; used most frequently in the installation of faucets for automatic washers and the riser pipe terminal for shower arms; also known as lug elbow, drop-eared elbow, or eared elbow

dug wells - wells dug by hand that are rarely deeper than 20 feet

duplex strainer - used in an enameled cast-iron, steel, or vitreous chain sink, and prevents objects larger than the drain can handle from entering the system

DWV - drainage, waste, and vent

eared elbow - has two lugs for fastening; used most frequently in the installation of faucets for automatic washers and the riser pipe terminal for shower arms; also known as lug elbow, drop-eared elbow; or drop elbow

effluent - the liquid part of sewage

elbow or ell - a fitting that changes the direction of pipelines

elbows - pipe fittings made to change the direction of a pipeline

elevation - a drawing of objects in a mostly vertical plane

end-to-center - the measurement to the ends of the pipe

end-to-end - the actual length measurement of the pipe to be cut

English style - measurements use fractions like 1/2, 1/4, 1/8, 1/16, and so on, and are based on feet and inches

F & M hanger - consists of two malleable iron halves hinged at the bottom and belted at the top

face feeding - when solder is applied to the face of the fitting

face of the fitting - any opening surface of a fitting

faucets - used to control the water at plumbing fixtures

faucet shank - machined on the inside and threaded on the outside

female end - the fitting opening that accepts the male or spigot end

fillet weld - used to join metal with faces that form an angle

finishing nails - nails that have small heads and are placed in exposed places such as the trim around an access door

fixture supports - used to hold the rear lip of the tub securely

flame soldering - type of soldering commonly used to join copper pipe and fittings

flange - a type of union joint using a gasket and bolts

flared fittings - used to join soft copper tubing in underground service pipes or where pipes must be cleaned often

flaring tools - used to provide a mechanical joint in copper tubing

flood rim - the surface of the fixture over which water would pour if the fixture's drain was plugged and the water supply was left on

floor drains - drains that carry away any leakage or spilled water in such places as elevator pits, garages, boiler rooms, and laundries

flow constrictor - slows down the flow through the coil, enabling it to have more time to pick up heat

flow pipe - the top, or hottest, pipe

flue gases - waste products of combustion

flush valve - designed to replace the flush tank in urinals and toilets in commercial applications

flux - indispensable to soft soldering, it enables the solder to flow on the material being soldered

fresh air inlet - admits air to ventilate the house drainage system

friction loss - the pressure loss due to the rubbing action of the moving water against the sides of the pipe

front elevation - what looking at a drawing of the front wall of a house might be described as

frost line - the depth below the surface to which the soil can be expected to freeze in winter

garage sand trap - prevents sand, gasoline, and oil from entering regular sewer drains

gas - fuel in a gaseous form

gas welding - mixes acetylene and oxygen to create temperatures high enough to melt steel

gate valves - used for liquid, pump, and main lines where maximum flow is required; also called a stop valve

globe valves - used on water, air, gas, oil, or steam lines

grade - refers to the slope of a pipeline

grease trap - a receptacle placed in the waste pipes of sinks to separate and retain grease from the water

ground joint - a connection between brass supply pipes and faucets

hacksaw - a tool for cutting pipe

hanger bolt - used to fasten a clevis hanger to a wood joist

hardness (water) - the total mineral content of the water and affects the scaling of the inside pipe surfaces

hard solder - a solder that contains silver

head - the height of the water

heat exchanger - extracts heat from the freeze-protected heating loop

heating elements - elements that either maintain the standby temperature of the tank or heat a small portion of the water to a higher temperature for immediate use

hexagon bushing - a type of bushing with a six-sided collar

hose bibs - valves that allow a hose to be attached

hydronic - using water to transport heat

hydrostatic test - filling a pipe with water and checking for leaks

immersion or tankless coil - a heat exchanger

indirect heater - a method to heat water; this appliance consists of a copper coil within a cast-iron jacket

inverted wye - a wye used only at the top of vent stacks and is made in sizes from 2 inches to 6 inches

inverts as they refer to septic tanks - the tees or pipe halves installed on the inlet and outlet of the septic tank so that incoming sewage will be released and admitted below the scum layer

isometric - drawing shows piping both horizontally and vertically in the same view

jet pump - uses a two-pipe system within the well to push and draw the water to the surface

jobbing - repairing, maintaining, and enlarging existing plumbing systems

laser beam - a small diameter beam of light of very high intensity

laying length - the amount of pipe which must be cut out of the pipeline to allow the fitting to be installed

lead wall shield - can be used along with a lag screw to fasten to concrete and block walls and floors

level - used by the plumber to make surfaces either vertical or horizontal

lever and weight relief valve - balances the pressure within a tank or system

long nipple - made in graduations of up to 12 inches in length

long-turn pattern tee - used on horizontal or vertical lines and at 90-degree turns

loop vent - used in single-story houses or on the top floor of a multistory building, in which the vent branch goes back into the stack vent at a point above all drain inlets

lug elbow, drop-eared elbow, eared elbow, drop elbow - has two lugs for fastening; used most frequently in the installation of faucets for automatic washers and the riser pipe terminal for shower arms

main (the) - the pipe that leads from the water meter or pressure tank

main house sewer - leads from the end of the house drain to the sanitary sewer

make up - to apply joint compound to the male threads and tighten the fitting onto the pipe

make-up water - water added when there is insufficient water

male end, spigot end - the end of a fitting or pipe that enters into a fitting opening

manhole - positioned above an invert so that a rod may be pushed through the sludge to measure its depth without disturbing the scum layer

manometer - a measuring device that will show small fluctuations in pressure

maximum fixture-unit load - the greatest number of fixture units that may be connected to a given size of a house drain, horizontal branch, vertical soil, or waste stack

metric style - measurements use fractions like 1.01, 21.362, 1.0 and are based on meters, centimeters, etc.

moly fastener - used to fasten to plasterboard and thin wall materials

monkey wrench - used when assembling fittings with flat sides

oakum - a material used for sealing cast-iron pipe joints

offset - can set over a pipeline without changing its direction

offset pipe wrench - used for work in confined areas

O-rings - made of neoprene, are used as seals to prevent leakage

orthographic - drawing as it applies to building plans, separates horizontal piping from vertical

patent overflow or P.O. plug - the waste connection used in a lavatory having an integral overflow

PE (polyethylene) - a flexible plastic pipe used for underground installations such as wells, sprinkler systems, and water supply systems

percolation test - a test that demonstrates the speed at which the soil will absorb water

perforated strap iron - used to support small pipes, such as water and heating pipes

pipe cutter - a tool with one to four cutting wheels

pipe dope - a compound for lubricating and sealing threaded pipe joints

pipe hangers - used to fasten to wood, stone, brick, concrete, and metal

piping - an installed system of pipe

pipe nipple - a short piece of pipe threaded at each end

pipe reamer - removes metal burrs that appear on the pipe after a pipe cutter has been used

pipe strap - used for small pipes on walls or ceilings

pipe threaders - may be divided into hand threaders and power threaders

pipe vise - a tool used to hold the pipe firmly during threading or assembly

pipe wrench - a special type of wrench with teeth that are set at an angle in one direction and is used when attaching fittings to threaded pipe

piston-type pump - uses a long pump rod extending to a piston that is submerged in the well water

plan view - a drawing of something that lies mostly horizontally

plug cock - the most common design for a valve used to stopper water, wine, and whiskey casks

plumb bob - a string with a weight on one end

polyethylene (PE) - a flexible plastic pipe used for underground installations such as wells, sprinkler systems, and water supply systems

polyvinyl chloride (PVC) - plastic pipe material used for higher-pressure applications

potable - drinkable

pressure loss method - one of two methods of pipe sizing

pressure-reducing valves - installed to protect water supply systems from excessive pressure; installed every tenth floor to avoid damage to the terminal outlets and piping from excess pressure

pressure switch - mounted on the tank or in the discharge pipe from the pump to activate an electrical switch

proving plug - a soil pipe testing tool

PVC - (polyvinyl chloride) plastic pipe material used for higher-pressure applications

quick closing faucet - operates on the principle of the inclined plane or the eccentric is used to raise the stem; also called self-closing faucet

quick opening faucet - operates on the principle of the screw but the threads have more pitch, enabling the faucet to be fully opened with a quarter turn of the handle; also called quick-pression faucet

quick-pression or quick opening faucet - operates on the principle of the screw but the threads have more pitch, enabling the faucet to be fully opened with a quarter turn of the handle; also called quick opening faucet

radiation - the passage of heat through space

rain leaders - carry rainwater from the roof to an underground or a ground level disposal conduit

reducers - used to reduce pipelines to smaller sizes

reducing elbow - changes the direction of the pipeline and also the size of the pipe used

relief vents - sometimes installed to relieve air pressure on branch waste near the stack

return elbow - a 180-degree elbow that returns the pipe to the same direction from which it started

return pipe - the bottom, or cooler, pipe

Reznor hook - a bent piece of wire with two pointed ends that are driven into a wooden joist

roof flanges - placed around stacks where they pass through the roof to prevent leakage

roofing nails - nails that are used to secure roof flashings in place

rosin - the dried resin from pine trees used to dust strap wrenches to prevent slipping

roughing-in - the installation of all of the pipelines and parts of the plumbing system that can be put in before the placement of the fixtures

roughing-in tests - tests for leaks on drainage systems by using air, water, or smoke

roundhead screw - used where appearance is not important and the head will not interfere with moving parts

run of an offset - the distance the offset occupies in the length of the pipeline

sacrificial rod - a magnesium rod that is installed to retard corrosion

safety or relief valve - a device that is placed on a closed water, air, steam, or oil system to prevent damage from excessive pressure

safety valve - placed between the pressure-reducing valve and the heater to guard against an explosion

sanitary sewer - a sewer line that carries human waste

sanitary systems - drainage, waste, and vent systems

sanitary tee - a fitting that is added to a soil or waste stack

sanitary tee - a tee used only on vertical stacks

scale - when used in blueprint reading, it is a reference to the size of the drawing compared to the size of the object or building, which the drawing represents

seal of a trap - the vertical distance that gas must travel down through the water in the trap before it can bubble up into the living space

self-closing or quick closing faucet - operates on the principle of the inclined plane or the eccentric is used to raise the stem; also called quick closing faucet

self-syphon - the tendency of the slug of draining water to pull the water, which should remain to maintain the seal, along with it

septic tank - a holding tank that allows raw sewage to settle and separate

serial distribution - a system that fills one pipe at a time with effluent and the excess will overflow to the next pipe, and so on

set of an offset - the distance the offset displaces the pipe without changing its direction

sewage - poisonous liquid waste that contains animal or vegetable matter

sewer saddle - a fitting complete with straps to go around the main sewer for an added outlet

shallow-well pump - a type of pump that uses atmospheric pressure (14.7 psi) to lift the water

sheet metal screws - screws that have parallel sides with the exception of the starting point

short-turn pattern tee - used on vertical pipes

shoulder nipple - has a short space between the threads

side outlet, three-way elbow - has an additional outlet 90 degrees from the others; used mostly on railings

single-lever faucet - uses one control handle to control the amount and also the temperature of the water

single-line technique - a technique in which the pipeline is represented by a single line

6-foot folding rule - a common measuring device used by most construction craftsmen

slants - outlets that are inserted for each building lot when a new sewer is laid; also known as Y-connections

slip coupling - used on traps and other easy-to-get-at locations

slip joint - used on brass tubing traps, connected waste and overflows, and flush elbows

sludge - soil that has been reduced to a nearly inert state by bacteria

smoke test - smoke is introduced into the system and then it is checked visually and by smell

soapstone - a flat marking tool somewhat like chalk

soft soldering - the most common operation used by plumbers to join fittings and tubing together

soft water - water that contains few minerals

soil - human waste

soil pipe - any pipe that carries human waste

soil stack - a vertical drainpipe or stack that receives discharge from water closets (toilets) or urinals

soldering - the process of joining two pieces of metal using heat and the application of another metal of a different composition

soldering iron- an iron that is first heated then applied to the metal to be soldered

solder or sweat fittings - a type of fitting for use with copper tubing

SOVENT drainage system - a self-venting system that uses only one stack

spigot end - the end of a fitting or pipe that enters into a fitting opening

split ring hanger - a commercial hanger that may be disassembled while remaining attached to the joists

spreader - installed after the water service pipe and the main house sewers are installed

spring cushion hanger - used to support pipes in buildings that have considerable vibration

spring relief valve - uses the tension of a coiled spring to withstand the internal pressure; sometimes known as a pop valve

square - forms an accurate 90-degree angle

stack - a vertical drainpipe

stack vents - vents that vent stacks

stadia rod - a measuring rule

steel tape - a light flexible blade with the measurements marked on it

stock - a device that holds the thread-cutting part, or die, of a pipe threader

stock and dies - another term for fixed-die-type pipe threader or ratchet-type pipe threader

storm sewer - a sewer line that carries runoff water

straight tee - a tee in which all three outlets are the same size

strainers - installed in the lines ahead of valves and equipment to prevent grit, rust, scale, and other harmful materials from getting caught in these valves and equipment

strap wrench - a tool that is useful for chrome-plated pipe because it does not mar the finish

street, service elbow - has one outside thread on one end and an inside thread on the other end; made for close work because it is shorter than a close nipple and elbow

stub pieces - should penetrate the walls and floors; help locate supply and drain lines after the floor and wall coverings are in place

stuffing box - an enclosure around a faucet's valve stem

submersible pump - a type of deep-well pump that is lowered into the water in the well

summer-winter hookup - a method of heating domestic water using the home heating system

sump - the depression in a basement floor where water collects

sump pump - a pump that removes water from the sump when the water reaches a preset level

swaging tool - used to join two lengths of tubing together

sweeps - another term for elbows

syphon - a bent tube with arms of unequal length

tack - a short weld that holds the fitting in line, while the weld is being made

temperature differential control box - a thermostatically controlled device that measures temperatures from two different points

tempering valve - a device that adds cold water to the hot water coming out of the immersion coil

test tee - a tee placed on the main drain just inside the wall of a building

thermocouple - a self-contained source of electrical power used to maintain the pilot flame in some gas-fired furnaces

thermopile - produces more electricity than the thermocouple because it is a number of thermocouples joined together in series

thermostat - a device which will automatically take some action that is desired in response to a rise or a fall in temperature, which can usually be set with a control on the device

thermosyphon systems - systems that depend on natural circulation

the run of - the centerline which is not on the branch of a tee or a wye

toggle bolt - used to fasten cabinet bases to walls

TPR valve (temperature pressure relief valve) a valve that opens on excessive temperature or pressure

transition fittings - the name given to one kind of adapter used

trap - a fitting or an assembly of fittings that holds a small amount of water

trap arms - the lateral distance between the trap and its vent

travel of an offset - the length of pipe and fittings used to make the offset

trip lever assembly - the assembly that goes directly against the bathtub; may be called waste and overflow

trouble or access door - placed at the head of the bathtub so that the plumber can get into the parts of the tub that may cause trouble

try handle - a metal handle that may be lifted to check for flow

tubing cutters - lightly constructed pipe cutters

tucker or slip-and-caulk - a special tee constructed with a bell on the top opening for caulking

union - used so that the fixture and fitting may be connected and disconnected without cutting the pipe

upright wye - a wye used at the bottom of vent lines

vacuum breaker - prevents back siphonage when installed at the outlet of the flush valve

velocity method - one of two methods of pipe sizing

vent boxes - vent fittings in the main house sewer to form the fresh air inlet

vent branch fitting - a fitting used at the top of the vent stacks where they reenter the soil stack

vent stacks - stacks that supply only air

washer joint - used to connect supply pipes to flush tanks and brass tailpieces to fixture wastes

waste and overflow - the assembly that goes directly against the bathtub; sometimes called a trip lever assembly

waste stack - a vertical drainpipe that receives the discharge of small fixtures

water-closet - toilet

water hammer - a sharp sound caused by a moving body of water suddenly stopping in a pipe

water meters - instruments used for measuring the quantity of water used in buildings

water service - an underground water supply pipe

water table - the level at which the saturated ground is found

wet vent - a pipe that serves as a drain and also as a vent for a fixture below

wood screws - used for fastening large hangers in place and for installing wall carriers for various fixtures

working pressure - the expected internal pressure when in use

wrought copper - copper that has been hammered or formed by pressure into a desired shape

Y-connections - outlets that are inserted for each building lot when a new sewer is laid; also known as slants

yoke vents - relief vents in tall buildings that are installed at 10-story intervals from the top floor to relieve air pressure

INDEX